Frank Wigglesworth Clarke

A Recalculation of the Atomic Weights

Frank Wigglesworth Clarke

A Recalculation of the Atomic Weights

ISBN/EAN: 9783337337988

Printed in Europe, USA, Canada, Australia, Japan

Cover: Foto ©berggeist007 / pixelio.de

More available books at **www.hansebooks.com**

SMITHSONIAN MISCELLANEOUS COLLECTIONS
— 1075 —

THE CONSTANTS OF NATURE

PART V

A RECALCULATION

OF

THE ATOMIC WEIGHTS

BY

FRANK WIGGLESWORTH CLARKE
Chief Chemist of the U. S. Geological Survey

NEW EDITION, REVISED AND ENLARGED

CITY OF WASHINGTON
PUBLISHED BY THE SMITHSONIAN INSTITUTION
1897

ADVERTISEMENT.

The present publication is one of a series devoted to the discussion and more precise determination of various "Constants of Nature;" and forms the Fifth contribution to that subject published by this Institution.

The First number of the series, embracing tables of "Specific Gravities" and of Melting and Boiling Points of Bodies, prepared by the same author, Prof. F. W. Clarke, was published in 1873. The Fourth part of the series, comprising a complete digest of the various "Atomic Weight" determinations of the chemical elements published since 1814, commencing with the well-known "Table of Equivalents" by Wollaston (given in the Philosophical Transactions for that year), compiled by Mr. George F. Becker, was published by the Institution in 1880. The present work comprises a very full discussion and recalculation of the "Atomic Weights" from all the existing data, and the assignment of the most probable value to each of the elements.

The first edition of this work was published in 1882, and this new edition, revised and enlarged by Professor Clarke, contains new information accumulated during the past fifteen years.

S. P. LANGLEY,
Secretary of the Smithsonian Institution.

WASHINGTON, *January, 1897.*

TABLE OF CONTENTS

	PAGE.
Introduction	1
Formulæ for the Calculation of Probable Error	7
1. Oxygen	8
2. Silver, Potassium, Sodium, Chlorine, Bromine, and Iodine	34
3. Nitrogen	58
4. Carbon	72
5. Sulphur	80
6. Lithium	84
7. Rubidium	87
8. Cæsium	89
9. Copper	91
10. Gold	101
11. Calcium	110
12. Strontium	113
13. Barium	118
14. Lead	127
15. Glucinum	132
16. Magnesium	135
17. Zinc	146
18. Cadmium	156
19. Mercury	166
20. Boron	171
21. Aluminum	176
22. Gallium	181
23. Indium	182
24. Thallium	184
25. Silicon	188
26. Titanium	190
27. Germanium	195
28. Zirconium	196
29. Tin	199
30. Thorium	204
31. Phosphorus	209
32. Vanadium	211
33. Arsenic	213
34. Antimony	216
35. Bismuth	229
36. Columbium	234
37. Tantalum	236
38. Chromium	238
39. Molybdenum	250
40. Tungsten	255
41. Uranium	263
42. Selenium	268
43. Tellurium	271
44. Fluorine	277

TABLE OF CONTENTS.

	PAGE.
45. Manganese	282
46. Iron	287
47. Nickel and Cobalt	291
48. Ruthenium	311
49. Rhodium	313
50. Palladium	315
51. Osmium	322
52. Iridium	325
53. Platinum	327
54. Cerium	335
55. Lanthanum	344
56. The Didymiums	351
57. Scandium	354
58. Yttrium	355
59. Samarium, Gadolinium, Erbium, and Ytterbium	359
60. Terbium, Thulium, Holmium, Dysprosium, etc.	362
61. Argon and Helium	363
Table of Atomic Weights	364
Index	367

A RECALCULATION OF THE ATOMIC WEIGHTS.

By FRANK WIGGLESWORTH CLARKE.

INTRODUCTION.

In the autumn of 1877 the writer began collecting data relative to determinations of atomic weight, with the purpose of preparing a complete résumé of the entire subject, and of recalculating all the estimations. The work was fairly under way, the material was collected and partly discussed, when I received from the Smithsonian Institution a manuscript by Professor George F. Becker, entitled "Atomic Weight Determinations: a Digest of the Investigations Published since 1814." This manuscript, which has since been issued as Part IV of the "Constants of Nature," covered much of the ground contemplated in my own undertaking. It brought together all the evidence, presenting it clearly and thoroughly in compact form; in short, that portion of the task could not well be improved upon. Accordingly, I decided to limit my own labors to a critical recalculation of the data; to combine all the figures upon a common mathematical basis, and to omit everything which could as well be found in Professor Becker's "Digest."

In due time my work was completed, and early in 1882 it was published. About a year later Meyer and Seubert's recalculation appeared, to be followed later still by the less elaborate discussions of Sebelien and of Ostwald. All of these works differed from one another in various essential particulars, presenting the subject from different points of view, and with different methods of calculation. Each one, therefore, has its own special points of merit, and, in a sense, reinforces the others. At the same time, the scientific activity which they represent shows how widespread was the interest in the subject of atomic weights, and how fundamentally important these constants undoubtedly are.

The immediate effect of all these publications was to render manifest the imperfections of many of the data, and to point out most emphatically in what directions new work needed to be done. Consequently, there has been since 1884 an extraordinary activity in the determination of atomic weights, and a great mass of new material has accumulated. The assimilation of this material, and its combination with the old data, is the object of the present volume.

At the very beginning of my work, certain fundamental questions confronted me. Should I treat the investigations of different individuals separately, or should I combine similar data together in a manner irrespective of persons? For example, ought I, in estimating the atomic weight of silver, to take Stas' work by itself, Marignac's work by itself, and so on, and then average the results together; or should I rather combine all series of figures relating to the composition of potassium chlorate into one mean value, and all the data concerning the composition of silver chloride into another mean, and, finally, compute from such general means the constant sought to be established? The latter plan was finally adopted; in fact, it was rendered necessary by the method of least squares, which, in a special, limited form, was chosen as the best method of dealing with the problem.

The mode of discussion and combination of results was briefly as follows. The formulæ employed are given in another chapter. I began with the ratio between oxygen and hydrogen; in other words, with the atomic weight of oxygen referred to hydrogen as unity. Each series of experiments was taken by itself, its arithmetical mean was found, and the probable error of that mean was computed. Then the several means were combined according to the appropriate formula, each receiving a weight dependent upon its probable error. The general mean thus established was taken as the most probable value for the atomic weight of oxygen, and, at the same time, its probable error was mathematically assigned.

Next in order came a group of elements which were best discussed together, namely, silver, chlorine, potassium, sodium, bromine, and iodine. For these elements there were data from many experimenters. All similar figures were first reduced to common standards, and then the means of individual series were combined into general means. Thus all the data were condensed into nineteen ratios, from which several independent values for the atomic weight of each element could be computed. The probable errors of these values, however, all involved the probable error of the atomic weight of oxygen, and were, therefore, higher than they would have been had the latter element not entered into consideration. Here, then, we have suggested a chief peculiarity of this whole revision. The atomic weight of each element involves the probable errors of all the other elements to which it is directly or indirectly referred. Accordingly, an atomic weight determined by reference to elements whose atomic weights have been defectively ascertained will receive a high probable error, and its weight, when combined with other values, will be relatively low. For example, an atomic weight ascertained by direct comparison with hydrogen will, other things being equal, have a lower probable error than one which is referred to hydrogen through the intervention of oxygen; and a metal whose equivalent involves only the probable error of oxygen should be more exactly known than one which depends upon the errors of silver and chlorine. These points will appear more clearly evident in the subsequent actual discussions.

But although the discussion of atomic weights is ostensibly mathematical, it cannot be purely so. Chemical considerations are necessarily involved at every turn. In assigning weights to mean values I have been, for the most part, rigidly guided by mathematical rules; but in some cases I have been compelled to reject altogether series of data which were mathematically excellent, but chemically worthless because of constant errors. In certain instances there were grave doubts as to whether particular figures should be included or rejected in the calculation of means, there having been legitimate reasons for either procedure. Probably many chemists would differ with me upon such points of judgment. In fact, it is doubtful whether any two chemists, working independently, would handle all the data in precisely the same way, or combine them so as to produce exactly the same final results. Neither would any two mathematicians follow identical rules or reach identical conclusions. In calculating the atomic weight of any element those values are assigned to other elements which have been determined in previous chapters. Hence a variation in the order of discussion might lead to slight differences in the final results.

As a matter of course the data herein combined are of very unequal value. In many series of experiments the weighings have been reduced to a vacuum standard; but in most cases chemists have neglected this correction altogether. In a majority of instances the errors thus introduced are slight; nevertheless they exist, and interfere more or less with all attempts at a theoretical consideration of the results.

Necessarily, this work omits many details relative to experimental methods, and particulars as to the arrangement of special forms of apparatus. For such details original memoirs must be consulted. Their inclusion here would have rendered the work unwarrantably bulky. There is such a thing as over-exhaustiveness of treatment, which is equally objectionable with under-thoroughness.

Of course, none of the results reached in this revision can be considered as final. Every one of them is liable to repeated corrections. To my mind the real value of the work, great or little, lies in another direction. The data have been brought together and reduced to common standards, and for each series of figures the probable error has been determined. Thus far, however much my methods of combination may be criticised, I feel that my labors will have been useful. The ground is cleared, in a measure, for future experimenters; it is possible to see more distinctly what remains to be done; some clues are furnished as to the relative merits of different series of results.

On the mathematical side my method of recalculation has obvious deficiencies. It is special, rather than general, and at some future time, when a sufficiently large mass of evidence has accumulated, it must give way to a more thorough mode of treatment. For example, the ratio $Ag_2 : BaBr_2$ has been used for computing the atomic weight of barium, the atomic weights of silver and bromine being supposed to be known. But these atomic weights are subject to small errors, and they are superimposed upon that of the ratio itself in the process of calculation. Obviously, the ratio should contribute to our knowledge of all three of the atomic weights involved in it, its error being distributed into three parts instead of appearing in one only. The errors may be in part compensatory; but that is not certainly known.

Suppose now that for every element we had a goodly number of atomic weight ratios, connecting it with at least a dozen other elements, and all measured with reasonable accuracy. These hundreds of ratios could then be treated as equations of observation, reduced to linear form, and combined by the general method of least squares into normal equations. All errors would thus be distributed, never becoming cumulative; and the normal equations, solved once for all, would give the atomic weights of all the elements simultaneously. The process would be laborious but the result would be the closest possible approach to accuracy. The data as yet are inadequate, although some small groups of ratios may be handled in that way; but in time the method is sure to be applied, and indeed to be the only general method applicable. Even if every ratio was subject to some small constant error, this, balanced against the similar errors of other ratios, would become accidental or unsystematic with reference to the entire mass of material, and would practically vanish from the final means.

Concerning this subject of constant and accidental errors, a word may be said here. My own method of discussion eliminates the latter, which are removable by ordinary averaging; but the constant errors, vicious and untractable, remain, at least partially. Still, where many ratios are considered, even the systematic errors may in part compensate each other, and do less harm than might be expected. They have, moreover, a peculiarity which deserves some attention.

In the discussion of instrumental observations, the systematic errors are commonly constant, both as to direction and as to magnitude. They are therefore independent of the accidental errors, and computation of means leaves them untouched. But in the measurement of chemical ratios the constant errors are most frequently due to an impurity in one of the materials investigated. If different samples of a substance are studied, although all may contain the same impurity, they are not likely to contain it in the same amount; and so the values found for the ratio will vary. In other words, such errors may be constant in direction but variable in magnitude. That variation appears in the probable error computed for the series of observations, diminishes its weight when combined with other series, and so, in part, corrects itself. It is not removed from the result, but it is self-mitigated. The constant errors familiar to the physicist and astronomer are obviously of a different order.

That all methods of averaging are open to objections, I am of course perfectly aware. I also know the doubts which attach to all questions of probable error, and to all combinations of data which depend upon them. I have, however, preferred to face these objections and to recognize these doubts rather than to adopt any arbitrary scheme which permits of a loose selection of data. After all, the use of probable error as a means of weighting is but a means of weighting, and perhaps more justifiable than any other method of attaining the same result. When observations are weighted empirically—that is, by individual judgment—far greater dangers arise. Almost unconsciously, the work of a famous man is given greater weight than that of some obscure chemist, although the latter may ultimately prove to be the best. But the probable error of a series of measurements is not affected by the glamor of great names; and the weight which it assigns to the observations is at least as safe as any other. In the long run, I believe it assigns weight more accurately, and therefore I have trusted to its indications, not as if it were a mathematical fetish, but regarding it as a safe guide, even though sometimes fallible.

In Meyer and Seubert's recalculation, weights are assigned in quite a novel manner. In each series of experiments the maximum and minimum results are given, but instead of the mean there is a value deduced from the sum of the weighings—that is, each experiment is weighted proportionally to the mass of the material handled in it. For this method I am unable to find any complete justification. Of course, the errors due to the operations of weighing become proportionally smaller as the quantity of material increases, but these errors, with modern apparatus, are relatively unimportant. The real errors in atomic weight determinations are much larger than these, and due to different causes. Hence an experiment upon ten grammes of material may be a little better than one made upon five grammes, but it is by no means necessarily twice as good. The ordinary mean of a series of observations, with its measure of concordance, the probable error, is a better value than one obtained in the manner just described. If only errors of weighing were to be considered, Meyer and Seubert's summation method would be valid, but in the presence of other and greater errors it seems to have but little real pertinency to the problem at hand.

In addition to the usual periodicals, the following works have been freely used by me in the preparation of this volume:

BERZELIUS, J. J. Lehrbuch der Chemie. 5 Auflage. Dritter Band. SS. 1147–1231. 1845.

VAN GEUNS, W. A. J. Prœve eener Geschiedenis van de Æquivalentgetallen der Scheikundige Grondstoffen en van hare Soortelijke Gewigten in Gasvorm, voornamelijk in Betrekking tot de vier Grondstoffen der Bewerktuigde Natuur. Amsterdam, 1853.

MULDER, E. Historisch-Kritisch Overzigt van de Bepalingen der Æquivalent-Gewigten van 13 Eenvoudige Ligchamen. Utrecht, 1853.

MULDER, L. Historisch-Kritisch Overzigt van de Bepalingen der Æquivalent-Gewigten van 24 Metalen. Utrecht, 1853.

OUDEMANS, A. C., Jr. Historisch-Kritisch Overzigt van de Bepaling der Æquivalent-Gewigten van Twee en Twintig Metalen. Leiden, 1853.

STAS, J. S. Untersuchungen über die Gesetze der Chemischen Proportionen über die Atomgewichte und ihre gegenseitigen Verhältnisse. Uebersetzt von Dr. L. Aronstein. Leipzig, 1867.
  See also his "Oeuvres Complètes," 3 vols., published at Bruxelles in 1894.

MEYER, L., and SEUBERT, K. Die Atomgewichte der Elemente, aus den Originalzahlen neu berechnet. Leipzig, 1883.

SEBELIEN, J. Beiträge zur Geschichte der Atomgewichte. Braunschweig, 1884.

OSTWALD, W. Lehrbuch der allgemeinen Chemie. Zweite Aufl. I Band. SS. 18–138. Leipzig, 1891.

The four Dutch monographs above cited are especially valuable. They represent a revision of all atomic weight data down to 1853, as divided between four writers.

For the sake of completeness the peculiar volume by Hinrichs [*] must also be cited, although the methods and criticisms embodied in it have not been generally endorsed. Hinrichs' point of view is so radically different from mine that I have been unable to make use of his discussions. His objections to the researches of Stas seem to be quite unfounded; and the rejoinders by Spring and by Van der Plaats are sufficiently thorough.

---

[*] The True Atomic Weight of the Chemical Elements and the Unity of Matter. St. Louis, 1894. Compare Spring, Chem. Zeitung, Feb. 22, 1893, and Van der Plaats, Compt. Rend., 116, 1362. See also a paper by Vogel, with adverse criticisms by Spring and L. Henry, in Bull. Acad. Bruxelles, (3), 26, 469.

FORMULÆ FOR THE CALCULATION OF PROBABLE ERROR.

The formula for the probable error of an arithmetical mean, familiar to all physicists, is as follows:

(1.) $$e = 0.6745 \sqrt{\frac{S}{n(n-1)}}$$

Here n represents the number of observations or experiments in the series, and S the sum of the squares of the variations of the individual results from the mean.

In combining several arithmetical means, representing several series, into one general mean, each receives a weight inversely proportional to the square of its probable error. Let A, B, C, etc., be such means, and a, b, c their probable errors respectively. Then the general mean is determined by the formula:

(2.) $$M = \frac{\frac{A}{a^2} + \frac{B}{b^2} + \frac{C}{c^2} \cdots}{\frac{1}{a^2} + \frac{1}{b^2} + \frac{1}{c^2} \cdots}$$

For the probable error of this general mean we have:

(3.) $$e = \frac{1}{\sqrt{\frac{1}{a^2} + \frac{1}{b^2} + \frac{1}{c^2} \cdots}}$$

In the calculation of atomic and molecular weights the following formulæ are used: Taking, as before, capital letters to represent known quantities, and small letters for their probable errors respectively, we have for the probable error of the sum or difference of two quantities, A and B:

(4.) $$e = \sqrt{a^2 + b^2}$$

For the product of A multiplied by B the probable error is

(5.) $$e = \sqrt{(Ab)^2 + (Ba)^2}$$

For the product of three quantities, ABC:

(6.) $$e = \sqrt{(BCa)^2 + (ACb)^2 + (ABc)^2}$$

For a quotient, $\frac{B}{A}$, the probable error becomes

(7.) $$e = \frac{\sqrt{\left(\frac{Ba}{A}\right)^2 + b^2}}{A}$$

Given a proportion, $A : B :: C : x$, the probable error of the fourth term is as follows:

$$(8.) \quad e = \frac{\sqrt{\left(\frac{BCa}{A}\right)^2 + (Cb)^2 + (Bc)^2}}{A}$$

This formula is used in nearly every atomic weight calculation, and is, therefore, exceptionally important. Rarely a more complicated case arises in a proportion of this kind:

$$A : B :: C + x : D + x$$

In this proportion the unknown quantity occurs in two terms. Its probable error is found by this expression, and is always large:

$$(9.) \quad e = \sqrt{\frac{(C-D)^2}{(A-B)^4}(B^2 a^2 + A^2 b^2) + \frac{B^2 c^2 + A^2 d^2}{(A-B)^2}}$$

When several independent values have been calculated for an atomic weight they are treated like means, and combined according to formulæ (2) and (3). Each final result is, therefore, to be regarded as the general or weighted mean of all trustworthy determinations. This method of combination is not theoretically perfect, but it seems to be the one most available in practice.

OXYGEN.

The ratio between oxygen and hydrogen is the foundation upon which the entire system of atomic weights is sustained. Hence, the accuracy of its determination has, from the beginning, been recognized as of extreme importance. A trifling error here may become cumulative when repeated through a moderate series of other ratios. But few of the elements have, so far, been compared directly with the unit, hydrogen; practically all of them are referred to it through the intervention of oxygen, and therefore the ratio in question requires discussion before any other can be profitably considered.

Leaving out of account the earliest researches, which now have only historical value, the first determinations to be noted are those of Dulong and Berzelius,[*] who, like some of their successors, effected the synthesis of water over heated oxide of copper. The essential features of the method are in all cases the same. Hydrogen gas is passed over the hot oxide, and the water thus formed is collected and weighed. From this weight and the loss of weight which the oxide undergoes, the exact com-

---

[*] Thomson's Annals of Philosophy, July, 1821, p. 50.

position of water is readily calculated. Dulong and Berzelius made but three experiments, with the following results for the percentages of oxygen and hydrogen in water:

O.	H.
88.942	11.058
88.809	11.191
88.954	11.046

From these figures we get, for the atomic weight of oxygen, the values—

16.124
15.863
16.106
—————
Mean, 16.031, ± .057.

As the weighings were not reduced to a vacuum, this correction was afterwards applied by Clark,* who showed that these syntheses really make $O = 15.894$; or, in Berzelian terms, if $O = 100$, $H = 12.583$. The value 15.894, ± .057 we may therefore take as the true result of Dulong and Berzelius' experiments, a result curiously close to that reached in the latest and best researches.

In 1842 Dumas† published his elaborate investigation upon the composition of water. The first point was to get pure hydrogen. This gas, evolved from zinc and sulphuric acid, might contain oxides of nitrogen, sulphur dioxide, hydrosulphuric acid, and arsenic hydride. These impurities were removed in a series of wash bottles; the H_2S by a solution of lead nitrate, the H_3As by silver sulphate, and the others by caustic potash. Finally, the gas was dried by passing through sulphuric acid, or, in some of the experiments, over phosphorus pentoxide. The copper oxide was thoroughly dried, and the bulb containing it was weighed. By a current of dry hydrogen all the air was expelled from the apparatus. and then, for ten or twelve hours, the oxide of copper was heated to dull redness in a constant stream of the gas. The reduced copper was allowed to cool in an atmosphere of hydrogen. The weighings were made with the bulbs exhausted of air. The following table gives the results:

Column A contains the symbol of the drying substance; B gives the weight of the bulb and copper oxide; C, the weight of bulb and reduced copper; D, the weight of the vessel used for collecting the water; E, the same, plus the water; F, the weight of oxygen; G, the weight of water formed; H, the crude equivalent of H when $O = 10,000$; I, the equivalent of H, corrected for the air contained in the sulphuric acid employed. This correction is not explained, and seems to be questionable.

---

* Philosophical Magazine, 3d series, 20, 341.
† Compt. Rend., 14, 537.

THE ATOMIC WEIGHTS.

A.	B.	C.	D.	E.	F.	G.	II.	I.
H₂SO₄	291.985	278.806	480.807	495.634	13.179	14.827	1250.5	1249.6
"	344.548	324.186	488.227	511.132	20.362	22.905	1249.0	1248.0
"	316.671	296.175	439.711	462.764	20.495	23.053	1248.1	1247.2
P₂O₅	625.829	568.825	884.190	948.323	57.004	64.044	1250.6	1249.0
H₂SO₄	804.546	728.182	887.331	973.291	76.364	85.960	1256.2	1254.6
"	533.726	490.155	867.159	916.206	43.571	49.047	1256.3	1255.0
"	661.915	627.104	839.304	878.482	34.811	39.178	1254.6	1253.3
P₂O₅	612.625	566.738	824.624	876.244	45.887	51.623	1250.0	1249.0
"	904.643	844.612	822.660	890.246	60.031	67.586	1258.3	1255.1
H₂SO₄	642.325	590.487	741.095	799.417	51.838	58.320	1250.4	1248.9
P₂O₅	587.645	535.137	874.832	933.910	52.508	59.078	1251.2	1249.0
"	673.280	613.492	931.487	998.700	59.789	67.282	1253.3	1250.8
H₂SO₄	660.855	598.765	682.374	752.273	62.090	69.899	1257.7	1254.8
"	642.325	590.487	741.097	799.455	51.838	58.360	1258.1	1256.2
"	937.845	881.362	1064.762	1128.319	56.483	63.577	1255.8	1252.2
P₂O₅	756.352	719.563	878.640	920.030	36.789	41.390	1250.6	1249.1
"	754.162	720.000	887.817	926.275	34.162	38.458	1257.3	1255.1
"	759.762	727.631	888.662	924.837	32.133	36.175	1257.5	1254.7
"	747.652	716.825	877.862	912.539	30.827	34.677	1248.8	1248.0
Means......							1253.3	1251.5

In the sum total of these nineteen experiments, 840.161 grammes of oxygen form 945.439 grammes of water. This gives, in percentages, for the composition of water—oxygen, 88.864; hydrogen, 11.136. Hence the atomic weight of oxygen, calculated in mass, is 15.9608. In the following column the values are deduced from the individual data given under the headings F and G:

15.994
16.014
16.024
15.992
15.916
15.916
15.943
16.000
15.892
15.995
15.984
15.958
15.902
15.987
15.926
15.992
15.904
15.900
16.015

Mean, 15.9607, with a probable error of ± .0070.

In calculating the above column several discrepancies were noted, probably due to misprints in the original memoir. On comparing columns B and C with F, or D and E with G, these anomalies chiefly appear. They were detected and carefully considered in the course of my own calculations; and, I believe, eliminated from the final result.

The investigation of Erdmann and Marchand * followed closely after that of Dumas. The method of procedure was essentially that of the latter chemist, differing from it only in points of detail. The hydrogen used was prepared from zinc and sulphuric acid, and the zinc, which contained traces of carbon, was proved to be free from arsenic and sulphur. The copper oxide was made partly from copper turnings and partly by the ignition of the nitrate. The results obtained are given in two series, in one of which the weighings were not actually made in vacuo, but were, nevertheless, reduced to a vacuum standard. In the second series the copper oxide and copper were weighed in vacuo. The following table contains the corrected weights of water obtained and of the oxygen in it, with the value found for the atomic weight of oxygen in a third column. The weights are given in grammes.

---

* Journ. für Prakt. Chem., 1842, bd. 26, s. 461.

THE ATOMIC WEIGHTS.

First Series.

Wt. Water.	Wt. O.	At. Wt. O.
62.980	55.950	15.917
95.612	84.924	15.891
94.523	84.007	15.977
35.401	31.461	15.970

Mean, 15.939, ± .014

Second Series.

Wt. Water.	Wt. O.	At. Wt. O.
41.664	37.034	15.996
44.089	39.195	16.018
53.232	47.321	16.011
55.636	49.460	16.017

Mean, 16.010, ± .0036

The effect of discussing these two series separately is somewhat startling. It gives to the four experiments in Erdmann and Marchand's second group a weight vastly greater than their other four and Dumas' nineteen taken together. For so great a superiority as this there is no adequate reason; and it is highly probable that it is due almost entirely to fortunate coincidences, rather than to greater accuracy of work. We will, therefore, treat Erdmann and Marchand's experiments as one series, giving all equal weight, the mean now becoming $O = 15.975, \pm .0113$. If we take the sum of the eight experiments, 483.137 grammes water and 429.352 grammes oxygen, and compute from these figures, then $O = 15.966$.

It would be easy to point out the sources of error in the foregoing sets of determinations, but it is hardly worth while to do so in detail. A few leading suggestions are enough for present purposes. First, there is an insignificant error due to the occlusion of hydrogen by metallic copper, rendering the apparent weight of the latter a trifle too high. Secondly, as shown by Dittmar and Henderson, hydrogen dried by passage through sulphuric acid becomes perceptibly contaminated with sulphur dioxide. In the third place, Morley* has found that hydrogen prepared from zinc always contains carbon compounds not removable by absorption and washing. Erdmann and Marchand themselves note that their zinc contained traces of carbon. Finally, copper oxide, especially when prepared by the ignition of the nitrate, is very apt to contain gaseous impurities, and particularly occluded nitrogen.† Any or all of these sources of error may have vitiated the three investigations so far considered, but it would be useless to speculate as to the extent of their influence. They

---

* Amer. Chem. Journ., 12, 469. 1890.
† See Richards' work cited in the chapter on copper.

amply account, however, for the differences between the older and the later determinations of the constant under discussion.

Leaving out of account all measurements of the relative densities of hydrogen and oxygen, to be considered separately later, the next determination to be noted is that published by J. Thomsen in 1870.* Unfortunately this chemist has not published the details of his work, but only the end results. Partly by the oxidation of hydrogen over heated copper oxide, and partly by its direct union with oxygen, Thomsen finds that at the latitude of Copenhagen, and at sea level, one litre of dry hydrogen at 0° and 760 mm. pressure will form .8041 gramme of water. According to Regnault, at this latitude, level, temperature, and pressure, a litre of hydrogen weighs .08954 gramme. From these data, $O = 15.9605$. It will be seen at once that Thomsen's work depends in great part upon that of Regnault, and is therefore subject to the corrections recently applied by Crafts and others to the latter. These corrections, which will be discussed further on, reduce the value of O from 15.9605 to 15.91. In order to combine this value with others, it is necessary to assign it weight arbitrarily, and as Thomsen made eight experiments, which are said to be concordant, it may be fair, to rank his determination with that of Erdmann and Marchand, and to assume for it the same probable error. The value 15.91, $\pm$ .0113 will therefore be taken as the outcome of Thomsen's research.

In 1887 Cooke and Richards published the results of their elaborate investigation.† These chemists weighed hydrogen, burned it over copper oxide, and weighed the water produced. The copper oxide was prepared from absolutely pure electrolytic copper, and the hydrogen was obtained from three distinct sources, as follows: First, from pure zinc and hydrochloric acid; second, by electrolysis, in a generator containing dilute hydrochloric acid and zinc-mercury amalgam; third, by the action of caustic potash solution upon sheet aluminum. The gas was dried and purified by passage through a system of tubes and towers containing potash, calcium chloride, glass beads drenched with sulphuric acid, and phosphorus pentoxide. No impurity could be discovered in it, and even nitrogen was sought for spectroscopically without being found.

The hydrogen was weighed in a glass globe holding nearly five litres and weighing 570.5 grammes, which was counterpoised by a second globe of exactly the same external volume. Before filling, the globe was exhausted to within 1 mm. of mercury and weighed. It was then filled with hydrogen and weighed again. The difference between the two weights gives the weight of hydrogen taken.

In burning, the hydrogen was swept from the globe into the combustion furnace by means of a stream of air which had previously been passed over hot reduced copper and hot cupric oxide, then through potash

---

* Berichte d. Deutsch. Chem. Gesell., 1870, s. 928.
† Proc. Amer. Acad., 23, 149. Am. Chem. Journ., 10, 81.

bulbs, and finally through a system of driers containing successively calcium chloride, sulphuric acid, and phosphorus pentoxide. The water formed by the combustion was collected in a condensing tube connected with a U tube containing phosphorus pentoxide. The latter was followed by a safety tube containing either calcium chloride or phosphorus pentoxide, added to the apparatus to prevent reflex diffusion. Full details as to the arrangement and construction of the apparatus are given. The final results appear in three series, representing the three sources from which the hydrogen was obtained. All weights are corrected to a vacuum.

First Series.—Hydrogen from Zinc and Acid.

Wt. of H.	Wt. H_2O.	At. Wt. O.
.4233	3.8048	15.977
.4136	3.7094	15.937
.4213	3.7834	15.960
.4163	3.7345	15.941
.4131	3.7085	15.954

Mean, 15.954, ± .0048

Second Series.—Electrolytic Hydrogen.

.4112	3.6930	15.962
.4089	3.6709	15.955
.4261	3.8253	15.955
.4197	3.7651	15.942
.4144	3.7197	15.953

Mean, 15.953, ± .0022

Third Series.—Hydrogen from Aluminum.

.42205	3.7865	15.943
.4284	3.8436	15.944
.4205	3.7776	15.967
.43205	3.8748	15.937
.4153	3.7281	15.954
.4167	3.7435	15.967

Mean, 15.952, ± .0035
Mean of all as one series, 15.953, ± .0020

Shortly after the appearance of this paper by Cooke and Richards Lord Rayleigh pointed out the fact, already noted by Agamennone, that a glass globe when exhausted is sensibly condensed by the pressure of the surrounding atmosphere. This fact involves a correction to the foregoing data, due to a change in the tare of the globe used, and this correction was promptly determined and applied by the authors.* By a

---

* Proc. Amer. Acad., 23, 182. Am. Chem. Journ., 10, 191.

careful series of measurements they found that the correction amounted to an average increase of 1.98 milligrammes to the weight of hydrogen taken in each experiment. Hence O equals not 15.953, but 15.869, the probable error remaining unchanged. The final result of Cooke and Richards' investigation, therefore, is

$$O = 15.869, \pm .0020.$$

Keiser's determinations of the atomic weight of oxygen were published almost simultaneously with Cooke and Richards'. He burned hydrogen occluded by palladium, and weighed the water so formed. In a preliminary paper * the following results are given:

$Wt.\ of\ H.$	$Wt.\ of\ H_2O.$	$At.\ Wt.\ O.$
.65100	5.81777	15.873
.60517	5.41540	15.897
.33733	3.00655	15.822

Mean, 15.864, ± .015

Not long after the publication of the foregoing data Keiser's full paper appeared.† Palladium foil, warmed to a temperature of 250°, was saturated with hydrogen prepared from dilute sulphuric acid and zinc free from arsenic. From 100 to 140 grammes of palladium were taken, and it was first proved that the metal did not absorb other gases which might contaminate the hydrogen. Before charging, the foil was heated to bright redness in vacuo. After charging, the tube containing the palladium hydride was exhausted by means of a Geissler pump to remove any nitrogen which might have been present. In the preliminary investigation cited above, the latter precaution was neglected, which may account for the low results.

Between the palladium tube and the combustion tube a U tube was interposed, containing phosphorus pentoxide. This was to determine the amount of moisture in the hydrogen. The combustion tube was filled with granular copper oxide, prepared by reducing the commercial oxide in hydrogen, heating the metal so obtained to bright redness in a vacuum, and then reoxidizing with pure oxygen.

Upon warming the palladium tube, which was first carefully weighed, hydrogen was given off and allowed to pass into the combustion tube. When the greater part of it had been burned, the tube was cut off by means of a stopcock and allowed to cool. Meanwhile a stream of nitrogen was passed through the combustion tube, sweeping hydrogen before it. This was followed by a current of oxygen, reoxidizing the reduced copper; and the copper oxide was finally cooled in a stream of dry air. The water produced by the combustion was collected in a weighed bulb tube, followed by a weighed U tube containing phosphorus pentoxide.

---

* Berichte, 20, 2323. 1887.
† Amer. Chem. Journ., 10, 249. 1888.

A second phosphorus pentoxide tube served to prevent the sucking back of moisture from the external air. The loss in weight of the palladium tube, corrected by the gain in weight of the first phosphorus pentoxide, gave the weight of hydrogen taken. The gain in weight of the two collecting tubes gave the weight of water formed. All weights in the following table of results are reduced to a vacuum:

Wt. of H.	Wt. H_2O.	At. Wt. O.
.34145	3.06338	15.943
.68394	6.14000	15.955
.65529	5.88200	15.952
.65295	5.86206	15.954
.66664	5.98116	15.944
.66647	5.98341	15.955
.57967	5.20493	15.958
.66254	5.94758	15.952
.87770	7.86775	15.950
.77215	6.93036	15.951

Mean, 15.9514, ± .0011.

In sum, 6.55880 grammes of hydrogen gave 52.30383 of water, whence O = 15.9492.

In March, 1889, Lord Rayleigh * published a few determinations of the atomic weight of oxygen obtained by still a new method. Pure hydrogen and pure oxygen were both weighed in glass globes. From these they passed into a mixing chamber, and thence into a eudiometer, where they were gradually exploded by a series of electric sparks. After explosion the residual gas remaining in the eudiometer was determined and measured. The results, given without weighings or explicit details, are as follows:

15.93
15.98
15.98
15.93
15.92

Mean, 15.948, ± .009

Correcting this result for shrinkage of the globes and consequent change of tare, it becomes O = 15.89, ± .009.

In the same month that Lord Rayleigh's paper appeared, Noyes † published his first series of determinations. His plan was to pass hydrogen into an apparatus containing hot copper oxide, condensing the water formed in the same apparatus, and from the gain in weight of the latter getting the weight of the hydrogen absorbed. The apparatus devised for

---

* Proc. Roy. Soc., 45, 425.
† Amer. Chem. Journ., 11, 155. 1889.

this purpose consisted essentially of a glass bulb of 30 to 50 cc. capacity, with a stopcock tube on one side and a sealed condensing tube on the other. In weighing, it was counterpoised by another apparatus of nearly the same volume but somewhat less weight, in order to obviate reductions to a vacuum. After filling the bulb with commercial copper oxide (90 to 150 grammes), the apparatus was heated in an airbath, exhausted by means of a Sprengel pump, cooled, and weighed. It was next replaced in the airbath, again heated, and connected with an apparatus delivering purified hydrogen. When a suitable amount of the latter had been admitted, the stopcock was closed, and the heating continued long enough to convert all gaseous hydrogen within it into water. The apparatus was then cooled and weighed, after which it was connected with a Sprengel pump, in order to extract the small quantity of nitrogen which was always present. The latter was pumped out into a eudiometer, where it was measured and examined. The gain in weight of the apparatus, less the weight of this very slight impurity, gave the weight of hydrogen oxidized.

The next step in the process consisted in heating the apparatus to expel water, and weighing again. After this, pure oxygen was admitted and the heating was resumed, so as to oxidize the traces of hydrogen which had been retained by the copper. Again the apparatus was cooled and weighed, and then reheated, when the water formed was received in a bulb filled with phosphorus pentoxide, and the gaseous contents were collected in a eudiometer. On cooling and weighing the apparatus, the loss of weight, less the weight of gases pumped out, gave the amount of water produced by the traces of residual hydrogen under consideration. This weight, added to the loss of weight when the original water was expelled, gives the weight of oxygen taken away from the copper oxide. Having thus the weight of hydrogen and the weight of oxygen, the atomic weight sought for follows. Six results are given, but as they are repeated, with corrections, in Noyes' second paper, they need not be considered now.

Noyes' methods were almost immediately criticised by Johnson,* who suggested several sources of error. This chemist had already shown in an earlier paper † that copper reduced in hydrogen persistently retains traces of the latter, and also that when the reduction is effected below 700°, water is retained too. The possible presence of sulphur in the copper oxide was furthermore mentioned. Errors from these sources would tend to make the apparent atomic weight of oxygen too low.

In his second paper ‡ Noyes replies to the foregoing criticisms, and shows that they carry no weight, at least so far as his work is concerned. He also describes a number of experiments in which oxides other than copper oxide were tried, but without distinct success, and he gives fuller

---

* Chem. News, 59, 272.
† Journ. Chem. Soc., May, 1879.
‡ Amer. Chem. Journ., 12, 441. 1890.

details as to manipulations and materials. His final results are in four series, as follows:

First Series.—Hydrogen from Zinc and Hydrochloric Acid.

Wt. of H.	Wt. of O.	At. Wt. O.
.9443	7.5000	15.885
.6744	5.3555	15.882
.7866	6.2569	15.909
.5521	4.3903	15.904
.4274	3.3997	15.909
.8265	6.5686	15.895

Mean, 15.8973, ± .0032.

This series appeared in the earlier paper, but with an error which is here corrected.

Second Series.—Electrolytic Hydrogen, Dried by Phosphorus Pentoxide.

Wt. of H.	Wt. of O.	At. Wt. O.
.5044	4.0095	15.898
.6325	5.0385	15.932
.6349	5.0517	15.913
.5564	4.4175	15.879
.7335	5.8224	15.876
.6696	5.3181	15.885

Mean, 15.8971, ± .0064.

Third Series.—Electrolytic Hydrogen, Dried by Passage Through a Tube Packed with Sodium Wire.

Wt. of H.	Wt. of O.	At. Wt. O.
.9323	7.4077	15.891
.9952	7.9045	15.885
.3268	2.5977	15.898
.7907	6.2798	15.884
.7762	6.1671	15.891
1.1221	8.9131	15.887

Mean, 15.8893, ± .0014

At the end of this series it was found that the hydrogen contained a trace of water, estimated to be equivalent to an excess of three milligrammes in the total hydrogen of the six experiments. Correcting for this, the mean becomes O = 15.899.

Fourth Series.—Electrolytic Hydrogen, Dried over Freshly Sublimed Phosphorus Pentoxide.

Wt. of H.	Wt. of O.	At. Wt. O.
1.0444	8.3017	15.898
.7704	6.1233	15.896
.8231	6.5421	15.896
.8872	7.0490	15.890
.9993	7.9403	15.892
1.1910	9.4595	15.885

Mean, 15.8929, ± .0013

OXYGEN.   19

The mean of all the twenty-four determinations, taken as one series, with the correction to the third series included, is O = 15.8966, ± .0017. In sum, there were consumed 18.5983 grammes of hydrogen and 147.8145 of oxygen; whence O = 15.8955.

Dittmar and Henderson,* who effected the synthesis of water over copper oxide by what was essentially the old method, begin their memoir with an exhaustive criticism of the work done by Dumas and by Erdmann and Marchand. They show, as I have already mentioned, that hydrogen dried by sulphuric acid becomes contaminated with sulphur dioxide, and also that a gas passed over calcium chloride may still retain as much as one milligramme of water per litre. Fused caustic potash they found to dry a gas quite completely.

In their first series of syntheses, Dittmar and Henderson generated their hydrogen from zinc and acid, sometimes hydrochloric and sometimes sulphuric, and dried it by passage, first through cotton wool, then through vitrioled pumice, then over red-hot metallic copper to remove oxygen. In later experiments it first traversed a column of fragments of caustic soda to remove antimony derived from the zinc. The oxide of copper used was prepared by heating chemically pure copper clippings in a muffle, and was practically free from sulphur. In weighing the several portions of apparatus it was tared with somewhat lighter similar pieces of as nearly as possible the same displacement. The results of this series of experiments, which are vitiated by the presence, unsuspected at first, of sulphur dioxide in the hydrogen, are stated in values of H when O = 16, but in the following table have been recalculated to the usual unit:

Wt. of Water.	Wt. of O.	At. Wt. O.
4.7980	4.26195	15.901
7.55025	6.71315	16.039
6.2372	5.53935	15.875
11.29325	10.03585	15.963
11.6728	10.3715	15.940
11.8433	10.5256	15.976
11.7317	10.4243	15.947
19.2404	17.0926	15.916
20.83435	18.5234	16.031
17.40235	15.4598	15.917
19.2631	17.11485	15.934

Mean, 15.949, ± .0103.

Reducing to a vacuum, this becomes 15.843, while a correction for the sulphur dioxide estimated to be present in the hydrogen brings the value

---

* Proc. Roy. Soc. Glasgow, 22, 33. Communicated Dec. 17, 1890.

up again to 15.865. Still another correction is suggested, namely, that as the reduced copper in the combustion tube, before weighing, was exposed to a long-continued current of dry air, it may have taken up traces of oxygen chemically, thereby increasing its weight. As this correction, however, is quantitatively uncertain, it may be neglected here, and the result of this series will be taken as O = 15.865, ± .0103. Its weight, relatively to some other series of experiments, is evidently small.

In their second and final series Dittmar and Henderson dried their hydrogen, after deoxidation by red-hot copper, over caustic potash and subsequently phosphorus pentoxide. The copper oxide and copper of the combustion tube were both weighed in vacuo. The results were as follows, vacuum weights being given:

Wt. Water.	Wt. O.	At. Wt. O.
19.2057	17.0530	15.843
19.5211	17.3342	[15.853]
19.4672	17.2882	15.868
22.9272	20.3540	15.820
23.0080	20.4421	[15.934]
23.4951	20.8639	15.859
23.5612	20.9226	[15.859]
23.7542	21.0957	15.870
23.6568	21.8994	15.884
23.6179	21.8593	15.848
24.6021	21.8499	15.878
24.3047	21.5788	15.832
23.6172	20.9709	15.849

Mean, 15.861, ± .0052.

The authors reject the three bracketed determinations, because of irregularities in the course of the experiments. The mean of the ten remaining determinations is 15.855, ± .0044. Both means, however, have to be corrected for the minute trace of hydrogen occluded by the reduced copper. This correction, experimentally measured, amounts to + .006. Hence the mean of all the experiments in the series becomes 15.867, ± .0052, and of the ten accepted experiments, 15.861, ± .0044. The authors themselves select out seven experiments, giving a corrected mean of 15.866, which they regard as the best value. Taking all their evidence, their two series combine thus:

First series	15.865,	± .0103
Second series	15.867,	± .0052
General mean	15.8667,	± .0046

Leduc,* who also effected the synthesis of water over copper oxide,

---

* Compt. Rend., 115, 41. 1892.

following Dumas' method with slight modifications, gives the results of only two experiments, as follows:

Wt. Water.	Wt. O.	At. Wt. O.
22.1632	19.6844	15.882
19.7403	17.5323	15.880
		Mean, 15.881

These experiments we may arbitrarily assign equal weight with two in Dittmar and Henderson's later series, when the result becomes 15.881, ± .0132, the value to be accepted. Leduc states that his copper oxide, which was reduced at as low a temperature as possible, was prepared by heating clippings of electrolytic copper in a stream of oxygen.

To E. W. Morley * we owe the first complete quantitative syntheses of water, in which both gases were weighed separately, and afterwards in combination. The hydrogen was weighed in palladium, as was done by Keiser, and the oxygen was weighed in compensated globes, after the manner of Regnault. The globes were contained in an artificial "cave," to protect them from moisture and from changes of temperature; being so arranged that they could be weighed by the method of reversals without opening either the "cave" or the balance case. For each weighing of hydrogen about 600 grammes of palladium were employed. After weighing, the gases were burnt by means of electric sparks in a suitable apparatus, from which the unburned residue could be withdrawn for examination. Finally, the apparatus containing the water produced was closed by fusion and also weighed. Rubber joints were avoided in the construction of the apparatus, and the connections were continuous throughout. The weights are as follows:

H taken	O taken.	H_2O formed.
3.2645	25.9176	29.1788
3.2559	25.8531	29.1052
3.8193	30.3210	34.1389
3.8450	30.5294	Lost
3.8382	30.4700	34.3151
3.8523	30.5818	34.4327
3.8298	30.4013	34.2284
3.8286	30.3966	34.2261
'3.8225	30.3497	34 1742
3.8220	30.3479	34.1743
3.7637	29.8865	33.6540
3.8211	30.3429	34.1559

* "On the Density of Oxygen and Hydrogen, and on the Ratio of their Atomic Weights," by Edward W. Morley. Smithsonian Contributions to Knowledge, 1895, 4to, xi + 117 pp., 40 cuts. Abstract in Am. Chem. Journ., 17, 267 (gravimetric), and Ztschr. Phys. Chem., 17, 87 (gaseous densities); also note in Am. Chem. Journ., 17, 396. Preliminary notice in Proc. Amer. Association, 1891, p. 185.

Hence we have—

$H:O$ Ratio	$H:H_2O$ Ratio
15.878	17.877
15.881	17.878
15.878	17.873
15.880	......
15.877	17.881
15.877	17.876
15.877	17.875
15.878	17.879
15.879	17.881
15.881	17.883
15.881	17.883
15.882	17.878
Mean, 15.8792, ± .00032	Mean, 17.8785, ± .00066

Combined, these data give:

From ratio $H_2:O$ .... $O = 15.8792, \pm .00032$
"   " $H_2:H_2O$ ...... $O = 15.8785, \pm .00066$
General mean........ $O = 15.8790, \pm .00028$

For details, Morley's full paper must be consulted. No abstract can do justice to the remarkable work therein recorded.

Two other series of determinations, by Julius Thomsen, remain to be noticed. In the earlier paper[*] he determined the ratio between HCl and NH_3, and thence, using Stas' values for Cl and N, fixed by reference to $O = 16$, computed the ratio $H:O$. This method was so indirect as to be of little importance, and gave for the atomic weight of oxygen approximately the round number 16. I shall use the data farther on in calculating the atomic weight of nitrogen. The paper has been sufficiently criticised by Meyer and Seubert,[†] who have discussed its sources of error.

In Thomsen's later paper [‡] a method of determination is described which is, like the preceding, quite novel, but more direct. First, aluminum, in weighed quantities, was dissolved in caustic potash solution. In one set of experiments the apparatus was so constructed that the hydrogen evolved was dried and then expelled. The loss of weight of the apparatus gave the weight of the hydrogen so liberated. In the second set of experiments the hydrogen passed into a combustion chamber in which it was burned with oxygen, the water being retained. The increase in weight of this apparatus gave the weight of oxygen so taken up. The two series, reduced to the standard of a unit weight of aluminum, gave the ratio between oxygen and hydrogen.

---

[*] Zeitsch. Physikal. Chem., 13, 398. 1894.
[†] Ber., 27, 2770.
[‡] Zeitsch. Anorg. Chem., 11, 14. 1895.

The results of the two series, reduced to a vacuum and stated as ratios, are as follows:

First. $\dfrac{\text{Weight of H}}{\text{Weight of Al}}$	Second. $\dfrac{\text{Weight of O}}{\text{Weight of Al}}$
0.11180	0.88788
0.11175	0.88799
0.11194	0.88774
0.11205	0.88779
0.11189	0.88785
0.11200	0.88789
0.11194	0.88798
0.11175	0.88787
0.11190	0.88773
0.11182	0.88798
0.11204	0.88785
0.11202	
0.11204	0.88787, ± 0.000018
0.11179	
0.11178	
0.11202	
0.11188	
0.11186	
0.11185	
0.11190	
0.11187	
0.11190, ± 0.000015	

Dividing the mean of the second column by the mean of the first, we have for the equivalent of oxygen:

$$\frac{0.88787, \pm 0.000018}{0.11190, \pm 0.000015} = 7.9345, \pm 0.0011$$

Hence $O = 15.8690, \pm 0.0022$.

The details of the investigation are somewhat complicated, and involve various corrections which need not be considered here. The result as stated includes all corrections and is evidently good. The ratios, however, cannot be reversed and used for measuring the atomic weight of aluminum, because the metal employed was not absolutely pure.

We have now before us, representing syntheses of water, thirteen series, as follows:

Dulong and Berzelius..........	$O = 15.894, \pm .057$
Dumas......................	15.9607, ± .0070
Erdmann and Marchand........	15.975, ± .0113
Thomsen, 1870............	15.91, ± .0113
Cooke and Richards...........	15.869, ± .0020
Keiser, 1887	15.864, ± .015
" 1888	15.9514, ± .0011

Rayleigh	15.89,	±.009
Noyes	15.8966,	±.0017
Dittmar and Henderson	15.8667,	±.0046
Leduc	15.881,	±.0132
Morley	15.8790,	±.00028
Thomsen, 1895	15.8690,	±.0022
General mean	O = 15.8837,	±.00026
Rejecting Keiser	15.8796,	±.00027

If we reject all except the determinations of Cooke and Richards, Rayleigh, Noyes, Dittmar and Henderson, Leduc, Thomsen, and Morley, the general mean of these becomes 15.8794, ±.00027. From this it is evident that Keiser's determinations alone, among the higher values for O, carry any appreciable weight; and it also seems clear that the rounded-off number, O = 15.88, ±.0003, cannot be very far from the truth; at least so far as the synthetic evidence goes.

In discussing the relative densities of oxygen and hydrogen gases we need consider only the more modern determinations, beginning with those of Dumas and Boussingault. As the older work has some historical value, I may in passing just cite its results. For the density of hydrogen we have .0769, Lavoisier; .0693, Thomson; .092, Cavendish; .0732, Biot and Arago; .0688, Dulong and Berzelius. For oxygen there are the following determinations: 1.087, Fourcroy, Vauquelin, and Séguin; 1.103, Kirwan; 1.128, Davy; 1,088, Allen and Pepys; 1.1036, Biot and Arago; 1.1117, Thomson; 1.1056, De Saussure; 1.1026, Dulong and Berzelius; 1.106, Buff; 1.1052, Wrede.*

In 1841 Dumas and Boussingault† published their determinations of gaseous densities. For hydrogen they obtained values ranging from .0691 to .0695; but beyond this mere statement they give no details. For oxygen three determinations were made, with the following results:

1.1055
1.1058
1.1057

Mean, 1.10567, ±.00006

If we take the two extreme values given above for hydrogen, and regard them as the entire series, they give us a mean of .0693, ±.00013. This mean hydrogen value, combined with the mean for oxygen, gives for the latter, when H = 1, the density ratio 15.9538, ±.031.

Regnault's researches, published four years later,‡ were much more

---
*For Wrede's work, see Berzelius' Jahresbericht for 1843. For Dulong and Berzelius, see the paper already cited. All the other determinations are taken from Gmelin's Handbook, Cavendish edition, v. 1, p. 279.
†Compt. Rend., 12, 1005. Compare also with Dumas, Compt. Rend., 14, 537.
‡Compt. Rend., 20, 975.

elaborately executed. Indeed, they have long stood among the classics of physical science, and it is only recently that they have been supplanted by other measurements.

For hydrogen three determinations of density gave the following results:

.06923
.06932
.06924
———
Mean, .069263, ± .000019

For oxygen four determinations were made, but in the first one the gas was contaminated by traces of hydrogen, and the value obtained, 1.10525, was, therefore, rejected by Regnault as too low. The other three are as follows:

1.10561
1.10564
1.10565
———
Mean, 1.105633, ± .000008

Now, combining the hydrogen and oxygen series, we have the ratio H : O : : 1 : 15.9628, ± .0044. According to Le Conte,* Regnault's reductions contain slight numerical errors, which, corrected, give for the density of oxygen, 1.105612, and for hydrogen, .069269. Ratio, 1 : 15.9611.

A much weightier correction to Regnault's data has already been indicated in the discussion of Cooke and Richards' work. He assumed that the globes in which the gases were weighed underwent no changes of volume, but Agamennone,† and after him, but independently,‡ Lord Rayleigh showed that an exhausted vessel was perceptibly compressed by atmospheric pressure. Hence its volume when empty was less than its volume when filled with gas. Crafts, having access to Regnault's original apparatus, has determined the magnitude of the correction indicated.§ Unfortunately, the globe actually used by Regnault had been destroyed, but another globe of the same lot was available. With this the amount of shrinkage during exhaustion was measured, and Regnault's densities were thereby changed to 1.10562 for oxygen, and .06949 for hydrogen. Corrected ratio, 1 : 15.9105. Doubtless Dumas and Boussingault's data are subject to a similar correction, and if we assume that it is proportionally the same in amount, the ratio derived from their experiments becomes 1 : 15.9015.

In the same paper, that which contained the discovery of this correction, Lord Rayleigh gives a short series of measurements of his own.

---
* Private communication. See also Phil. Mag. (4), 27, 29, 1864, and Smithsonian Report, 1878, p. 428.
† Atti Rendiconti Acad. Lincei, 1885.
‡ Proc. Roy. Soc., 43, 356. Feb., 1888.
§ Compt. Rend., 106, 1662.

His hydrogen was prepared from zinc and sulphuric acid, and was purified by passage over liquid potash, then through powdered mercuric chloride, and then through pulverized solid potash. It was dried by means of phosphorus pentoxide. His oxygen was derived partly from potassium chlorate, and partly from the mixed chlorates of sodium and potassium. Equal volumes of the two gases weighed as follows:

H.	*O.*
.15811	2.5186, ± .00061*
.15807	
.15798	
.15792	

Mean, .15802, ± .000029.

Corrected for shrinkage of the exhausted globe these become—H, 0.15860; O, 2.5192. Hence the ratio 1 : 15.884, ± .0048.

In 1892 Rayleigh published a much more elaborate determination of this ratio.† The gases were prepared electrolytically from caustic potash, and dried by means of solid potash and phosphorus pentoxide. The hydrogen was previously passed over hot copper. The experiments, stated like the previous series, are in five groups; two for oxygen and three for hydrogen; but for present purposes the similar sets may be regarded as equal in weight, and so discussable together. The weights of equal volumes are as follows:

	H.		*O.*	
First set Mean, .15808	.15807 .15816 .15811 .15803 .15801 .15809		2.5182 2.5173 2.5172 2.5193 2.5174 2.5177	First set. Mean, 2.51785.
Second set Mean, .15797	.15800 .15820 .15792 .15788 .15783		2.5183 2.5168 2 5172 2.5181 2.5156	Second set. Mean, 2.5172.
			Mean, 2.5176, ± .00019.	
Third set Mean, .15804	.15807 .15801 .15817 .15790 .15810 .15798 .15802 .15807			

Mean, .15804, ± .000019.

---

*Arbitrarily assigned the probable error of a single experiment in Rayleigh's paper of 1892.
† Proc. Roy. Soc., 50, 448, Feb. 18, 1892.

OXYGEN. 27

These weights with various corrections relative to temperatures and pressures, and also for the compression of the exhausted globe, ultimately become for H, .158531; and for O, 2.51777. Hence the ratio 1 : 15.882, ± .0023. For details relative to corrections the original memoir should be consulted.

In his paper "On a new method of determining gas densities,"* Cooke gives three measurements for hydrogen, referred to air as unity. They are:

.06957
.06951
.06966

Mean, .06958, ± .000029

Combining this with Regnault's density for oxygen, as corrected by Crafts, 1.10562, ± .000008, we get the ratio H : O : : 1 : 15.890, ± .0067.

Leduc, working by Regnault's method, somewhat modified, and correcting for shrinkage of exhausted globes, gives the following densities: †

$H.$	$O.$
.06947	1.10501
.06949	1.10516
.06947	

Mean, .06948, ± .00006745

The two oxygen measurements are the extremes of three, the mean being 1.10506, ± .0000337. Hence the ratio 1 : 15.905, ± .0154.

The first two hydrogen determinations were made with gas produced by the electrolysis of caustic potash, while the third sample was derived from zinc and sulphuric acid. The oxygen was electrolytic. Both gases were passed over red-hot platinum sponge, and dried by phosphorus pentoxide.

Much more elaborate determinations of the two gaseous densities are those made by Morley.‡ For oxygen he gives three series of data; two with oxygen from potassium chlorate, and one with gas partly from the same source and partly electrolytic. In the first series, temperature and pressure were measured with a mercurial thermometer and a mano-barometer. In the second series they were not determined for each experiment, but were fixed by comparison with a standard volume of hydrogen by means of a differential manometer. In the third series the gas was kept at the temperature of melting ice, and the mano-barometer

---
* Proc. Amer. Acad., 24. 202. 1889. Also Am. Chem. Journ., 11, 509.
† Compt. Rend., 113, 186. 1891.
‡ Paper already cited, under the gravimetric portion of this chapter.

alone was read. The results for the weight in grammes, at latitude 45°, of one litre of oxygen are as follows:

First Series.	Second Series.	Third Series.
1.42864	1.42952	1.42920
1.42849	1.42900	1.42860
1.42838	1.42863	1.42906
1.42900	1.42853	1.42957
1.42907	1.42858	1.42910
1.42887	1.42873	1.42930
1.42871	1.42913	1.42945
1.42872	1.42905	1.42932
1.42883	1.42896	1.42908
	1.42880	1.42910
Mean, 1.42875, ± .000051	1.42874	1.42951
Corrected,* 1.42879, ± .000051	1.42878	1.42933
	1.42872	1.42905
	1.42859	1.42914
	1.42851	1.42849
		1.42894
	Mean, 1.42882, ± .000048	1.42886
	Corrected, 1.42887, ± .000048	
		Mean, 1.42912, ± .000048
		Corrected, 1.42917, ± .000048

General mean of all three series, 1.42896, ± .000028.

Morley himself, for experimental reasons, prefers the last series, and gives it double weight, getting a mean density of 1.42900. The difference between this mean and that given above is insignificant with reference to the atomic weight problem.

In the case of hydrogen, Morley's determinations fall into two groups, but in both the gas was prepared by the electrolysis of pure dilute sulphuric acid, and was most elaborately purified. In the first group there are two series of measurements. Of these, the first involved the reading of temperature and pressure by means of a mercurial thermometer and mano-barometer. In the second series, the gas was delivered into the weighing globes after occlusion in palladium; it was then kept at the temperature of melting ice, and only the syphon barometer was read. In this group the hydrogen was possibly contaminated with mercurial vapor, and the results are discarded by Morley in his final summing up. For present purposes, however, it is unnecessary to reject them, for they have confirmatory value, and do not appreciably affect the final mean. The weight of one litre of hydrogen at 45° latitude, as found in these two sets of determinations, is as follows:

---

*Correction applied by Morley to all his series, for a slight error, $\frac{1}{50000}$, in the length of his standard metre bar.

OXYGEN.

First Series.	Second Series.
.089904	.089977
.089936	.089894
.089945	.089987
.089993	.089948
.089974	.089951
.089941	.089960
.089979	.090018
.089936	.089909
.089904	.089953
.089863	.089974
.089878	.089922
.089920	.090093
.089990	.090007
.089926	.089899
.089928	.089974
	.089900
Mean, .089934, ± .000007	.089869
Corrected, .089938, ± .000007	.090144
	.089984

Mean, .089967, ± .000011
Corrected, .089970, ± .000011

In the second group of experiments, the hydrogen was weighed in palladium before transfer to the calibrated globe; and in weighing, the palladium tube was tared by a similar apparatus of nearly equal volume and weight. After transfer, which was effected without the intervention of stopcocks, the volume and pressure of the gas were taken at the temperature of melting ice. A preliminary set of measurements was made, followed by three regular series; of these, the first and second were with the same apparatus, and are different only in point of time, a vacation falling between them. The last series was with a different apparatus. The data are as follows, with the means as usual:

Preliminary.	Third Series.	Fourth Series.	Fifth Series.
.089946	.089874	.089972	.089861
.089915	.089891	.089877	.089877
.089881	.089886	.089867	.089870
.089901	.089866	.089916	.089867
.089945	.089911	.089770	.089839
	.089856	.089846	.089874
Mean, .089918,	.089912		.089864
± .0000271	.089872	Mean, .089875,	.089883
Corrected, .089921		± .0000187	.089830
	Mean, .089883,	Corrected, .089880	.089877
	± .0000049		.089851
	Corrected, .089886		

Mean, .089863,
± .0000034
Corrected, .089866

Now, rejecting nothing, we may combine all the series into a general mean, giving the weight of one litre of hydrogen as follows:

First series.............................	.089938, ± .000007
Second series	.089970, ± .000011
Preliminary series, second method.......	.089921, ± .0000271
Third series...........................	.089886, ± .0000049
Fourth "	.089880, ± .0000187
Fifth "	.089866, ± .0000034
General mean..................	.089897, ± .0000025
Rejecting the first three...............	.089872, ± .0000028

This last mean value for hydrogen will be used in succeeding chapters of this work for reducing volumes of the gas to weights. Combining the general mean of all with the value found for the weight of a litre of oxygen, 1.42896, ± .000028, we get for the ratio H : O,

$$O = 15.8955, \pm .0005$$

If we take only the second mean for H, excluding the first three series, we have—

$$O = 15.9001, \pm .0005$$

This value is undoubtedly nearest the truth, and is preferable to all other determinations of this ratio. Its probable error, however, is given too low; for some of the oxygen weighings involved reductions for temperature and pressure. These reductions involve, again, the coefficient of expansion of the gas, and its probable error should be included. Since, however, that factor has been disregarded elsewhere, it would be an over-refinement of calculation to include it here.

In a memoir of this kind it is impossible to do full justice to so elaborate an investigation as that of Morley. The details are so numerous, the corrections so thorough, the methods for overcoming difficulties so ingenious, that many pages would be needed in order to present anything like a satisfactory abstract. Hardly more than the actual results can be cited here; for all else the original memoir must be consulted.

Still more recently, by a novel method, J. Thomsen has measured the two densities in question.* In his gravimetric research, already cited, he ascertained the weights of hydrogen and of oxygen equivalent to a unit weight of aluminum. In his later paper he describes a method of measuring the corresponding volumes of both gases during the same reactions. Then, having already the weights of the gases, the volume-weight ratio, or density, is in each case easily computable. From 1.0171 to 2.3932 grammes of aluminum were used in each experiment. Omitting details, the volume of hydrogen in litres, equivalent to one gramme of the metal, is as follows:

---

* Zeitschr. Anorg. Chem., 12, 4. 1896.

OXYGEN. 31

1.24297
1.24303
1.24286
1.24271
1.24283
1.24260
1.24314
1.24294

Mean, 1.24289, ± .00004

The weight of hydrogen evolved from one gramme of aluminum was found in Thomsen's gravimetric research to be 0.11190, ± .000015. Hence the weight of one litre at 0°, 760 mm., and 10.6 meters above sea level at Copenhagen is:

.090032, ± .000012;

or at sea level in latitude 45°,

.089947, ± .000012 gramme.

The data for oxygen are given in somewhat different form, namely, for the volume of one gramme of the gas at 0°, 760, and at Copenhagen. The values are, in litres:

.69902
.69923
.69912
.69917
.69903
.69900
.69901
.69921
.69901
.69922

Mean, .69910, ± .00002
At sea level in latitude 45°, .69976, ± .00002

Hence one litre weighs 1.42906, ± .00004 grammes.

Dividing this by the weight found for hydrogen, 0.089947, ± .000012 we have for the ratio H : O,

15.8878, ± .0022.

The density ratios, H : O, now combine as follows:

Dumas and Boussingault, corrected........ 15.9015, ± .031
Regnault, corrected..................... 15.9105, ± .0044
Rayleigh, 1888......................... 15.884, ± .0048
   "   1892......................... 15.882, ± .0023
Cooke.................................. 15.890, ± .0067
Leduc .... 15.905, ± .0154
Morley, including all the data.......... 15.8955, ± .0005
Thomsen................................ 15.8878, ± .0022

General mean................... 15.8948, ± .0048

If we reject all of Morley's data for the density of hydrogen except his third, fourth, and fifth series, the mean becomes

$$O = 15.8991, \pm .00048.$$

In either case Morley's data vastly outweigh all others.

If oxygen and hydrogen were perfect gases, uniting by volume to form water exactly in the ratio of one to two, then the density of the first in terms of the second would also express its atomic weight. But in fact, the two gases vary from Boyle's law in opposite directions, and the true composition of water by volume diverges from the theoretical ratio to a measurable extent. Hence, in order to deduce the atomic weight of oxygen from its density, a small correction must be applied to the latter, dependent upon the amount of this divergence. Until recently, our knowledge of the volumetric composition of water rested entirely upon the determinations made by Humboldt and Gay-Lussac * early in this century, which gave a ratio between H and O of a little less than 2 : 1, but their data need no farther consideration here.

In 1887 Scott † published his first series of experiments, 21 in number, finding as the most probable result a value for the ratio of 1.994 : 1. In March, 1888,‡ he gave four more determinations, ranging from 1.9962 to 1.998 : 1; and later in the same year § another four, with values from 1.995 to 2.001. In 1893, || however, by the use of improved apparatus, he was able to show that his previous work was vitiated by errors, and to give a series of measurements of far greater value. Of these, twelve were especially good, being made with hydrogen from palladium hydride, and with oxygen from silver oxide. In mean the value found is 2.00245, ± .00007, with a range from 2.0017 to 2.0030.

In 1891 an elaborate paper by Morley ¶ appeared, in which twenty concordant determinations of the volumetric ratio gave a mean value of 2.00023, ± .000015. These measurements were made in eudiometer tubes, and were afterwards practically discarded by the author. In his later and larger paper, ** however, he redetermined the ratio from the density of the mixed electrolytic gases, and found it to be, after applying all corrections, 2.00274. The probable error, roughly estimated, is .00005. Morley also reduces Scott's determinations, which were made at the temperature of the laboratory, to 0°, when the value becomes 2.00285. The mean value of both series may therefore be put at 2.0028, ± .00004, with sufficient accuracy for present purposes. Leduc's †† single determination,

---

* Journ. de Phys., 60, 129.
† Proc. Roy. Soc., 42, 396.
‡ Nature, 37, 439.
§ British Assoc. Report, 1888, 631.
|| Proc. Roy. Soc., 53, 130. In full in Philosophical Transactions, 184, 543. 1893.
¶ Amer. Journ. Sci. (3), 46, 220, and 276.
** Already cited with reference to syntheses of water.
†† Compt. Rend., 115, 311. 1892.

OXYGEN.

based upon the density of the mixed gases obtained by the electrolysis of water, gave 2.0037; but Morley shows that some corrections were neglected. This determination, therefore, may be left out of account.

Now, including all data, we have a mean value for the density ratio:

(A.) $\quad\quad\quad\quad H:O::1:15.8948, \pm .00048;$

or, omitting Morley's rejected series,

(B.) $\quad\quad\quad\quad H:O::1:15.8991, \pm .00048.$

Correcting these by the volume ratio, $2.0028, \pm .00004$, the final result for the atomic weight of oxygen as determined by gaseous densities becomes:

$$\text{From A}\ldots\ldots\ldots\ldots O = 15.8726, \pm .00058$$
$$\text{From B}\ldots\ldots\ldots\ldots O = 15.8769, \pm .00058$$

Combining these with the result obtained from the syntheses of water, rejecting nothing, we have—

$$\text{By synthesis of water}\ldots\ldots O = 15.8837, \pm .00026$$
$$\text{By gaseous densities}\ldots\ldots O = 15.8726, \pm .00058$$
$$\text{General mean}\ldots\ldots\ldots O = 15.8821, \pm .00024$$

If we reject Keiser's work under the first heading, and omit Morley's defective hydrogen series under the second, we get—

$$\text{By synthesis of water}\ldots\ldots O = 15.8796, \pm .00027$$
$$\text{By gaseous densities}\ldots\ldots O = 15.8769, \pm .00058$$
$$\text{General mean}\ldots\ldots\ldots O = 15.8794, \pm .00025$$

Morley, discussing his own data, gets a final value of $O = 15.8790, \pm .00026$, a result sensibly identical with the second of the means given above. These results cannot be far from the truth; and accordingly, rounding off the last decimals, the value

$$O = 15.879, \pm .0003,$$

will be used in computation throughout this work.

NOTE.—A useful "short bibliography" upon the composition of water, by T. C. Warrington, may be found in the Chemical News, vol. 73, pp. 137, 145, 156, 170, and 184.

SILVER, POTASSIUM, SODIUM, CHLORINE, BROMINE, AND IODINE.

The atomic weights of these six elements depend upon each other to so great an extent that they can hardly be considered independently. Indeed, chlorine, potassium, and silver have always been mutually determined. From the ratio between silver and chlorine, the ratio between silver and potassium chloride, and the composition of potassium chlorate, these three atomic weights were first accurately fixed. Similar ratios, more recently worked out by Stas and others, have rendered it desirable to include bromine, iodine, and sodium in the same general discussion.

Several methods of determination will be left altogether out of account. For example, in 1842 Marignac* sought to fix the atomic weight of chlorine by estimating the quantity of water formed when hydrochloric acid gas is passed over heated oxide of copper. His results were wholly inaccurate, and need no further mention here. A little later Laurent † redetermined the same constant from the analysis of a chlorinated derivative of naphthalene. This method did not admit of extreme accuracy, and it presupposed a knowledge of the atomic weight of carbon; hence it may be properly disregarded. Maumené's ‡ analyses of the oxalate and acetate of silver gave good results for the atomic weight of that metal; but they also depend for their value upon our knowledge of carbon, and will, therefore, be discussed farther on with reference to that element. Hardin's § work also, relating to the nitrate, acetate, and benzoate of silver, will be found in the chapters upon nitrogen and carbon.

Let us now consider the ratios upon which we must rely for ascertaining the atomic weights of the six elements in question. After we have properly arranged our data we may then discuss their meaning. First in order we may conveniently take up the percentage of potassium chloride obtainable from the chlorate.

The first reliable series of experiments to determine this percentage was made by Berzelius. || All the earlier estimations were vitiated by the fact that when potassium chlorate is ignited under ordinary circumstances a little solid material is mechanically carried away with the oxygen gas. Minute portions of the substance may even be actually volatilized. These sources of loss were avoided by Berzelius, who devised means for collecting and weighing this trace of potassium chloride.

---

* Compt. Rend., 14, 570. Also, Journ. f. Prakt. Chem., 26, 304.
† Compt. Rend., 14, 456. Journ. f. Prakt. Chem., 26, 307.
‡ Ann. d. Chim et d. Phys. (3), 18, 41. 1846.
§ Journ. Amer. Chem. Soc. 18, 990. 1896.
|| Poggend. Annalen, 8, 1. 1826.

All the successors of Berzelius in this work have benefited by his example, although for the methods by which loss has been prevented we must refer to the original papers of the several investigators. In short, then, Berzelius ignited potassium chlorate, and determined the percentage of chloride which remained. Four experiments gave the following results:

$$60.854$$
$$60.850$$
$$60.850$$
$$60.851$$

Mean, 60.851, ± .0006

The next series was made by Penny,* in England, who worked after a somewhat different method. He treated potassium chlorate with strong hydrochloric acid in a weighed flask, evaporated to dryness over a sand bath, and then found the weight of the chloride thus obtained. His results are as follows, in six trials:

$$60.825$$
$$60.822$$
$$60.815$$
$$60.820$$
$$60.823$$
$$60.830$$

Mean, 60.8225, ± .0014

In 1842 Pelouze† made three estimations by the ignition of the chlorate, with these results:

$$60.843$$
$$60.857$$
$$60.830$$

Mean, 60.843, ± .0053

Marignac, in 1842,‡ worked with several different recrystallizations of the commercial chlorate. He ignited the salt, with the usual precautions for collecting the material carried off mechanically, and also examined the gas which was evolved. He found that the oxygen from 50 grammes of chlorate contained chlorine enough to form .003 gramme of silver chloride. Here are the percentages found by Marignac:

In chlorate once crystallized	60.845
In chlorate once crystallized	60.835
In chlorate twice crystallized	60.833
In chlorate twice crystallized	60.844
In chlorate three times crystallized	60.839
In chlorate four times crystallized	60.839

Mean, 60.8392, ± .0013

---

* Phil. Transactions, 1839, p. 20.
† Compt. Rend., 15, 959.
‡ Ann. d. Chem. u. Pharm., 44, 18.

In the same paper Marignac describes a similar series of experiments made upon potassium perchlorate, $KClO_4$. In three experiments it was found that the salt was not quite free from chlorate, and in three more it contained traces of iron. A single determination upon very pure material gave 46.187 per cent. of oxygen and 53.813 of residue.

In 1845 two series of experiments were published by Gerhardt.* The first, made in the usual way, gave these results:

$$60.871$$
$$60.881$$
$$60.875$$

Mean, $60.8757, \pm .0020$

In the second series the oxygen was passed through a weighed tube containing moist cotton, and another filled with pumice stone and sulphuric acid. Particles were thus collected which in the earlier series escaped. From these experiments we get—

$$60.947$$
$$60.947$$
$$60.952$$

Mean, $60.9487, \pm .0011$

These last results were afterwards sharply criticised by Marignac,† and their value seriously questioned.

The next series, in order of time, is due to Maumené.‡ This chemist supposed that particles of chlorate, mechanically carried away, might continue to exist as chlorate, undecomposed; and hence that all previous series of experiments might give too high a value to the residual chloride. In his determinations, therefore, the ignition tube, after expulsion of the oxygen, was uniformly heated in all its parts. Here are his percentages of residue:

$$60.788$$
$$60.790$$
$$60.793$$
$$60.791$$
$$60.785$$
$$60.795$$
$$60.795$$

Mean, $60.791, \pm .0009$

The question which most naturally arises in connection with these results is, whether portions of chloride may not have been volatilized, and so lost.

---

* Compt. Rend., 21, 1280.
† Supp. Bibl. Univ. de Genéve, Vol. I.
‡ Ann. d. Chim. et d. Phys. (3), 18, 71. 1846.

SILVER, POTASSIUM, ETC.

Closely following Maumené's paper, there is a short note by Faget,* giving certain mean results. According to this chemist, when potassium chlorate is ignited slowly, we get 60.847 per cent. of residue. When the ignition is rapid, we get 60.942. As no detailed experiments are given, these figures can have no part in our discussion.

Last of all we have two series determined by Stas.† In the first series are the results obtained by igniting the chlorate. In the second series the chlorate was reduced by strong hydrochloric acid, after the method followed by Penny:

First Series.
60.8380
60.8395
60.8440
60.8473
60.8450

Mean, 60.84276, ± .0012

Second Series.
60.850
60.853
60.844

Mean, 60.849, ± .0017

In these experiments every conceivable precaution was taken to avoid error and insure accuracy. All weighings were reduced to a vacuum standard; from 70 to 142 grammes of chlorate were used in each experiment; and the chlorine carried away with the oxygen in the first series was absorbed by finely divided silver and estimated. It is difficult to see how any error could have occurred.

Now, to combine these different series of experiments.

Berzelius, mean result		60.851,	± .0006
Penny,	"	60.8225,	± .0014
Pelouze,	"	60.843,	± .0053
Marignac,	"	60.8392,	± .0013
Gerhardt, 1st	"	60.8757,	± .0020
" 2d	"	60.9487,	± .0011
Maumené,	"	60.791,	± .0009
Stas, 1st	"	60.8428,	± .0012
" 2d	"	60.849,	± .0017

General mean from all nine series, representing forty experiments..... 60.846, ± .00038

This value is exactly that which Stas deduced from both of his own series combined, and gives great emphasis to his wonderfully accurate

---
* Ann. d. Chim. et d. Phys. (3), 18, 80. 1846.
† See Aronstein's translation, p. 249.

work. It also finely illustrates the compensation of errors which occurs in combining the figures of different experimenters.

Similar analyses of silver chlorate have been made by Marignac and by Stas. Marignac's data are as follows:* The third column gives the percentage of O in $AgClO_3$:

24.510 grm. $AgClO_3$ gave	18.3616	AgCl.	25 103
25.809	"	19.3345	"	25.086
30.306	"	22.7072	"	25.074
28.358	"	21.2453	"	25.082
28.287	"	21.1833	"	25.113
57.170	"	42.8366	"	25.072

Mean, 25.088, ± .0044

Stas † found the following percentages in two experiments only:

25.081
25.078

Mean, 25.0795, ± .0010

Combined with Marignac's mean this gives a general mean of 25,080, ± .0010; that is, Marignac's series practically vanishes.

For the direct ratio between silver and chlorine there are seven available series of experiments. Here, as in many other ratios, the first reliable work was done by Berzelius. ‡

He made three estimations, using each time twenty grammes of pure silver. This was dissolved in nitric acid. In the first experiment the silver chloride was precipitated and collected on a filter. In the second and third experiments the solution was mixed with hydrochloric acid in a flask, evaporated to dryness, and the residue then fused and weighed without transfer. One hundred parts of silver formed of chloride:

132.700
132.780
132.790

Mean, 132.757, ± .019

Turner's work § closely resembles that of Berzelius. Silver was dissolved in nitric acid and precipitated as chloride. In experiments one, two, and three the mixture was evaporated and the residue fused. In experiment four the chloride was collected on a filter. A fifth experiment was made, but has been rejected as worthless.

The results were as follows: In a third column I put the quantity of AgCl proportional to 100 parts of Ag.

---

* Bibl. Univ. de Genóve, 46, 356. 1843.
† Aronstein's translation, p. 214.
‡ Thomson's Annals of Philosophy, 1820, v. 15, 89.
§ Phil. Transactions, 1829, 291.

28.407 grains Ag gave	37.737 AgCl.	132.844
41.917 "	55.678 "	132.829
40.006 "	53.143 "	132.837
30.922 "	41.070 "	132.818

Mean, 132.832, ± .0038

The same general method of dissolving silver in nitric acid, precipitating, evaporating, and fusing without transfer of material was also adopted by Penny.* His results for 100 parts of silver are as follows, in parts of chloride:

132.836
132.840
132.830
132.840
132.840
132.830
132.838

Mean, 132.8363, ± .0012

In 1842 Marignac† found that 100 parts of silver formed 132.74 of chloride, but gave no available details. Later,‡ in another series of determinations, he is more explicit, and gives the following data. The weighings were reduced to a vacuum standard:

79.853 grm. Ag gave	106.080 AgCl.	Ratio, 132.844
69.905 "	92.864 "	132.843
64.905 "	86.210 "	132.825
92.362 "	122.693 "	132.839
99.653 "	132.383 "	132.844

Mean, 132.839, ± .0024

The above series all represent the synthesis of silver chloride. Maumené§ made analyses of the compound, reducing it to metal in a current of hydrogen. His experiments make 100 parts of silver equivalent to chloride:

132.734
132.754
132.724
132.729
132.741

Mean, 132.7364, ± .0077

By Dumas ‖ we have the following estimations:

9.954 Ag gave	13.227 AgCl.	Ratio, 132.882
19.976 "	26.542 "	132.869

Mean, 132.8755, ± .0044

---

*Phil. Transactions, 1839, 28.
†Ann. Chem. Pharm., 44, 21.
‡ See Berzelius' Lehrbuch, 5th Ed., Vol. 3, pp. 1192, 1193.
‡ Ann. d. Chim. et d. Phys. (3), 18, 49. 1846.
‖ Ann. Chem. Pharm., 113, 21. 1860.

Finally, there are seven determinations by Stas,* made with his usual accuracy and with every precaution against error. In the first, second, and third, silver was heated in chlorine gas, and the synthesis of silver chloride thus effected directly. In the fourth and fifth silver was dissolved in nitric acid, and the chloride thrown down by passing hydrochloric acid gas over the surface of the solution. The whole was then evaporated in the same vessel, and the chloride fused, first in an atmosphere of hydrochloric acid, and then in a stream of air. The sixth synthesis was similar to these, only the nitric solution was precipitated by hydrochloric acid in slight excess, and the chloride thrown down was washed by repeated decantation. All the decanted liquids were afterwards evaporated to dryness, and the trace of chloride thus recovered was estimated in addition to the main mass. The latter was fused in an atmosphere of HCl. The seventh experiment was like the sixth, only ammonium chloride was used instead of hydrochloric acid. From 98.3 to 399.7 grammes of silver were used in each experiment, the operations were performed chiefly in the dark, and all weighings were reduced to vacuum. In every case the chloride obtained was beautifully white. The following are the results in chloride for 100 of silver:

$$132.841$$
$$132.843$$
$$132.843$$
$$132.849$$
$$132.846$$
$$132.848$$
$$122.8417$$

Mean, $132.8445, \pm .0008$

We may now combine the means of these seven series, representing in all thirty-three experiments. One hundred parts of silver are equivalent to chlorine, as follows:

Berzelius............................. $32.757, \pm .0190$
Turner................................ $32.832, \pm .0038$
Penny................................. $32.8363, \pm .0012$
Marignac.............................. $32.839, \pm .0024$
Maumené............................... $32.7364, \pm .0077$
Dumas................................. $32.8755, \pm .0044$
Stas.................................. $32.8445, \pm .0008$

General mean.................... $32.8418, \pm .0006$

Here, again, we have a fine example of the evident compensation of errors among different series of experiments. We have also another tribute to the accuracy of Stas, since this general mean varies from the mean of his results only within the limits of his own variations.

---
*Aronstein's translation, p. 171.

The ratio between silver and potassium chloride, or, in other words, the weight of silver in nitric acid solution which can be precipitated by a known weight of KCl, has been fixed by Marignac and by Stas. Marignac,* reducing all weighings to vacuum, obtained these results. In the third column I give the weight of KCl proportional to 100 parts of Ag:

4.7238 grm. Ag	=	3.2626 KCl.			69.067
22.725	"	15.001	"		69.050
21.759	"	15.028	"		69.066
21.909	"	15.131	"		69.063
22.032	"	15.216	"		69.063
25.122	"	17.350	"		69.063

Mean, 69.062, ± .0017

The work of Stas falls into several series, widely separated in point of time. His earlier experiments† upon this ratio may be divided into two sets, as follows: In the first set the silver was slightly impure, but the impurity was of known quantity, and corrections could therefore be applied. In the second series pure silver was employed. The potassium chloride was from several different sources, and in every case was purified with the utmost care. From 10.8 to 32.4 grammes of silver were taken in each experiment, and the weighings were reduced to vacuum. The method of operation was, in brief, as follows: A definite weight of potassium chloride was taken, and the exact quantity of silver necessary, according to Prout's hypothesis, to balance it was also weighed out. The metal, with suitable precautions, was dissolved in nitric acid, and the solution mixed with that of the chloride. After double decomposition the trifling excess of silver remaining in the liquid was determined by titration with a normal solution of potassium chloride. One hundred parts of silver required the following of KCl:

First Series.
69.105
69.104
69.103
69.104
69.102

Mean, 69.1036, ± .0003

Second Series.
69.105
69.099
69.107
69.103
69.103
69.105
69.104

---

* See Berzelius' Lehrbuch, 5th Ed., Vol. 3, pp. 1192–3.
† Aronstein's translation, pp. 250–257.

69.099
69.1034
69.104
69.103
69.102
69.104
69.104
69.105
69.103
69.101
69.105
69.103

Mean, 69.1033, ± .0003

In these determinations Stas did not take into account the slight solubility of precipitated silver chloride in the menstrua employed in the experiments. Accordingly, in 1882* he published a new series, in which by two methods he remeasured the ratio, guarding against the indicated error, and finding the following values:

69.1198
69.11965
69.121
69.123

Mean, 69.1209, ± .0003

Corrected for a minute trace of silica contained in the potassium chloride, this mean becomes

69.11903, ± .0003.†

Still later, in order to establish the absolute constancy of the ratio in question, Stas made yet another series of determinations,‡ in which he employed potassium chloride prepared from four different sources. One lot of silver was used throughout. The values obtained were as follows:

69.1227
69.1236
69.1234
69.1244
69.1235
69.1228
69.1222
69.1211
69.1219
69.1249
69.1238
69.1225
69.1211

---

* Mémoires Acad. Roy. de Belge, t. 43. 1882.
† See Van der Plaats, Ann. Chim. Phys. (6), 7, 15.
‡ Oeuvres Posthumes, edited by W. Spring.

A series was also begun in which one sample of potassium chloride was to be balanced against silver from various sources, but only one result is given, namely, 69.1240. This, with the previous series, gives a mean of 69.1230, ± .0002.

Five series of determinations are now at hand for the ratio Ag : KCl. They combine as follows:

Marignac	69.062, ± .0017
Stas, 1st series	69.1036, ± .0003
" 2d "	69.1033, ± .0003
" 3d "	69.1190, ± .0003
" 4th "	69.1230, ± .0002
General mean	69.1143, ± .00013

The difference between the highest and the lowest of Stas' series corresponds to a difference of 0.021 in the atomic weight of potassium. The rejection of the earlier work might be quite justifiable, but would exert a very slight influence upon our final result.

The quantity of silver chloride which can be formed from a known weight of potassium chloride has also been determined by Berzelius, Marignac and Maumené. Berzelius[*] found that 100 parts of KCl were equivalent to 194.2 of AgCl; a value which, corrected for weighings in air, becomes 192.32. This experiment will not be included in our discussion.

In 1842 Marignac[†] published two determinations, with these results from 100 KCl:

$$192.33$$
$$192.34$$

Mean, corrected for weighing in air, 192.26, ± .003

In 1846 Marignac[‡] published another set of results, as follows. The weighings were reduced to vacuum. The usual ratio is in the third column:

17.034 grm. KCl gave	32.761 AgCl.		192.327
14.427 "	27.749 "		192.341
15.028 "	28.910 "		192.374
15.131 "	29.102 "		192.334
15.216 "	29.271 "		192.370

Mean, 192.349, ± .006

Three estimations of the same ratio were also made by Maumené§ as follows:

---

[*] Poggend. Annal., 8, 1. 1826.
[†] Ann. Chem. Pharm., 44, 21, 1842.
[‡] Berzelius' Lehrbuch, 5th Ed., Vol. 3, pp. 1192, 1193.
[§] Ann. d. Chim. et d. Phys. (3), 18, 41. 1846.

10.700 grm. KCl gave 20.627 AgCl.   192.776
10.5195   "   20.273   "   192.716
 8.587   "   16.556   "   192.803
                                     ─────────
                              Mean, 192.765, ± .017

The three series of ten experiments in all foot up thus:

Marignac, 1842........................ 192.260, ± .003
    "   1846........................ 192.349, ± .006
Maumené............................... 192 765, ± .017
                                      ─────────
        General mean.................. 192.294, ± .0029

These figures show clearly that the ratio which they represent is not of very high importance. It might be rejected altogether without impropriety, and is only retained for the sake of completeness. It will obviously receive but little weight in our final discussion.

In estimating the atomic weight of bromine the earlier experiments of Balard, Berzelius, Liebig, and Löwig may all be rejected. Their results were all far too low, probably because chlorine was present as an impurity in the materials employed. Wallace's determinations, based upon the analysis of arsenic tribromide, are tolerably good, but need not be considered here. In the present state of our knowledge, Wallace's analyses are better fitted for fixing the atomic weight of arsenic, and will, therefore, be discussed with reference to that element.

The ratios with which we now have to deal are closely similar to those involving chlorine. In the first place, there are the analyses of silver bromate by Stas.* In two careful experiments he found in this salt the following percentages of oxygen:

                20.351
                20.347
               ────────
          Mean, 20.349, ± .0014

There are also four analyses of potassium bromate by Marignac.† The salt was heated, and the percentage loss of oxygen determined. The residual bromide was feebly alkaline. We cannot place much reliance upon this series. The results are as follows:

                28.7016
                28.6496
                28.6050
                28.7460
               ────────
          Mean, 28.6755, ± .0207

───────────────────────────────
* Aronstein's translation, pp. 200–206.
† See E. Mulder's Overzigt, p. 117; or Berzelius' Jahresbericht, 24, 72.

SILVER, POTASSIUM, ETC.   45

When silver bromide is heated in chlorine gas, silver chloride is formed. In 1860 Dumas* employed this method for estimating the atomic weight of bromine. His results are as follows. In the third column I give the weight of AgBr equivalent to 100 parts of AgCl:

$$
\begin{array}{lll}
2.028 \text{ grm. AgBr gave } 1.547 \text{ AgCl.} & & 131.092 \\
4.237 \quad\quad\quad\quad\quad\quad\quad\quad\quad 3.235 \quad\quad & & 130.974 \\
5.769 \quad\quad\quad\quad\quad\quad\quad\quad\quad 4.403 \quad\quad & & 131.024 \\
\end{array}
$$

Mean, 131.030, ± .023

This series is evidently of but little value.

The two ratios upon which, in connection with Stas' analyses of silver bromate, the atomic weight of bromine chiefly depends, are those which connect silver with the latter element directly and silver with potassium bromide.

Marignac,† to effect the synthesis of silver bromide, dissolved the metal in nitric acid, precipitated the solution with potassium bromide, washed, dried, fused, and weighed the product. The following quantities of bromine were found proportional to 100 parts of silver:

74.072
74.055
74.066

Mean, reduced to a vacuum standard, 74.077, ± .003

Much more elaborate determinations of this ratio are due to Stas.‡ In one experiment a known weight of silver was converted into nitrate, and precipitated in the same vessel by pure hydrobromic acid. The resulting bromide was washed thoroughly, dried, and weighed. In four other estimations the silver was converted into sulphate. Then a known quantity of pure bromine, as nearly as possible the exact amount necessary to precipitate the silver, was transformed into hydrobromic acid. This was added to the dilute solution of the sulphate, and, after precipitation was complete, the minute trace of an excess of silver in the clear supernatant fluid was determined. All weighings were reduced to a vacuum. From these experiments, taking both series as one, we get the following quantities of bromine corresponding to 100 parts of silver:

74.0830
74.0790
74.0795
74.0805
74.0830

Mean, 74.081, ± .0006

---

* Ann. Chem. Pharm., 113, 20.
† R. Mulder's Overzigt, p. 116. Berzelius' Jahresbericht, 24, 72.
‡ Aronstein's translation, pp. 154–170.

In his paper on the atomic weight of cadmium,* Huntington gives three syntheses and three analyses of silver bromide. The data are as follows, with the usual ratio given in the last column:

1.4852 grm. Ag gave	2.5855 AgBr.		74.084
1.4080 "	2.4510 "		74.077
1.4449 "	2.5150 "		74.060
4.1450 grm. AgBr gave	2.3817 Ag.		74.035
1.8172 "	1.0437 "		74.111
4.9601 "	2.8497 "		74.057

Mean, 74.071, ± .0072

Similar synthetic data are also given by Richards, incidentally to his work on copper.† There are two sets of three experiments each, which can here be treated as one series, thus:

1.11235 grm. Ag gave	1.93630 AgBr.		74.073
1.57620 "	2.74335 "		74.044
2.16670 "	3.77170 "		74.076
.9664 "	1.68205 "		74.053
.9645 "	1.6789 "		74.069
.9639 "	1.6779 "		74.074

Mean, 74.065, ± .0035

Another set of data by Richards appears in his research upon the atomic weight of barium;‡ in which $BaBr_2$ was balanced against silver, and the AgBr was also weighed. Richards gives from these data the percentage of Ag in AgBr, which figures are easily restated in the usual form as follows:

Percentage.	Ratio.
57.460	74.034
57.455	74.049
57.447	74 073
57.445	74.074
57.448	74.070
57.442	74.089
57.451	74.061
57.455	74.049
57.443	74.086
57.445	74.074
57.445	74.074

Mean, 74.067, ± .0034

The same ratio can also be computed indirectly from Cooke's experiments upon $SbBr_3$, Huntington's on $CdBr_2$, Thorpe's on $TiBr_4$, and

---

* Proc. Amer. Acad., 1881.
† Proc. Amer. Acad., 25, pp. 199, 210, 211. 1890.
‡ Proc. Amer. Acad., vol. 28. 1893.

Thorpe and Laurie's on gold. The values so obtained all confirm the results already given, varying within their limits, but having probable errors so high that their use would not affect the final mean. The latter is obtained as follows:

$$\begin{array}{lr}\text{Marignac,} & 74.077, \pm .0030 \\ \text{Stas,} & 74.081, \pm .0006 \\ \text{Huntington} & 74.071, \pm .0072 \\ \text{Richards, 1st series} & 74.065, \pm .0035 \\ \text{`` 2d ``} & 74.067, \pm .0034 \\ \\ \text{General mean} & 74.080, \pm .00057\end{array}$$

In this case again, as in so many others, Stas' work alone appears at the end, the remaining data having only corroborative value.

The ratio between silver and potassium bromide was first accurately determined by Marignac.* I give, with his weighings, the quantity of KBr proportional to 100 parts of Ag:

2.131 grm. Ag =	2.351 KBr.	110.324
2.559 ``	2.823 ``	110.316
2.447 ``	2.700 ``	110.339
3.025 ``	3.336 ``	110.283
3.946 ``	4.353 ``	110.314
11.569 ``	12.763 ``	110.321
20.120 ``	22.191 ``	110.293

Mean, corrected for weighing in air, 110.343, ± .005

Stas,† working in essentially the same manner, as when he fixed the ratio between potassium chloride and silver, obtained the following results:

110.361
110.360
110.360
110.342
110.346
110.338
110.360
110.336
110.344
110.332
110.343
110.357
110.334
110.335

Mean, 110.3463, ± .0020

Combining this with Marignac's mean result, 110.343, ± .005, we get a general mean of 110.3459, ± .0019.

---

* Berzelius' Jahresbericht, 24, 72.
† Aronstein's translation, pp. 334-347.

The ratios upon which we must depend for the atomic weight of iodine are exactly parallel to those used for the determination of bromine.

To begin with, the percentage of oxygen in potassium iodate has been determined by Millon.* In three experiments he found:

     22.46
     22.49
     22.47
    Mean, 22.473, ± .005

Millon also estimated the oxygen in silver iodate, getting the following percentages:

     17.05
     17.03
     17.06
    Mean, 17.047, ± .005

The analysis of silver iodate has also been performed with extreme care by Stas.† From 76 to 157 grammes were used in each experiment, the weights being reduced to a vacuum standard. As the salt could not be prepared in an absolutely anhydrous condition, the water expelled in each analysis was accurately estimated and the necessary corrections applied. In two of the experiments the iodate was decomposed by heat, and the oxygen given off was fixed upon a weighed quantity of copper heated to redness. Thus the actual weights, both of the oxygen and the residual iodide, were obtained. In a third experiment the iodate was reduced to iodide by a solution of sulphurous acid, and the oxygen was estimated only by difference. In the three percentages of oxygen given below, the result of this analysis comes last. The figures for oxygen are as follows:

     16.976
     16.972
     16.9761
    Mean, 16.9747, ± .0009

This, combined with Millon's series above cited, gives us a general mean of 16.9771, ± .0009.

The ratio between silver and potassium iodide seems to have been determined only by Marignac,‡ and without remarkable accuracy. In five experiments 100 parts of silver were found equivalent to potassium iodide as follows:

---

* Ann. Chim. Phys. (3), 9, 400. 1843.
† Aronstein's translation, pp. 170-200.
‡ Berzelius' Lehrbuch, 5th ed., 3, 1196.

1.616 grm. Ag =	2.483 KI.		Ratio,	153.651
2.503 "	3.846 "		"	153.665
3.427 "	5.268 "		"	153.720
2.141 "	3.290 "		"	153.667
10.821 "	16.642 "		"	153.794

Mean, 153.6994, ± .0178

The synthesis of silver iodide has been effected by both Marignac and Stas. Marignac, in the paper above cited, gives these weighings. In the last column I add the ratio between iodine and 100 parts of silver:

15.000 grm. Ag gave 31.625 AgI.	117.500
14.790 " 32.170 "	117.512
18.545 " 40.339 "	117.519

Mean, corrected for weighing in air, 117.5335, ± .0036

Stas* in his experiments worked after two methods, which gave, however, results concordant with each other and with those of Marignac.

In the first series of experiments Stas converted a known weight of silver into nitrate, and then precipitated with pure hydriodic acid. The iodide thus thrown down was washed, dried, and weighed without transfer. By this method 100 parts of silver were found to require of iodine:

117.529
117.536

Mean, 117.5325, ± .0024

In the second series a complete synthesis of silver iodide from known weights of iodine and metal was performed. The iodine was dissolved in a solution of ammonium sulphite, and thus converted into ammonium iodide. The silver was transformed into sulphate and the two solutions were mixed. When the precipitate of silver iodide was completely deposited the supernatant liquid was titrated for the trifling excess of iodine which it always contained. As the two elements were weighed out in the ratio of 127 to 108, while the atomic weight of iodine is probably a little under 127, this excess is easily explained. From these experiments two sets of values were deduced; one from the weights of silver and iodine actually employed, the other from the quantity of iodide of silver collected. From the first set we have of iodine for 100 parts of silver:

117.5390
117.5380
117.5318
117.5430
117.5420
117.5300

Mean, 117.5373, ± .0015

---

*Aronstein's translation, pp. 136, 152.

From the weight of silver iodide actually collected we get as follows. For experiment number three in the above column there is no equivalent here:

$$117.529$$
$$117.531$$
$$117.539$$
$$117.538$$
$$117.530$$

Mean, $117.5334, \pm .0014$

Now, combining these several sets of results, we have the following general mean:

Marignac	117.5335,	± .0036
Stas, 1st series	117.5325,	± .0024
" 2d "	117.5373,	± .0015
" 3d "	117.5334,	± .0014
General mean	117.5345,	± .0009

One other comparatively unimportant iodine ratio remains for us to notice. Silver iodide, heated in a stream of chlorine, becomes converted into chloride; and the ratio between these two salts has been thus determined by Berzelius and by Dumas.

From Berzelius[*] we have the following data. In the third column I give the ratio between AgI and 100 parts of AgCl:

5.000 grm. AgI gave	3.062 AgCl.	163.292
12.212 "	7.4755 "	163.360

Mean, $163.326, \pm .023$

Dumas'[†] results were as follows:

3.520 grm. AgI gave	2.149 AgCl.	163.793
7.011 "	4.281 "	163.770

Mean, $163.782, \pm .008$

General mean from the combination of both series, $163.733, \pm .0076$.

For sodium there are but four ratios of any value for present purposes.

The early work of Berzelius we may disregard entirely, and confine ourselves to the consideration of the results obtained by Penny, Pelouze, Dumas, and Stas, together with a single ratio measured incidentally by Ramsay and Aston.

The percentage of oxygen in sodium chlorate has been determined only by Penny[‡], who used the same method which he applied to the potassium salt. Four experiments gave the following results:

---

[*] Ann. Chim. Phys. (2), 40, 430. 1829.
[†] Ann. Chem. Pharm., 113, 28. 1860.
[‡] Phil. Transactions, 1839, p. 25.

45.060
45.075
45.080
45.067

Mean, 45.0705, ± .0029.

The ratio between silver and sodium chloride has been fixed by Pelouze, Dumas, and Stas. Pelouze* dissolved a weighed quantity of silver in nitric acid, and then titrated with sodium chloride. Equivalent to 100 parts of silver he found of chloride:

54.158
54.125
54.139

Mean, 54.141, ± .0063

By Dumas† we have seven experiments, with results as follows. The third column gives the ratio between 100 of silver and NaCl:

2.0535 grm. NaCl =	3.788 grm. Ag.	54.211
2.169 "	4.0095 "	54.097
4.3554 "	8.0425 "	54.155
6.509 "	12.0140 "	54.178
6.413 "	11.8375 "	54.175
2.1746 "	4.012 "	54.202
5.113 "	9.434 "	54.187

Mean, 54.172, ± .0096

Stas,‡ applying the method used in establishing the similar ratio for potassium chloride, and working with salt from six different sources, found of sodium chloride equivalent to 100 parts of silver:

54.2093
54.2088
54.2070
54.2070
54.2070
54.2060
54.2076
54.2081
54.2083
54.2089

Mean, 54.2078, ± .0002

As in the case of the corresponding ratio for potassium chloride, these data needed to be checked by others which took into account the solu-

---

*Compt. Rend., 20, 1047. 1845.
†Ann. Chem. Pharm., 113. 31. 1860.
‡ Aronstein's translation, p. 274.

bility of silver chloride. Such data are given in Stas' paper of 1882,* and four results are as follows :

$$54.2065$$
$$54.20676$$
$$54.2091$$
$$54.2054$$

Mean, 54.20694, ± .00045

Corrected for a trace of silica in the sodium chloride, this mean becomes 54.2046, ± .00045.† Combining all four series, we have for the NaCl equivalent to 100 parts of Ag—

Pelouze	54.141,	± .0063
Dumas	54.172,	± .0096
Stas, early series	54.2078,	± .0002
Stas, late "	54.2046,	± .00045
General mean	54.2071,	± .00018

Here the work of Stas is of such superior excellence that the other determinations might be completely rejected without appreciably affecting our final results.

In their research upon the atomic weight of boron, Ramsay and Aston ‡ converted borax into sodium chloride. In the latter the chlorine was afterwards estimated gravimetrically by weighing as silver chloride on a Gooch filter. Hence the ratio, AgCl : NaCl : : 100 : x, as follows :

3.0761 grm. NaCl gave	7.5259 AgCl.	Ratio,	40.874
2.7700 "	6.7794 "	"	40.859
2.8930 "	7.0804 "	"	40.859
2.7360 "	6.6960 "	"	40.860
1.9187 "	4.6931 "	"	40.863

Mean, 40.867, ± .0033

Finally, for the ratios between silver and sodium bromide we have one set of measurements by Stas.§ The bromide was prepared by saturating Na_2CO_3 with HBr. The NaBr proportional to 100 parts of silver was—

$$95.4420$$
$$95.4383$$
$$95.4426$$
$$95.4392$$

Mean, 95.4405, ± .0007

We have now before us the data for computing, with greater or less accuracy, the atomic weights of the six elements under discussion. In

---

* Mémoires Acad. Roy. de Belge., 43. 1882.
† See Van der Plaats, Ann. Chim. Phys. (6), 7, 16. 1886.
‡ Chem. News, 66, 92. 1892.
§ Mémoires Acad. Roy. Belge., 43. 1882.

SILVER, POTASSIUM, ETC.   53

all there are nineteen ratios, involving about two hundred and fifty separate experiments. These ratios may now be tabulated and numbered for reference, it being understood that the probable error in each case is that of the last term in the proportion.

(1.) Percentage of O in $KClO_3$...... 39.154, ± .00038
(2.)   "   " $KBrO_3$...... 28.6755, ± .0207
(3.)   "   " KIO_3....... 22.473, ± .0050
(4.)   "   " $NaClO_3$..... 45.0705, ± .0029
(5.)   "   " $AgClO_3$..... 25.080, ± .0010
(6.)   "   " $AgBrO_3$..... 20.349, ± .0014
(7.)   "   " $AgIO_3$...... 16.9771, ± .0009
(8.) Ag : NaCl :: 100 : 54.2071, ± .00018
(9.) Ag : NaBr :: 100 : 95.4405, ± .0007
(10.) Ag : KCl :: 100 : 69.1143, ± .00013
(11.) Ag : KBr :: 100 : 110.3459, ± .0019
(12.) Ag : KI :: 100 : 153.6994, ± .0178
(13.) Ag : Cl :: 100 : 32.8418, ± .0006
(14.) Ag : Br :: 100 : 74.080, ± .00057
(15.) Ag : I :: 100 : 117.5345, ± .0009
(16.) AgCl : NaCl :: 100 : 40.867, ± .0033
(17.) KCl : AgCl :: 100 : 192.294, ± .0029
(18.) AgCl : AgBr :: 100 : 131.030, ± .023
(19.) AgCl : AgI :: 100 : 163.733, ± .0076

Now, from ratios 1 to 7, inclusive, we can at once, by applying the known atomic weight of oxygen, deduce the molecular weights of seven haloid salts. Let us consider the first calculation somewhat in detail.

Potassium chlorate yields 39.154 per cent. of oxygen and 60.846 per cent. of residual chloride. For each of these quantities the probable error is ± .00038. The atomic weight of oxygen is 15.879, ± .0003, so that the value for three atoms becomes 47.637, ± .0009. We have now the following simple proportion:

$$39.154 : 60.846 :: 47.637 : x,$$

whence the molecular weight of potassium chloride becomes = 74.029.

The probable error being known for the first, second, and third term of this proportion, we can easily find that of the fourth term by the formula given in our introduction. It is ± .0073. By this method we obtain the following series of values, which may conveniently be numbered consecutively with the foregoing ratios:

(20) KCl, from (1) = 74.029, ± .0073
(21) KBr, "   (2) = 118.487, ± .0923
(22) KI,   "   (3) = 164.337, ± .0382
(23) NaCl, "   (4) = 58.057, ± .0050
(24) AgCl, "   (5) = 142.303, ± .0066
(25) AgBr, "   (6) = 186.463, ± .0137
(26) AgI, "   (7) = 232.959, ± .0134

With the help of these molecular weights, we are now able to compute seven independent values for the atomic weight of silver.

First,	from (10)	and	(20).....	Ag	= 107.111,	± .0106
Second,	"	(11), "	(21)........	"	= 107.378,	± .0837
Third,	"	(12) "	(22)........	"	= 106.921,	± .0278
Fourth,	"	(8) "	(23)........	"	= 107.102,	± .0092
Fifth,	"	(13) "	(24)........	"	= 107.122,	± .0050
Sixth,	"	(14) "	(25)........	"	= 107.113,	± .0079
Seventh,	"	(15) "	(26)........	"	= 107.091,	± .0062
			General mean..............	Ag	= 107.108,	± .0031

It is noticeable that five of these values agree very well. The second and third, however, diverge widely from the average, but in opposite directions; they have, moreover, high probable errors, and consequently little weight. Of these two, one represents little and the other none of Stas' work. Their trifling influence upon our final results becomes curiously apparent in the series of silver values given a little further along.

When we consider closely, in all of its bearings, any one of the values just given, we shall see that for certain purposes it must be excluded from our general mean. For example, the first is derived partly from the ratio between silver and potassium chloride. From this ratio, the atomic weight of one substance being known, we can deduce that of the other. We have already used it in ascertaining the atomic weight of silver, and the value thus obtained is included in our general mean. But if from it we are to determine the molecular weight of potassium chloride, we must use a silver value derived from other sources only, or we should be assuming a part of our result in advance. In other words, we must now use a general mean for silver from which this ratio with reference to silver has been rejected. Hence the following series of silver values, which are lettered for reference:

A.	General mean from all eight..........		107.108,	± .0031
B.	"	excluding the first........	107.108,	± .0032
C.	"	" second.....	107.107,	± .0031
D.	"	" third.......	107.110,	± .0031
E.	"	" fourth......	107.109,	± .0033
F.	"	" fifth........	107.099,	± .0039
G.	"	" sixth.......	107.106,	± .0034
H.	"	" seventh	107.113,	± .0036

We are now in a position to determine more closely the molecular weights of the haloid salts which we have already been considering.

For silver chloride, still employing the formula for the probable error of the last term of a proportion, we get the following values:

From (5).................... AgCl = 142.303, ± .0066
From (13) and (F)............ " = 142.276, ± .0052
From (16) " (23) .... " = 142.063, ± .0168
From (17) " (20) .... " = 142.353, ± .0156
From (18) " (25) .... " = 142.306, ± .0271
From (19) " (26) .... " = 142.278, ± .0105

General mean............. AgCl = 142.277, ± .0036

The third of these values is certainly too low, and although it reduces the atomic weight of chlorine by only 0.01, it ought to be rejected. The general mean of the other five values is AgCl = 142.287, ± .0037. Subtracting from this the atomic weight of silver, 107.108, ± .0031, we have for the atomic weight of chlorine—

$$Cl = 35.179, \pm .0048.$$

For silver bromide three ratios are available:

From (6)..................... AgBr = 186.463, ± .0137
From (14) and (G)............. " = 186.450, ± .0050
From (18) " (24) .... " = 186.459, ± .0339

General mean............. AgBr = 186.452, ± .0054

Hence, applying the atomic weight of silver as before—

$$Br = 79.344, \pm .0062.$$

For silver iodide we have—

From (7)..................... AgI = 232.959, ± .0134
From (15) and (H)............. " = 233.008, ± .0079
From (19) " (24)............. " = 232.997, ± .0153

General mean............. AgI = 232.996, ± .0062

Hence,
$$I = 125.888, \pm .0069.$$

For the molecular weight of sodium chloride three values appear, as follows:

From (4)..................... NaCl = 58.057, ± .0050
From (8) and (E)............. " = 58.061, ± .0018
From (16) " AgCl............. " = 58.148, ± .0049

General mean............. NaCl = 58.069, ± .0016

Rejecting the third value, which corresponds to the rejected value for AgCl and throws out ratio (16) entirely, the mean becomes

$$NaCl = 58.060, \pm .0017$$

From (9) and (A)................ NaBr = 102.224, ± .0031

Deducting from these molecular weights the values already found for Cl and Br, two measurements of the atomic weight of sodium are obtained, thus:

From NaCl.......................... Na = 22.881, ± .0051
From NaBr.......................... " = 22.880, ± .0112

General mean............... Na = 22.881, ± .0046

The rejection of ratio (16) in connection with the atomic weights of sodium and chlorine is fully justified by the fact that the data which it represents were never intended for use in such computations. They were obtained incidentally in connection with work upon boron, and their consideration here may have some bearing later upon the discussion of the last-named element.

For potassium, the ratios available give molecular weights for the chloride, bromide, and iodide. For the chloride,

From (1)......................... KCl = 74.029, ± .0073
From (10) and (B)................ "   = 74.027, ± .0022
From (17) " (24)................. "   = 74.003, ± .0049

General mean............... KCl = 74.025, ± .0019

For the bromide we have—

From (2)......................... KBr = 118.487, ± .0923
From (11) and (C)................ "   = 118.188, ± .0073

General mean............... KBr = 118.200, ± .0073

And for the iodide—

From (3)......................... KI = 164.337, ± .0382
From (12) and (D)................ " = 164.627, ± .0052

General mean............... KI = 164.622, ± .0051

Combining these values with those found for chlorine, bromine, and iodine, we have three values for the atomic weight of potassium, as follows:

From KCl........................ K = 38.846, ± .0078
From KBr........................ " = 38.856, ± .0096
From KI......................... " = 38.734, ± .0086

General mean............... K = 38.817, ± .0051

To sum up, the six atomic weights under discussion may be tabulated as follows, both for the standard chosen, and with O = 16 as the base of the system:

SILVER, POTASSIUM, ETC.   57

	$H = 1$.	$O = 16$.
Ag	107.108, ± .0031	107.924
K	38.817, ± .0051	39.112
Na	22.881, ± .0046	23.048
Cl	35.179, ± .0048	35.447
Br	79.344, ± .0062	79.949
I	125.888, ± .0069	126.847

It must be remembered that these values represent the summing up of work done by many investigators. Stas' ratios, taken by themselves, give various results, according to the method of combining them. This computation has been made by Stas himself, with his older determinations, and more recently by Ostwald,* Van der Plaats,† and Thomsen,‡ all with the standard of $O = 16$. By Van der Plaats two sets of results are given: one with Stas' ratios assigned equal weight (A), and the other with each ratio given weight inversely proportional to the square of its mean error (B). The results of these several computations may well be tabulated in comparison with the values obtained in my own general discussion, thus:

	Clarke.	Stas.	Ostwald.	V. der P., A.	V. der P., B.	Thomsen.
Ag	107.924	107.930	107.9376	107.9202	107.9244	107.9299
K	39.112	39.137	39.1361	39.1414	39.1403	39.1507
Na	23.048	23.043	23.0575	23.0453	23.0443	23.0543
Cl	35.447	35.457	35.4529	35.4516	35.4565	35.4494
Br	79.949	79.952	79.9628	79.9407	79.9548	79.9510
I	126.847	126.850	126.8640	126.8445	126.8494	126.8556

The agreement between the new values and the others is highly satisfactory, and gives a strong emphasis to the magnificent accuracy of Stas' determinations. No severer test could be applied to them.

---

* Lehrbuch der allgemeinen Chemie, 1, 41. 1885.
† Compt. Rend., 116, 1362. 1893.
‡ Zeitsch. Physikal. Chem., 13, 726. 1894.

NITROGEN.

The atomic weight of nitrogen has been determined from the density of the gas, and from a considerable variety of purely chemical ratios.

Upon the density of nitrogen a great many experiments have been made. In early times this constant was determined by Biot and Arago, Thomson, Dulong and Berzelius, Lavoisier, and others. But all of these investigations may be disregarded as of insufficient accuracy; and, as in the case of oxygen, we need consider only the results obtained by Dumas and Boussingault, by Regnault, and by recent investigators.

Taking air as unity, Dumas and Boussingault* found the density of nitrogen to be—

.970
.972
.974

Mean, .972, ± .00078

For hydrogen, as was seen in our discussion of the atomic weight of oxygen, the same investigators found a mean of .0693, ± .00013. Upon combining this with the above nitrogen mean, we find for the atomic weight of the latter element, $N = 14.026$, ± .0295.

By Regnault † much closer work was done. He found the density of nitrogen to be as follows:

.97148
.97148
.97154
.97155
.97108
.97108

Mean, .97137, ± .000062

For hydrogen, Regnault's mean value is .069263, ± .000019. Hence, combining as before, $N = 14.0244 ± .0039$.

Both of the preceding values are affected by a correction for the difference in volume between the weighing globes when full and when empty. This correction, in the case of Regnault's data, has been measured by Crafts,‡ who gives .06949 for the density of H, and .97138 for N. Corrected ratio, $N = 13.9787$. If we assume the same proportional correction for the determination by Dumas and Boussingault, that becomes $N = 13.9771$.

---

* Compt. Rend., 12, 1005. 1841.
† Compt. Rend., 20, 975. 1845.
‡ Compt. Rend., 106, 1664.

NITROGEN. 59

Von Jolly,* working with electrolytic oxygen and with nitrogen prepared by passing air over hot copper, but not with hydrogen, compared the weights of equal volumes of the two gases, with results as follows:

Oxygen.	Nitrogen.
1.442470	1.269609
1.442579	1.269389
1.442489	1.269307
1.442570	1.269449
1.442571	1.269515
1.442562	1.269443
1.442478	1.269478
Mean, 1.442545, ± .000013	Mean, 1.269455, ± .000024

The ratio, when $O = 16$, is $N = 14.0802$, ± .0003. Corrected by Rayleigh, the ratio between the weights becomes 14.0805. If $O = 15.879$, ± .0003, the final value for N, deducible from Von Jolly's data, is $N = 13.974$, ± .0004.

The next determination in order of time is Leduc's.† He made nine measurements of the density of nitrogen, giving a mean of .97203, with extremes of .9719 and .9721; but he neglects to cite the intermediate values. Taking the three figures given as representative, and assuming a fair distribution of the other values between the indicated limits, the probable error of the mean is not far from 0.00002. For hydrogen he found .06948, ± .00006745. The ratio between the two densities gives $N = 13.9901$, ± .0138.

Lord Rayleigh,‡ preparing nitrogen by passing air over hot copper, and weighing in a standard globe, obtained the following weights:

2.31035
2.31026
2.31024
2.31012
2.31027

Mean, 2.31025, ± 000025

With corrections for temperature, shrinkage of the globe when exhausted, etc., this becomes 2.30883, as against 2.37512 for the same volume of air. Hence the density of $N = .97209$, ± .00001. His former work on hydrogen gives .06960, ± .0000084, for the density of that gas. The ratio is $N = 13.9678$, ± .0017.

The foregoing data, however, all apply to nitrogen derived from the atmosphere. In a later memoir Rayleigh § found that nitrogen from

---

* Poggend. Annalen (2), 6, 529–530. 1879.
† Compt. Rend., 113, 186. 1891.
‡ Proc. Roy. Soc., 53, 134. 1894.
§ Chem. News, 69, 231. 1894.

chemical sources, such as oxides of nitrogen, ammonium nitrate, etc., was perceptibly lighter; and not long afterwards the discrepancy was explained by the astonishing discovery of argon. The densities given, therefore, are all too high, and unavailable for any discussion of atomic weight. As, however, the reductions had been completed in nearly all their details before the existence of argon was announced, they may be allowed to remain here as part of the record. Summing up, the ratios found between hydrogen and atmospheric "nitrogen" are as follows:

Dumas and Boussingault, corrected		13.977
Regnault,	"	13.979
Von Jolly,	"	13.974
Leduc,	"	13.990
Rayleigh,	"	13.968

Perhaps at some future time, when the density of argon is accurately known and its amount in the atmosphere has been precisely determined, these figures may be so corrected as to be useful for atomic weight calculations.

In discussing the more purely chemical ratios for establishing the atomic weight of nitrogen, we may ignore, for the present, the researches of Berzelius and of Anderson. These chemists experimented chiefly upon lead nitrate, and their work is consequently now of greater value for fixing the atomic weight of lead. Their results will be duly considered in the proper connection further on.

The ratio between ammonium chloride and silver has been determined by Pelouze, by Marignac, and by Stas. The method of working is essentially that adopted in the similar experiments with the chlorides of sodium and potassium.

For the ammonium chloride equivalent to 100 parts of silver, Pelouze* found:

$$49.556$$
$$49.517$$

Mean, $49.5365, \pm .013$

Marignac† obtained the following results. The usual ratio for 100 parts of silver is given also:

8.063 grm. Ag	=	3.992 grm. NH₄Cl.			49.510
9.402	"	4.656	"		49.521
10.339	"	5.120	"		49.521
12.497	"	6.191	"		49.540
11.337	"	5.617	"		49.546
11.307	"	5.595	"		49.483
4.326	"	2.143	"		49.538

Mean, $49.523, \pm .0055$

---

*Compt. Rend., 20, 1047. 1845.
†Berzelius' Lehrbuch, 5th ed., vol. 3, 1184, 1185.

But neither of these series can for a moment compare with that of Stas.* He used from 12.5 to 80 grammes of silver in each experiment, reduced his weighings to a vacuum standard, and adopted a great variety of precautions to insure accuracy. He found for every 100 parts of silver the following quantities of NH_4Cl:

$$\begin{aligned}
&49.600\\
&49.599\\
&49.597\\
&49.598\\
&49.597\\
&49.593\\
&49.597\\
&49.5974\\
&49.602\\
&49.597\\
&49.598\\
&49.592
\end{aligned}$$

Mean, 49.5973, ± .0005

In this work, as with the similar ratios for potassium and sodium chloride, the solubility of silver chloride was not guarded against so fully as is needful. Accordingly Stas published a new series of determinations in 1882,† carefully checked in this particular, with the subjoined values for the ratio:

$$\begin{aligned}
&49.60001\\
&49.59999\\
&49.599\\
&49.600\\
&49.597
\end{aligned}$$

Mean, 49.5992, ± .00039

Combining all four series, we have—

Pelouze	49.5365,	± .013
Marignac	49.523,	± .0055
Stas, early series	49.5973,	± .0005
Stas, later "	49.5992,	± .00039
General mean	49.5983,	± .00031

In the paper last cited Stas also gives a similar series of determinations for the ratio $Ag : NH_4Br :: 100 : x$. The results are as follows, with reduction to vacuum:

---

* Aronstein's translation, pp. 56-58.
† Mémoires Acad. Roy. de Belge., 43. 1882.

THE ATOMIC WEIGHTS.

$$
\begin{array}{r}
90.831 \\
90.831 \\
90.8297 \\
90.823 \\
90.8317 \\
90.8311 \\
90.832 \\
\hline
\end{array}
$$

Mean, 90.8299, ± .0008

The quantity of silver nitrate which can be formed from a known weight of metallic silver has been determined by Penny, by Marignac, and by Stas. Penny* dissolved silver in nitric acid in a flask, evaporated to dryness without transfer, and weighed. One hundred parts of silver thus gave of nitrate:

$$
\begin{array}{r}
157.430 \\
157.437 \\
157.458 \\
157.440 \\
157.430 \\
157.455 \\
\hline
\end{array}
$$

Mean, 157.4417, ± .0033

Marignac's† results were as follows. In the third column they are reduced to the common standard of 100 parts of silver:

68.987 grm. Ag gave	108.608 grm. $AgNO_3$.				157.433
57.844	"	91.047	"		157.401
66.436	"	104.592	"		157.433
70.340	"	110.718	"		157.404
200.000	"	314.894	"		157.447

Mean, 157.4236, ± .0061

Stas,‡ employing from 77 to 405 grammes of silver in each experiment, made two different series of determinations at two different times. The silver was dissolved with all the usual precautions against loss and against impurity, and the resulting nitrate was weighed, first after long drying without fusion, just below its melting point; and again, fused. Between the fused and the unfused salt there was in every case a slight difference in weight, the latter giving a maximum and the former a minimum value.

In Stas' first series there are eight experiments; but the seventh he himself rejects as inexact. The values obtained for the nitrate from 100

---

* Phil. Trans., 1839.
† Berzelius' Lehrbuch, 5th ed., 3, pp. 1184, 1185.
‡ Aronstein's translation, pp. 305 and 315.

parts of silver are given below in two columns, representing the two conditions in which the salt was weighed. The general mean given at the end I have deduced from the means of the two columns considered separately:

Unfused.	Fused.
157.492	157.474
157.510	157.481
157.485	157.477
157.476	157.471
157.478	157.470
157.471	157.463
157.488	157.469
Mean, 157.4857	Mean, 157.472

General mean, 157.474, ± .0014

In the later series there are but two experiments, as follows:

Unfused.	Fused.
157.4964	157.488
157.4940	157.480
Mean, 157.4952	Mean, 157.484

General mean, 157.486, ± .0003

The reverse ratio, namely, the amount of silver obtainable from a weighed quantity of nitrate, has been determined electrolytically by Hardin.* The data obtained, however, are reducible to the same form as in the preceding series, and all are properly combinable together. Pure silver was dissolved in pure aqueous nitric acid, and the crystalline salt thus formed was dried, fused, and used for the determinations. The silver nitrate, mixed with an excess of pure potassium cyanide solution, was electrolyzed in a platinum dish. The results obtained, reduced to vacuum weights, were as follows:

.31202 $AgNO_3$ gave	.19812 Ag.	Ratio,	157.490
.47832 "	.30370 "	"	157.498
.56742 "	.36030 "	"	157.485
.57728 "	.36655 "	"	157.490
.69409 "	.44075 "	"	157.479
.86367 "	.54843 "	"	157.479
.86811 "	.55130 "	"	157.466
.93716 "	.59508 "	"	157.485
1.06170 "	.67412 "	"	157.494
1.19849 "	.76104 "	"	157.477

Mean, 157.484, ± .0020

---

* Journ. Amer. Chem. Soc., 18, 995. 1896.

Now, to combine all five sets of results:

Penny	157.4417,	± .0033
Marignac	157.4236,	± .0061
Stas, 1st series	157.4740,	± .0014
Stas, 2d "	157.4860,	± .0003
Hardin	157.484,	± .0020
General mean	157.479,	± .0003

For the direct ratio between silver nitrate and silver chloride there are two series of estimations. A weighed quantity of nitrate is easily converted into chloride, and the weight of the latter ascertained. In two experiments Turner* found of chloride from 100 parts of nitrate:

84.357
84.389

Mean, 84.373, ± .011

Penny,† in five determinations, found the following percentages:

84.370
84.388
84.377
84.367
84.370

Mean, 84.3744, ± .0025

The general mean from both series is 84.3743, ± .0025.

The ratio directly connecting silver nitrate with ammonium chloride has been determined only by Stas. ‡ The usual method of working was followed, namely, nearly equivalent quantities of the two salts were weighed out, the solutions mixed, and the slight excess of one estimated by titration. In four experiments 100 parts of silver nitrate were found equivalent to chloride of ammonium, as follows:

31.489
31.490
31.487
31.486

Mean, 31.488, ± .0006

The similar ratio between potassium chloride and silver nitrate has been determined by both Marignac and Stas.

---

* Phil. Trans., 1833, 537.
† Phil. Trans., 1839.
‡ Aronstein's translation, p. 309.

NITROGEN.

Marignac* gives the following weights. I add the quantity of KCl proportional to 100 parts of $AgNO_3$:

1.849 grm. KCl =	4.218 grm. $AgNO_3$.	43.836
2.473 "	5.640 "	43.848
3.317 "	7.565 "	43.847
2.926 "	6.670 "	43.868
6.191 "	14.110 "	43.877
4.351 "	9.918 "	43.870

Mean, 43.858, ± .0044

Stas'† results are given in three series, representing silver nitrate from three different sources. In the third series the nitrate was weighed in vacuo, while for the other series this correction was applied in the usual way. For the KCl equivalent to 100 parts of $AgNO_3$ Stas found:

First Series.
43.878
43.875
43.875
43.874

Mean, 43.8755, ± .0005.

Second Series.
43.864
43.869
43.876

Mean, 43.8697, ± .0023.

Third Series.
43.894
43.878
43.885

Mean, 43.8857, ± .0031

Combining all four series we have:

Marignac	43.858,	± .0044
Stas, 1st series	43.8755,	± .0005
Stas, 2d "	43.8697,	± .0023
Stas, 3d "	43.8857,	± .0031
General mean	43.8715,	± .0004

There have also been determined by Penny, by Stas, and by Hibbs a series of ratios connecting the alkaline chlorides and chlorates with the corresponding nitrates. One of these, relating to the lithium salts, will be studied farther on with reference to that metal.

---

* Berzelius' Lehrbuch, 5th ed., 3d vol., 1184, 1185.
† Aronstein's translation, p. 308.

The general method of working upon these ratios is due to Penny.* Applied to the ratio between the chloride and nitrate of potassium, it is as follows: A weighed quantity of the chloride is introduced into a flask which is placed upon its side and connected with a receiver. An excess of pure nitric acid is added, and the transformation is gradually brought about by the aid of heat. Then, upon evaporating to dryness over a sand bath, the nitrate is brought into weighable form. The liquid in the receiver is also evaporated, and the trace of solid matter which had been mechanically carried over is recovered and also taken into account. In another series of experiments the nitrate was taken, and by pure hydrochloric acid converted into chloride, the process being the same. In the following columns of figures I have reduced both series to one standard, namely, so as to express the number of parts of nitrate corresponding to 100 of chloride:

First Series.—KCl treated with HNO_3.

135.639
135.637
135.640
135.635
135.630
135.640
135.630

Mean, 135.636, ± .0011

Second Series.—KNO_3 treated with HCl.

135.628
135.635
135.630
135.641
135.630
135.635
135.630

Mean, 135.633, ± .0011

Stas'† results are as follows:

135.643
135.638
135.647
135.649
135.640
135.645
135.655

Mean, 135.6453, ± .0014

---
* Phil. Trans., 1839.
† Aronstein's translation, p. 270.

These figures by Stas represent weighings in the air. Reduced to a vacuum standard, this mean becomes 135.6423.

The determinations made by Hibbs* differ slightly in method from those of Penny and Stas. He converted the nitrate into the chloride by heating in a stream of gaseous hydrochloric acid. His results were as follows, vacuum weights being given·

Weight KNO_3	Weight KCl.	Ratio.
.11090	.08177	135.624
.14871	.10965	135.622
.21067	.15533	135.627
.23360	.17225	135.620
.24284	.17903	135.642
		Mean, 135.627, ± .0026

Now, combining, we have:

Penny, 1st series	135.636,	± .0011
Penny, 2d "	135.633,	± .0011
Stas	135.6423,	± .0014
Hibbs	135.627,	± .0026
General mean.	135.636,	± .0007

By the same general process Penny † determined how much potassium nitrate could be formed from 100 parts of chlorate. He found as follows:

82.505
82.497
82.498
82.500

Mean, 82.500, ± .0012

For 100 parts of sodium chlorate he found of nitrate:

79.875
79.882
79.890

Mean, 79.8823, ± .0029

For the ratio between the chloride and nitrate of sodium Penny made two sets of estimations, as in the case of potassium salts. The subjoined figures give the amount of nitrate equivalent to 100 parts of chloride:

---

* Thesis for Doctor's degree, University of Pennsylvania, 1896. Work done under the direction of Professor E. F. Smith.
† Phil. Trans., 1839.

First Series.—NaCl treated with HNO_3.

$$145.415$$
$$145.408$$
$$145.420$$
$$145.424$$
$$145.410$$
$$145.418$$
$$145.420$$

Mean, 145.4164, ± .0015

Second Series.—$NaNO_3$ treated with HCl.

$$145.419$$
$$145.391$$
$$145.412$$
$$145.415$$
$$145.412$$
$$145.412$$

Mean, 145.410, ± .0026

Stas[*] gives the following series:

$$145.453$$
$$145.468$$
$$145.465$$
$$145.469$$
$$145.443$$

Mean, after reducing to vacuum standard, 145.4526, ± .0030

Hibbs'[†] data, obtained by the method employed in the case of the potassium compounds, are as follows, vacuum weights being stated:

Weight $NaNO_3$.	Weight NaCl.	Ratio.
.01550	.01066	145.403
.20976	.14426	145.404
.26229	.18038	145.410
.66645	.45829	145.429
.93718	.64456	145.399

Mean, 145.407, ± .0026

Combining, we have as follows:

Penny, 1st series........................	145.4164, ± .0015
Penny, 2d "	145.410, ± .0026
Stas.....................................	145.4526, ± .0030
Hibbs...................................	145.407, ± .0026
General mean...................	145.418, ± .0012

[*] Aronstein's translation, p. 278.
[†] Thesis, University of Pennsylvania, 1896.

Julius Thomsen,* for the purpose of fixing indirectly the ratio H : O, has made a valuable series of determinations of the ratio $HCl : NH_3$, which may properly be used toward establishing the atomic weight of nitrogen. First, pure, dry, gaseous hydrochloric acid is passed into a weighed absorption apparatus containing pure distilled water. After noting the increase in weight, pure ammonia gas is passed in until a very slight excess is present, and the apparatus is weighed again. The excess of NH_3, which is always minute, is measured by titration with standard hydrochloric acid. In weighing, the apparatus is tared by one of similar form, and containing about the same amount of water. Three series of determinations were made, differing only in the size of the absorption apparatus; so that for present purposes the three may be taken as one. Thomsen considers them separately, and so gives greatest weight to the experiments involving the largest masses of material. I give his weighings, and also, as computed by him, the ratio $\frac{HCl}{NH_3}$.

	HCl.	NH_3.	Ratio.
First series,	5.1624	2.4120	2.1403
	3.9425	1.8409	2.1416
	4.6544	2.1739	2.1411
	3.9840	1.8609	2.1409
	5.3295	2.4898	2.1406
	4.2517	1.9863	2.1405
	4.8287	2.2550	2.1414
	6.4377	3.0068	2.1411
	4.1804	1.9528	2.1407
	5.0363	2.3523	2.1410
	4.6408	2.1685	2.1411
Second series,	11.8418	5.5302	2.14130
	14.3018	6.6808	2.14073
	12.1502	5.6759	2.14067
	11.5443	5.3927	2.14073
	12.3617	5.7733	2.14118
Third series,	19.3455	9.0360	2.14094
	19.4578	9.0890	2.14081

Mean of all, 2.14093, ± .000053
Reduced to vacuo, 2.1394

From the sums of the weights Thomsen finds the ratio to be 2.14087, or 2.13934 in vacuo. From this, using Ostwald's reductions of Stas' data for the atomic weights of N and Cl, he finds the atomic weight of H = 0.99946, when O = 16.

We have now, apart from the determinations of gaseous density, eleven ratios, representing one hundred and sixty-four experiments, from which

---

* Zeitsch. Physikal. Chem., 13, 398. 1894.

to calculate the atomic weight of nitrogen. Let us first collect and number these ratios:

(1.) Ag : AgNO₃ : : 100 : 157.479, ± .0003
(2.) AgNO₃ : AgCl : : 100 : 84.3743, ± .0025
(3.) AgNO₃ : KCl : : 100 : 43.8715, ± .0004
(4.) AgNO₃ : NH₄Cl : : 100 : 31.488, ± .0006
(5.) Ag : NH₄Cl : : 100 : 49.5983, ± .00031
(6.) Ag : NH₄Br : : 100 : 90.8299, ± .0008
(7.) KCl : KNO₃ : : 100 : 135.636, ± .0007
(8.) KClO₃ : KNO₃ : : 100 : 82.500, ± .0012
(9.) NaCl : NaNO₃ : : 100 : 145 418, ± .0011
(10.) NaClO₃ : NaNO₃ : : 100 : 79.8823, ± .0029
(11.) NH₃ : HCl : : 1.00 : 2.1394, ± .000053

From these ratios we are now able to deduce the molecular weight of ammonium chloride, ammonium bromide, and three nitrates. For these calculations we must use the already ascertained atomic weights of oxygen, silver, chlorine, bromine, sodium and potassium, and the molecular weights of sodium chloride, potassium chloride, and silver chloride. The following are the antecedent values to be employed:

Ag   = 107.108, ± .0031
K   = 38.817, ± .0051
Na   = 22.881, ± .0046
Cl   = 35.179, ± .0048
Br   = 79.344, ± .0062
O₃   = 47.637, ± .0009
AgCl = 142.287, ± .0037
KCl = 74.025, ± .0019
NaCl = 58.060, ± .0017

Now, from ratio number five we get the molecular weight of NH₄Cl = 53.124, ± .0016, and N = 13.945, ± .0051.

From ratio number six, NH₄Br = 97.286, ± .0029, and N = 13.942, ± .0077.

From ratio number eleven, NH₃ = 16.911, ± .0048, and N = 13.911, ± .0048.

From ratio number four, which involves an expression of the type $A : B :: C + x : D + x$, an independent value is deducible, N = 13.935, ± .0073.

For the molecular weight of silver nitrate there are three values, namely :

From (1)............... AgNO₃ = 168.673, ± .0049
From (2)...............   "   = 168.634, ± .0066
From (3) ....   "   = 168.731, ± .0046
General mean...........AgNO₃ = 168.690, ± .0030

Hence N = 13.945, ± .0044.

The molecular weight of potassium nitrate is twice calculable, as follows:

From (7).................... $KNO_3 = 100.405, \pm .0026$
From (8).................... " $= 100.371, \pm .0059$

General mean........... $KNO_3 = 100.401, \pm .0024$

Hence $N = 13.947, \pm .0057$.
And for sodium nitrate we have:

From (9).................... $NaNO_3 = 84.430, \pm .0026$
From (10)................... " $= 84.433, \pm .0053$

General mean........... $NaNO_3 = 84.431, \pm .0023$

Hence $N = 13.913, \pm .0052$.

There are now seven estimates of the atomic weight of nitrogen, to be combined by means of the usual formula.

1. From NH_4Cl................... $N = 13.945, \pm .0051$
2. " NH_4Br................... " $= 13.942, \pm .0077$
3. " ratio (4).................. " $= 13.935, \pm .0073$
4. " " (11).................. " $= 13.911, \pm .0048$
5. " $AgNO_3$................... " $= 13.945, \pm .0044$
6. " KNO_3.................... " $= 13.947, \pm .0057$
7. " $NaNO_3$................... " $= 13.913, \pm .0052$

General mean............... $N = 13.935, \pm .0021$

If oxygen is 16, this becomes 14.041. From Stas' data alone, Stas finds 14.044; Ostwald, 14.0410; Van der Plaats, 14.0421 (A), and 14.0519 (B); and Thomsen, 14.0396. The new value, representing all available data, falls between these limits of variation.

CARBON.

Although there is a large mass of material relating to the atomic weight of carbon, much of it may be summarily set aside as having no value for present purposes. The density of carbon dioxide, which has been scrupulously determined by many investigators,* leads to no safe estimate of the constant under consideration. The numerous analyses of hydrocarbons, like the analyses of naphthalene by Mitscherlich, Woskresensky, Fownes, and Dumas, give results scarcely more satisfactory. In short, all the work done upon the atomic weight of carbon before the year 1840 may be safely rejected as unsuited to the present requirements of exact science. As for methods of estimation we need consider but four, as follows:

First. The analysis of organic salts of silver.
Second. The determination of the weight of carbon dioxide formed by the combustion of a known weight of carbon.
Third. The method of Stas, by the combustion of carbon monoxide.
Fourth. From the density of carbon monoxide.

The first of these methods, which is probably the least accurate, was employed by Liebig and Redtenbacher † in 1840. They worked with the acetate, tartrate, racemate, and malate of silver, making five ignitions of each salt, and determining the percentage of metal. From one to nine grammes of material were used in each experiment.

In the acetate the following percentages of silver were found:

$$64.615$$
$$64.624$$
$$64.623$$
$$64.614$$
$$64.610$$

Mean, $64.6172, \pm .0018$

After applying corrections for weighing in air, this mean becomes 64.6065.

In the tartrate the silver came out as follows:

$$59.297$$
$$59.299$$
$$59.287$$
$$59.293$$
$$59.293$$

Mean, $59.2938, \pm .0014$
Or, reduced to a vacuum, 59.2806

---

* Notably by Lavoisier, Biot and Arago, De Saussure, Dulong and Berzelius, Buff, Von Wrede, Regnault, and Marchand. For details, Van Geun's monograph may be consulted.
† Ann. Chem. Pharm., 38, 137. Mem. Chem. Soc., 1, 9. Phil. Mag. (3), 19, 210.

In the racemate we have:

59.290
59.292
59.287
59.283
59.284

Mean, 59.2872, ± .0012
Or, corrected, 59.2769

And from the malate:

61.996
61.972
62.015
62.059
62.011

Mean, 62.0106, ± .0096
Or, corrected, 62.0016

Now, applying to these mean results the atomic weights already found for oxygen and silver, we get the following values for carbon:

From the acetate.................. $C = 11.959$, ± .0021
From the tartrate .... " $= 11.967$, ± .0019
From the racemate................. " $= 11.973$, ± .0017
From the malate................... " $= 11.972$, ± .0098

Now these results, although remarkably concordant, are by no means unimpeachable. They involve two possible sources of constant error, namely, impurity of material and the volatility of the silver. These objections have both been raised by Stas, who found that the silver tartrate, prepared as Liebig and Redtenbacher prepared it, always carried traces of the nitrate, and that he, by the ignition of that salt, could not get results at all agreeing with theirs. In the case of the acetate a similar impurity would lower the percentage of silver, and thus both sources of error would reinforce each other and make the atomic weight of carbon come out too high. With the three other salts the two sources of error act in opposite directions, although the volatility of the silver is probably far greater in its influence than the impurity. Even if we had no other data relating to the atomic weight of carbon, it would be clear from these facts that the results obtained by Liebig and Redtenbacher must be decidedly in excess of the true figure.

Strecker,* however, discussed the data given by Liebig and Redtenbacher by the method of least squares, using the Berzelian scale, and assuming $H = 12.51$. Thus treated, they gave $C = 75.415$, and $Ag = 1348.79$; or, with $O = 16$, $C = 12.066$ and $Ag = 107.903$. These values

---

* Ann. Chem. Pharm., 59, 280. 1846.

of course would change somewhat upon adoption of the modern ratio between O and H.

Observations upon silver acetate, like those of Liebig and Redtenbacher, were also made by Marignac.* The salt was prepared by dissolving silver carbonate in acetic acid, and repeatedly recrystallizing. Two experiments gave as follows:

3.3359 grm. acetate gave 2.1561 Ag.	64.633 per cent.
3.0527 " 1.9727 "	64.621 "

Mean, 64.627, ± .0040

Reduced to a vacuum, this becomes 64.609.

In a second series, conducted with special precautions to avoid mechanical loss by spurting, Marignac found:

24.717 grm. acetate gave 15.983 Ag.	64.665 per cent.
21.202 " 13.709 "	64.661 "
31.734 " 20.521 "	64.666 "

Mean, 64.664, ± .0010
Or, reduced to a vacuum, 64.646

Other experiments, comparable with the preceding series, have recently been published by Hardin,† who sought to redetermine the atomic weight of silver. Silver acetate and silver benzoate, carefully purified, were subjected to electrolysis in a platinum dish, and the percentage of silver so determined. For the acetate, using vacuum weights, he gives the following data, the percentage column being added by myself:

.32470 grm. acetate gave .20987 Ag.	64.635 per cent.
.40566 " .26223 "	64.643 "
.52736 " .34086 "	64.635 "
.60300 " .38976 "	64.637 "
.67235 " .43455 "	64.631 "
.72452 " .46830 "	64.636 "
.78232 " .50563 "	64.632 "
.79804 " .51590 "	64.646 "
.92101 " .59532 "	64.638 "
1.02495 " .66250 "	64.637 "

Mean, 64.637, ± .0011

Combining this series with those of the earlier investigators we have:

Liebig and Redtenbacher	64.6065,	± .0018
Marignac, 1st series	64.609,	± .0040
Marignac, 2d "	64.646,	± .0010
Hardin	64.637,	± .0011
General mean	64.636,	± .0007

---

* Ann. Chem. Pharm., 59, 287. 1846.
† Journ. Amer. Chem. Soc., 18, 990. 1896.

With silver benzoate, $C_7H_5AgO_2$, Hardin's results are as follows:

.40858 grm. benzoate gave	.19255 Ag.	47.127 per cent.
.46674 "	.21999 "	47.133 "
.48419 "	.22815 "	47.120 "
.62432 "	.29418 "	47.120 "
.66496 "	.31340 "	47.131 "
.75853 "	.35745 "	47.124 "
.76918 "	.36247 "	47.124 "
.81254 "	.38286 "	47.119 "
.95673 "	.45079 "	47.118 "
1.00840 "	.47526 "	47.130 "

Mean, 47.125, ± .0012

A different method of dealing with organic silver salts was adopted by Maumené,* in 1846, for the purpose of establishing by reference to carbon the atomic weight of silver. We will simply reverse his results and apply them to the atomic weight of carbon. He effected the combustion of the acetate and the oxalate of silver, and, by weighing both the residual metal and the carbon dioxide formed, he fixed the ratio between these two substances. In the case of the acetate his weighings show that for every gramme of metallic silver the weights of CO_2 were produced which are shown in the third column:

8.083 grm. Ag =	6.585 grm. CO_2.	.8147
11.215 "	9.135 "	.8136
14.351 "	11.6935 "	.8148
9.030 "	7.358 "	.8148
20.227 "	16.475 "	.8145

Mean, .81448

The oxalate of silver, ignited by itself, decomposes too violently to give good results; and for this reason it was not used by Liebig and Redtenbacher. Maumené, however, found that when the salt was mixed with sand the combustion could be tranquilly effected. The oxalate employed, however, with the exception of the sample represented in the last experiment of the series, contained traces of nitrate, so that these results involve slight errors. For each gramme of silver the appended weights of CO_2 were obtained:

14.299 grm. Ag. =	5.835 grm. CO_2.	.4081
17.754 "	7.217 "	.4059
11.550 "	4.703 "	.4072
10.771 "	4.387 "	.4073
8.674 "	3.533 "	.4073
11.4355 "	4.658 "	.4073

Mean, .40718

---

*Ann. Chim. Phys. (3), 18, 41. 1846.

Now, one of these salts being formed by a bivalent and the other by a univalent acid, we have to reduce both to a common standard. Doing this, we have the following results for the ratio between the atomic weight of silver and the molecular weight of CO_2; if $Ag = 1.00$:

From the acetate.................$CO_2 = .40724, \pm .000076$
From the oxalate................." $= .40718, \pm .000185$

General mean..............$CO_2 = .40723, \pm .000071$

Here the slight error due to the impurity of the oxalate becomes of such trifling weight that it practically vanishes.

As has already been said, the volatility of silver renders all the foregoing results more or less uncertain. Far better figures are furnished by the combustion of carbon directly, as carried out by Dumas and Stas[*] in 1840 and by Erdmann and Marchand[†] in 1841. In both investigations weighed quantities of diamond, of natural graphite, and of artificial graphite were burned in oxygen, and the amount of dioxide produced was estimated by the usual methods. The graphite employed was purified with extreme care by treatment with strong nitric acid and by fusion with caustic alkali. I have reduced all the published weighings to a common standard, so as to show in the third column the amount of oxygen which combines with a unit weight (say one gramme) of carbon. Taking Dumas and Stas' results first in order, we have from natural graphite:

1.000 grm. C gave	3.671 grm. CO_2.		2.6710
.998 "	3.660 "		2.6673
.994 "	3.645 "		2.6670
1.216 "	4.461 "		2.6686
1.471 "	5.395 "		2.6676

Mean, $2.6683, \pm .0005$

With artificial graphite :

.992 grm. C gave	3.642 grm. CO_2.		2.6714
.998 "	3.662 "		2.6682
1.660 "	6.085 "		2.6654
1.465 "	5.365 "		2.6744

Mean, $2.66985, \pm .0013$

And with diamond:

.708 grm. C gave	2.598 grm. CO_2.		2.6695
.864 "	3.1675 "		2.6661
1.219 "	4.465 "		2.6628
1.232 "	4.519 "		2.6680
1.375 "	5.041 "		2.6662

Mean, $5.6665 \pm .0007$

---

[*] Compt. Rend., 11, 991-1008. Ann. Chim. Phys. (3), 1, 1.
[†] Jour. f Prakt. Chem., 23, 159.

CARBON. 77

Erdmann and Marchand's figures for natural graphite give the following results:

1.5376 grm.	gave 5.6367 grm. CO_2.	2.6659
1.6494	" 6.0384 "	2.6609
1.4505	" 5.31575 "	2.6647

In one experiment 1.8935 grm. of artificial graphite gave 6.9355 grm. CO_2. Ratio for O, 2.6628. This, combined with the foregoing series, gives a mean of 2.6636, ± .0007.
With the diamond they found:

.8052 grm.	gave 2.9467 grm. CO_2.	2.6596
1.0858	" 3.9875 "	2.6632
1.3557	" 4.9659 "	2.6629
1.6305	" 5.97945 "	2.6673
.7500	" 2.7490 "	2.6653

Mean, 2.6637, ± .0009

In more recent years the ratio under consideration has been carefully redetermined by Roscoe, by Friedel, and by Van der Plaats. Roscoe* made use of transparent Cape diamonds, and in a sixth experiment he burned carbonado. The combustions were effected in a platinum boat, contained in a tube of glazed Berlin porcelain; and in each case the ash was weighed and its weight deducted from that of the diamond. The results were as follows, with the ratios stated as in the preceding series:

1.2820 grm.	C gave 4.7006 CO_2.	2.6666
1.1254	" 4.1245 "	2.6649
1.5287	" 5.6050 "	2.6665
.7112	" 2.6070 "	2.6656
1.3842	" 5.0765 "	2.6675
.4091	" 1.4978 "	2.6612

Mean, 2.6654, ± .0006

Friedel's work,† also upon Cape diamond, was in all essential particulars like Roscoe's. The data, after deduction of ash, were as follows:

.4705 grm.	C gave 1.7208 CO_2.	2.6628
.8616	" 3.1577 "	2.6640

Mean, 2.6634, ± .0004

By Van der Plaats‡ we have six experiments, numbers one to three on graphite, numbers four and five on sugar charcoal, and number six on charcoal made from purified filter paper. Each variety of carbon was submitted to elaborate processes of purification, and all weights were

* Ann. Chim. Phys. (5), 26, 136. Zeit. Anal. Chem., 22, 306. 1883. Compt. Rend., 94, 1180. 1882.
† Bull. Soc. Chim., 42, 100, 1884.
‡ Compt. Rend., 100, 52. 1885.

reduced to vacuum standards. The data, with ash deducted, are subjoined:

1.	5.1217 grm. C gave	18.7780	CO$_2$.		2.6664
2.	9.0532	"	33.1931	"	2.6664
3.	13.0285	"	47.7661	"	2.6663
4.	11.7352	"	43.0210	"	2.6660
5.	19.1335	"	70.1336	"	2.6655
6.	4.4017	"	16.1352	"	2.6657

Mean, 2.6660, ± .0001

This combines with the previous series thus:

Dumas and Stas, first set............	2.6683,	± .0005
Dumas and Stas, second set............	2.66985,	± .0013
Dumas and Stas, third set............	2.6665,	± .0007
Erdmann and Marchand, first set........	2.6636,	± .0007
Erdmann and Marchand, second set.......	2.6637,	± .0009
Roscoe.................................	2.6654,	± .0006
Friedel................................	2.6634,	± .0004
Van der Plaats.........................	2.6660,	± .0001
General mean..........................	2.6659,	± .0001

Another very exact method for determining the atomic weight of carbon was employed by Stas* in 1849. Carefully purified carbon monoxide was passed over a known weight of copper oxide at a red heat, and both the residual metal and the carbon dioxide formed were weighed. The weighings were reduced to a vacuum standard, and in each experiment a quantity of copper oxide was taken representing from eight to twenty-four grammes of oxygen. The method, as will at once be seen, is in all essential features similar to that usually employed for determining the composition of water. The figures in the third column, deduced from the weights given by Stas, represent the quantity of carbon monoxide corresponding to one gramme of oxygen:

9.265 grm. O =	25.483	CO$_2$.		1.75046
8.327	"	22.900	"	1.75010
13.9438	"	38.351	"	1.75040
11.6124	"	31.935	"	1.75008
18.763	"	51.6055	"	1.75039
19.581	"	53.8465	"	1.74994
22.515	"	61.926	"	1.75043
24.360	"	67.003	"	1.75053

Mean, 1.75029, ± .00005

For the density of carbon monoxide the determinations made by Leduc† are available. The globe used contained 2.9440 grm. of air.

---

* Bull. Acad. Bruxelles, 1849 (1), 31.
† Compt. Rend., 115, 1072. 1893.

Filled with CO, it held the following weights, which give the accompanying densities:

Wt. CO.	Density.
2.8470	.96705
2.8468	.96698
2.8469	.96702

Mean, .96702, ± .000015

Combining this density with Leduc's determination of the density of hydrogen, 0.6948, ± .00006745, it gives for the atomic weight of carbon:

$$C = 11.957, \pm .0270.$$

Leduc himself combines the data with the density of oxygen, taken as 1.10503, and finds $C = 11.913$. In either case, however, the probable error of the result is so high that it can carry little weight in the final combination.

For carbon, including all the foregoing series, we now have the subjoined ratios:

(1.) Per cent. Ag in silver acetate.... 64.636, ± .0007
(2.) " " tartrate.... 59.2806, ± .0014
(3.) " " racemate.. 59.2769, ± .0012
(4.) " " malate.... 62.0016, ± .0096
(5.) " " benzoate... 47.125, ± .0012
(6.) Ag : CO_2 :: 1.00 : 0.40723, ± .000071
(7.) C : O_2 :: 1.00 : 2.6659, ± .0001
(8.) O : CO :: 1.00 : 1.75029, ± .00005
(9.) Density of CO (air = 1), 0.96702, ± .000015

Now, computing with $O = 15.879$, ± .0003, and $Ag = 107.108$, ± .0031, we get nine values for the atomic weight of carbon, as follows:

From (1)............................ $C = 11.921$, ± .0012
From (2)............................ " $= 11.967$, ± .0019
From (3)............................ " $= 11.973$, ± .0017
From (4)............................ " $= 11.972$, ± .0098
From (5)............................ " $= 11.917$, ± .0008
From (6)............................ " $= 11.860$, ± .0077
From (7)............................ " $= 11.913$, ± .0006
From (8)............................ " $= 11.914$, ± .0010
From (9)............................ " $= 11.957$, ± .0270

General mean................ $C = 11.920$, ± .0004

If $O = 16$, this becomes $C = 12.011$.

SULPHUR.

The atomic weight of sulphur has been determined by means of four ratios connecting it with silver, chlorine, oxygen, sodium and carbon. Other ratios have also been considered, but they are hardly applicable here. The earlier results of Berzelius were wholly inaccurate, and his later experiments upon the synthesis of lead sulphate will be used in discussing the atomic weight of lead. Erdmann and Marchand determined the amount of calcium sulphate which could be formed from a known weight of pure Iceland spar; and later they made analyses of cinnabar, in order to fix the value of sulphur by reference to calcium and to mercury. Their results will be applied in this discussion toward ascertaining the atomic weights of the metals just named.

First in order let us take up the composition of silver sulphide, as directly determined by Dumas, Stas, and Cooke. Dumas'[*] experiments were made with sulphur which had been thrice distilled and twice crystallized from carbon disulphide. A known weight of silver was heated in a tube in the vapor of the sulphur, the excess of the latter was distilled away in a current of carbon dioxide, and the resulting silver sulphide was weighed.

I subjoin Dumas' weighings, and also the quantity of Ag_2S proportional to 100 parts of Ag, as deduced from them:

9.9393 grm. Ag = 1.473 S.	Ratio, 114.820
9.962 "	1.4755 "	" 114.811
30.637 "	4.546 "	" 114.838
30.936 "	4.586 "	" 114.824
30.720 "	4.554 "	" 114.824

Mean, 114.8234, ± .0029

Dumas used from ten to thirty grammes of silver in each experiment. Stas,[†] however, in his work employed from sixty to two hundred and fifty grammes at a time. Three of Stas' determinations were made by Dumas' method, while in the other two the sulphur was replaced by pure sulphuretted hydrogen. In all cases the excess of sulphur was expelled by carbon dioxide, purified with scrupulous care. Impurities in the dioxide may cause serious error. The five results come out as follows for 100 parts of silver:

114.854
114.853
114.854
114.851
114.849

Mean, 114.8522, ± .0007

---

[*] Ann. Chem. Pharm., 113, 24. 1860.
[†] Aronstein's translation, p. 179.

The experiments made by Professor Cooke* with reference to this ratio were only incidental to his elaborate researches upon the atomic weight of antimony. They are interesting, however, for two reasons: they serve to illustrate the volatility of silver, and they represent, not syntheses, but reductions of the sulphide by hydrogen. Cooke gives three series of results. In the first the silver sulphide was long heated to full redness in a current of hydrogen. Highly concordant and at the same time plainly erroneous figures were obtained, the error being eventually traced to the fact that some of the reduced silver, although not heated to its melting point, was actually volatilized and lost. The second series, from reductions at low redness, are decidedly better. In the third series the sulphide was fully reduced below a visible red heat. Rejecting the first series, we have from Cooke's figures in the other two the subjoined quantities of sulphide corresponding to 100 parts of silver:

7.5411 grm. Ag_2S lost .9773 grm. S.	Ratio,	114.889
5.0364 " .6524 "	"	114.882
2.5815 " .3345 "	"	114.886
2.6130 " .3387 "	"	114.892
2.5724 " .3334 "	"	114.891

Mean, 114.888, ± .0012

1.1357 grm. Ag_2S lost .1465 S.	Ratio,	114.810
1.2936 " .1670 "	"	114.823

Mean, 114.8165, ± .0044

Now, combining all four series, we get the following results:

Dumas	114.8234, ± .0029
Stas	114.8522, ± .0007
Cooke's 2d	114.888, ± .0012
Cooke's 3d	114.8165, ± .0044
General mean	114.8581, ± .0006

Here again we encounter a curious and instructive compensation of errors, and another evidence of the accuracy of Stas.

The percentage of silver in silver sulphate has been determined by Struve and by Stas. Struve† reduced the sulphate by heating in a current of hydrogen, and obtained these results:

5.1860 grm. Ag_2SO_4 gave 3.5910 grm. Ag.	69.244 per cent.
6.0543 " 4.1922 "	69.243 "
8.6465 " 5.9858 "	69.228 "
11.6460 " 8.0608 "	69.215 "
9.1090 " 6.3045 "	69.212 "
9.0669 " 6.2778 "	69.239 "

Mean, 69.230, ± .004

---

* Proc. Amer. Acad. of Arts of Sciences, vol. 12. 1877.
† Ann. Chem. Pharm., 80, 203. 1851.

Stas,* working by essentially the same method, with from 56 to 83 grammes of sulphate at a time, found these percentages:

$$69.200$$
$$69.197$$
$$69.204$$
$$69.209$$
$$69.207$$
$$69.202$$

Mean, 69.203, ± .0012

Combining this mean with that from Struve's series, we get a general mean of 69.205, ± .0011.

The third sulphur ratio with which we have now to deal is one of minor importance. When silver chloride is heated in a current of sulphuretted hydrogen the sulphide is formed. This reaction was applied by Berzelius † to determining the atomic weight of sulphur. He gives the results of four experiments; but the fourth varies so widely from the others that I have rejected it. I have reason to believe that the variation is due, not to error in experiment, but to error in printing; nevertheless, as I am unable to track out the cause of the mistake, I must exclude the figures involving it entirely from our discussion.

The three available experiments, however, give the following results: The last column contains the ratio of silver sulphide to 100 parts of chloride.

6.6075 grm. AgCl gave	5.715 grm. Ag$_2$S.	86.478
9.2323 "	7.98325 "	86.471
10.1775 "	8.80075 "	86.472

Mean, 86.4737, ± .0015

We have also a single determination of this value by Svanberg and Struve.‡ After converting the chloride into sulphide they dissolved the latter in nitric acid. A trifling residue of chloride, which had been enclosed in sulphide, and so protected against change, was left undissolved. Hence a slight constant error probably affects this whole ratio. The experiment of Svanberg and Struve gave 86.472 per cent. of silver sulphide derived from 100 of chloride. If we assign this figure equal weight with the results of Berzelius, and combine, we get a general mean of 86.4733, ± .0011.

The work done by Richards § relative to the atomic weight of sulphur is of a different order from any of the preceding determinations. Sodium carbonate was converted into sodium sulphate, fixing the ratio Na_2CO_3 : Na_2SO_4 :: $100 : x$. The data are as follows, with vacuum weights:

---

* Aronstein's translation, pp. 214-218.
† Berzelius' Lehrbuch, 5th ed., vol. 3, p. 1187.
‡ Journ. Prakt. Chem., 44, 320. 1848.
§ Proc. Amer. Acad., 26, 268. 1891.

SULPHUR. 83

Na_2CO_3.	Na_2SO_4.	Ratio.
1.29930	1.74113	134.005
3.18620	4.26790	133.950
1.01750	1.36330	133.985
2.07680	2.78260	133.985
1.22427	1.63994	133.952
1.77953	2.38465	134.005
2.04412	2.73920	134.004
3.06140	4.10220	133.997

Mean, 133.985, ± .0055

The available ratios for sulphur are now as follows:

(1.) $Ag_2 : Ag_2S :: 100 : 114.8581$, ± .0006
(2.) Per cent. Ag in Ag_2SO_4, 69.205, ± .0011
(3.) $2\ AgCl : Ag_2S :: 100 : 86.4733$, ± .0011
(4.) $Na_2CO_3 : Na_2SO_4 :: 100 : 133.985$, ± .0055

From these ratios, four values for the atomic weight of sulphur are deducible. Calculating with—

$$O = 15.879, \pm .0003$$
$$Ag = 107.108, \pm .0031$$
$$Cl = 35.179, \pm .0048$$
$$Na = 22.881, \pm .0046$$
$$C = 11.920, \pm .0004$$
$$AgCl = 142.287, \pm .0037,$$

we have:

From (1)...... $S = 31.828$, ± .0016
From (2)...... " $= 31.806$, ± .0048
From (3)...... " $= 31.864$, ± .0086
From (4).... " $= 31.835$, ± .0191

General mean............ $S = 31.828$, ± .0015

If $O = 16$, $S = 32.070$. From Stas' ratios alone, Stas found 32.074; Ostwald, 32.0626; Van der Plaats, (A) 32.0576, (B) 32.0590, and Thomsen, 32.0606. Here again Stas' determinations far outweigh all others.

LITHIUM.

The earlier determinations of the atomic weight of lithium by Arfvedson, Stromeyer, C. G. Gmelin, and Kralovanzky were all erroneous, because of the presence of sodium compounds in the material employed. The results of Berzelius, Hagen, and Hermann were also incorrect, and need no further notice here. The only investigations which we need to consider are those of Mallet, Diehl, Troost, Stas, and Dittmar.

Mallet's experiments[*] were conducted upon lithium chloride, which had been purified as completely as possible. In two trials the chloride was precipitated by nitrate of silver, which was collected upon a filter and estimated in the ordinary way. The figures in the third column represent the LiCl proportional to 100 parts of AgCl:

```
7.1885 grm. LiCl gave 24.3086 grm. AgCl.    29.606
8.5947        "         29.0621       "     29.574
```

In a third experiment the LiCl was titrated with a standard solution of silver. 3.9942 grm. LiCl balanced 10.1702 grm. Ag, equivalent to 13.511 grm. AgCl. Hence 100 AgCl = 29.563 LiCl. Mean of all three experiments, 29.581, ± .0087.

Diehl,[†] whose paper begins with a good résumé of all the earlier determinations, describes experiments made with lithium carbonate. This salt, which was spectroscopically pure, was dried at 130° before weighing. It was then placed in an apparatus from which the carbon dioxide generated by the action of pure sulphuric acid upon it could be expelled, and the loss of weight determined. From this loss the following percentages of CO_2 in Li_2CO_3 were determined:

```
59.422
59.404
59.440
59.401
```
Mean, 59.417, ± .006

Diehl's investigation was quickly followed by a confirmation from Troost.[‡] This chemist, in an earlier paper,[§] had sought to fix the atomic weight of lithium by an analysis of the sulphate, and had found a value not far from 6.5, thus confirming the results of Berzelius and of Hagen, who had employed the same method. But Diehl showed that the $BaSO_4$ precipitated from Li_2SO_4 always retained traces of Li, which were recog-

---

[*] Silliman's Amer. Journal, November, 1856. Chem. Gazette, 15, 7.
[†] Ann. Chem. Pharm., 121, 93.
[‡] Zeit. Anal. Chem., 1, 402.
[§] Annales d. Chim. et d. Phys., 51, 108.

nizable by spectral analysis, and which accounted for the error. In the later paper Troost made use of the chloride and the carbonate of lithium, both spectroscopically pure. The carbonate was strongly ignited with pure quartz powder, thus losing carbon dioxide, which loss was easily estimated. The subjoined results were obtained:

.970 grm. Li_2CO_3 lost .577 grm. CO_2.   59.485 per cent.
1.782   "   1.059   "   59.427   "

Mean, 59.456, ± .020

The lithium chloride employed by Troost was heated in a stream of dry hydrochloric acid gas, of which the excess, after cooling, was expelled by a current of dry air. The salt was weighed in the same tube in which the foregoing operations had been performed, and the chlorine was then estimated as silver chloride. The usual ratio between LiCl and 100 parts of AgCl is given in the third column:

1.309 grm. LiCl gave 4.420 grm. AgCl.   29.615
2.750   "   9.300   "   29.570

Mean, 29.5925, ± .0145

This, combined with Mallet's mean, 29.581, ± .0087, gives a general mean of 59.584, ± .0075.

Next in order is the work of Stas,* which was executed with his usual wonderful accuracy. In three titrations, in which all the weights were reduced to a vacuum standard, the following quantities of LiCl balanced 100 parts of pure silver:

39.356
·39.357
39.361

Mean, 39.358, ± .001

In a second series of experiments, intended for determining the atomic weight of nitrogen, LiCl was converted into $LiNO_3$. The method was that employed for a similar purpose with the chlorides of sodium and of potassium. One hundred parts of LiCl gave of $LiNO_3$:

162.588
162.600
162.598

Mean, 162.5953, ± .0025

The determinations of Dittmar† resemble those of Diehl; but the lithium carbonate used was dehydrated by fusion in an atmosphere of carbon dioxide. The carbonate was treated with sulphuric acid, and

---
* Aronstein's translation, 279-302.
† Trans. Roy. Soc. Edinburgh, 35, II, 429. 1889.

the CO_2 was collected and weighed in an absorption apparatus, which was tared by a similar apparatus after the method of Regnault. The following percentages of CO_2 in Li_2CO_3 were found:

> 59.601
> 59.645
> 59.529—rejected.
> 59.655
> 59.683
> 59.604
> 59.517
> 59.663
> 60.143—rejected.
> 59.794
> 59.584
>
> Mean of all, 59.674

Rejecting the two experiments which Dittmar regards as untrustworthy, the mean of the remaining nine becomes 59.638, ± .0173. This combines with the work of Diehl and Troost, as follows:

> Diehl............................. 59.417, ± .0060
> Troost............................. 59.456, ± .0200
> Dittmar............................ 59.638, ± .0173
>
> General mean..................... 59.442, ± .0054

Dittmar's determinations give a much lower value for the atomic weight of lithium than any of the others, and therefore seem to be questionable. As, however, they carry little weight in the general combination, it is not necessary to speculate upon their possible sources of error.

The ratios for lithium are now as follows:

> (1.) AgCl : LiCl : : 100 : 29.584, ± .0075
> (2.) Ag : LiCl : : 100 : 39.358, ± .001.
> (3.) LiCl : $LiNO_3$: : 100 : 162.5953, ± .0025
> (4.) Per cent. of CO_2 in Li_2CO_3, 59.442, ± .0054

And the data to use in their reduction are—

> O = 15.879, ± .0003   N = 13.935, ± .0015
> Ag = 107.108, ± .0031   C = 11.920, ± .0004
> Cl = 35.179, ± .0048   AgCl = 142.287, ± .0037

These factors give two values for the molecular weight of lithium chloride, thus:

> From (1)...................... LiCl = 42.0942, ± .0110
> From (2)...................... " = 42.1556, ± .0016
>
> General mean .... LiCl = 42.1542, ± .0016

For lithium itself there are three values:

From molecular weight LiCl.......... Li = 6.9752, ± .0051
From (3)........................... " = 6.9855, ± .0129
From (4)........................... " = 6.9628, ± .0077

General mean............... Li = 6.9729, ± .0040

If O = 16, Li = 7.026. From Stas' ratios, Stas found Li = 7.022; Ostwald, 7.0303; Van der Plaats (A), 7.0273; (B), 7.0235; and Thomsen, 7.0307.

RUBIDIUM.

The atomic weight of rubidium has been determined by Bunsen, Piccard, Godeffroy, and Heycock from analyses of the chloride and bromide.

Bunsen,* employing ordinary gravimetric methods, estimated the ratio between AgCl and RbCl. His rubidium chloride was purified by fractional crystallization of the chloroplatinate. He obtained the following results, to which, in a third column, I add the ratio between RbCl and 100 parts of AgCl:

One grm. RbCl gave 1.1873 grm. AgCl.   84.225
"   1.1873 "   84.225
"   1.1850 "   84.388
"   1.1880 "   84.175

Mean, 84.253, ± .031

The work of Piccard† was similar to that of Bunsen. In weighing, the crucible containing the silver chloride was balanced by a precisely similar crucible, in order to avoid the correction for displacement of air. The filter was burned separately from the AgCl, as usual; but the small amount of material adhering to the ash was reckoned as metallic silver. The rubidium chloride was purified by Bunsen's method. The results, expressed according to the foregoing standard, are as follows:

1.1587 grm. RbCl = 1.372 AgCl + .0019 Ag.   84.300
1.4055   "   1.6632 "   .0030 "   84.303
1.001   "   1.1850 "   .0024 "   84.245
1.5141   "   1.7934 "   .0018 "   84.313

Mean, 84.290, ± .0105

Godeffroy,‡ starting with material containing both rubidium and

---
* Zeit. Anal. Chem., 1, 136. Poggend. Annual., 113, 339. 1861.
† Journ. für Prakt. Chem., 86, 454. 1862. Zeit. Anal. Chem., 1, 518.
‡ Ann. Chem. Pharm., 181, 185. 1876.

cæsium, separated the two metals by fractional crystallization of their alums, and obtained salts of each spectroscopically pure. The nitric acid employed was tested for chlorine and found to be free from that impurity, and the weights used were especially verified. In two of his analyses of RbCl the AgCl was handled by the ordinary process of filtration. In the other two it was washed by decantation, dried, and weighed in a glass dish. The usual ratio is appended in the third column:

1.4055 grm.	RbCl gave	1.6665 grm.	AgCl.		84.338
1.8096	"	2.1461	"		84.320
2.2473	"	2.665	"		84.326
2.273	"	2.6946	"		84.354

Mean, 84.3345, ± .0051

Combining the three series, we get the following result:

Bunsen...............	84.253, ± .031	Rb =	84.702
Piccard...............	84.290, ± .0105	" =	84.754
Godeffroy.............	84.3345, ± .0051	" =	84.817
General mean......	84.324, ± .0045		

Heycock [*] worked by two methods, but unfortunately his results are given only in abstract, without details. First, silver solution was added in slight deficiency to a solution of rubidium chloride, and the excess of the latter was measured by titration. The mean of seven experiments gave—

Ag : RbCl : : 107.93 : 120.801

Hence Rb = 84.702.

Two similar experiments with the bromide gave—

Ag : RbBr : : 107.93 : 165.437
Ag : RbBr : : 107.93 : 165.342

Mean, 165.3895, ± .0320

There are now three ratios for the metal rubidium, as follows:

(1.) AgCl : RbCl : : 100 : 84.324, ± .0045
(2.) Ag : RbCl : : 107.93 : 120.801
(3.) Ag : RbBr : : 107.93 : 165.3895, ± .0320

To reduce these ratios we have—

Ag = 107.108, ± .0031
Br = 79.344, ± .0062
Cl = 35.179, ± .0048
AgCl = 142.287, ± .0037

---

[*] British Association Report, 1882, p. 499.

For the molecular weight of RbCl, two values are calculable:

From (1).................... RbCl = 119.981, ± .0109
From (2)....................   "   = 119.881, ± .0218

General mean............ RbCl = 119.961, ± .0097

To the value from ratio (2) I have arbitrarily assigned a weight represented by the probable error as written above. The data for systematic weighting are deficient, and no other course of procedure seemed advisable.

From RbCl.................... Rb = 84.782, ± .0109
From RbBr, ratio (3) ....   " = 84.786, ± .0329

General mean............... Rb = 84.783, ± .0103

If $O = 16$, $Rb = 85.429$.

CÆSIUM.

The atomic weight of cæsium, like that of rubidium, has been determined from the analysis of the chloride. The earliest determination, by Bunsen,[*] was incorrect, because of impurity in the material employed.

In 1863 Johnson and Allen published their results.[†] Their material was extracted from the lepidolite of Hebron, Maine, and the cæsium was separated from the rubidium as bitartrate. From the pure cæsium bitartrate cæsium chloride was prepared, and in this the chlorine was estimated as silver chloride by the usual gravimetric method. Reducing their results to the convenient standard adopted in preceding chapters, we have, in a third column, the quantities of CsCl equivalent to 100 parts of AgCl:

1.8371 grm.	CsCl gave	1.5634 grm.	AgCl.		117.507
2.1295	"	1.8111	"		117.580
2.7018	"	2.2992	"		117.511
1.56165	"	1.3302	"		117.399

Mean, 117.499, ± .025

Shortly after the results of Johnson and Allen appeared a new series of estimations was published by Bunsen.[‡] His cæsium chloride was purified by repeated crystallizations of the chloroplatinate, and the ordi-

---
[*] Zeit. Anal. Chem., I, 137.
[†] Amer. Journ. Sci. and Arts (2), 35, 94.
[‡] Poggend. Annalen, 119, 1. 1863.

nary gravimetric process was employed. The following results represent, respectively, material thrice, four times, and five times purified:

1.3835 grm. CsCl gave	1.1781 grm. AgCl.	Ratio,	117.435
1.3682 "	1.1644 "	"	117.503
1.2478 "	1.0623 "	"	117.462

Mean, 117.467, ± .013

Godeffroy's work* was, in its details of manipulation, sufficiently described under rubidium. In three of the experiments upon cæsium the silver chloride was washed by decantation, and in one it was collected upon a filter. The results are subjoined:

1.5825 grm. CsCl gave	1.351 grm. AgCl.	Ratio,	117.135
1.3487 "	1.1501 "	"	117.265
1.1880 "	1.0141 "	"	117.148
1.2309 "	1.051 "	"	117.107

Mean, 117.164, ± .023

We may now combine the three series to form a general mean:

Johnson and Allen	117.499, ± .025	Cs = 132.007
Bunsen...............	117.467, ± .013	" = 131.961
Godeffroy...........	117.164, ± .023	" = 131.560
General mean...	117.413, ± .010	

Hence, if AgCl = 142.287, ± .0037, and Cl = 35.179, ± .0048, Cs = 131.885, ± .0142.

If O = 16, Cs = 132.890.

---

* Ann. Chem. Pharm., 181, 185. 1876.

COPPER.

The atomic weight of copper has been chiefly determined by means of the oxide, the sulphate, and the bromide, and by direct comparison of the metal with silver.

In dealing with the first-named compound all experimenters have agreed in reducing it with a current of hydrogen, and weighing the metal thus set free.

The earliest experiments of any value were those of Berzelius,[*] whose results were as follows:

7.68075 grm. CuO lost 1.55 grm. O.   79.820 per cent. Cu in CuO.
9.6115   "   1.939 " .   79.826   "   "

Mean, 79.823, ± .002

Erdmann and Marchand,[†] who come next in chronological order, corrected their results for weighing in air. Their weighings, thus corrected, give us the subjoined percentages of metal in CuO:

63.8962 grm. CuO gave 51.0391 grm. Cu.   79.878 per cent.
65.1590   "   52.0363   "   79.860   "
60.2878   "   48.1540   "   79.874   "
46.2700   "   36.9449   "   79.846   "

Mean, 79.8645, ± .0038

Still later we find a few analyses by Millon and Commaille.[‡] These chemists not only reduced the oxide by hydrogen, but they also weighed, in addition to the metallic copper, the water formed in the experiments. In three determinations the results were as follows:

6.7145 grm. CuO gave 5.3565 grm. Cu and 1.5325 grm. H_2O.   79.775 per cent.
3.3915   "   2.7085   "   .7680   "   79.791   "
2.7880   "   2.2240   "   79.770   "

Mean, 79.7787, ± .0043

For the third of these analyses the water estimation was not made, but for the other two it yielded results which, in the mean, would make the atomic weight of copper 62.680. This figure has so high a probable error that we need not consider it further.

The results obtained by Dumas [§] are wholly unavailable. Indeed, he does not even publish them in detail. He merely says that he reduced copper oxide, and also effected the synthesis of the subsulphide, but without getting figures which were wholly concordant. He puts $Cu = 63.5$.

---

[*] Poggend. Annal., 8, 177. 1826.
[†] Journ. für Prakt. Chem., 31, 380. 1844.
[‡] Fresenius' Zeitschrift, 2, 475. 1863.
[§] Ann. Chim. et Phys. (3), 55, 129. 1859.

In 1873 Hampe* published his careful determinations, which were for many years almost unqualifiedly accepted. First, he attempted to estimate the atomic weight of copper by the quantity of silver which the pure metal could precipitate from its solutions. This attempt failed to give satisfactory results, and he fell back upon the old method of reducing the oxide. From ten to twenty grammes of material were taken in each experiment, and the weights were reduced to a vacuum standard:

20.3260 grm. CuO gave 16.2279 grm. Cu. 79.838 per cent.
20.68851 " 16.51669 " 79.835 "
10.10793 " 8.06926 " 79.831 "

Mean, 79.8347, ± .0013

Hampe also determined the quantity of copper in the anhydrous sulphate, $CuSO_4$. From 40 to 45 grammes of the salt were taken at a time, the metal was thrown down by electrolysis, and the weights were all corrected. I subjoin the results:

40.40300 grm. $CuSO_4$ gave 16.04958 grm. Cu. 39.724 per cent.
44.64280 " 17.73466 " 39.726 "

Mean, 39.725, ± .0007

The last series of data gives $Cu = 62.839, \pm .0035$, and is interesting for comparison with results obtained by Richards later.

In all of the foregoing experiments with copper oxide, that compound was obtained by ignition of the basic nitrate. But, as was shown in the chapter upon oxygen, copper oxide so prepared always carries occluded gases, which are not wholly expelled by heat. This point was thoroughly worked up by Richards† in his fourth memoir upon the atomic weight of copper, and it vitiates all the determinations previously made by this method.

By a series of experiments with copper oxide ignited at varying temperatures, and with different degrees of heat during the process of reduction, Richards obtained values for Cu ranging from 63.20 to 63.62, when $O = 16$. In two cases selected from this series he measured the amount of gaseous impurity, and corrected the results previously obtained. The results were as follows, with vacuum standards:

1.06253 grm. CuO gave. .84831 grm. Cu. 79.802 per cent.
1.91656 " 1.5298 " 79.820 "

Mean, 79.811, ± .0061

Correcting for the occluded gases in the oxide, the sum of the two experiments gives 79.901 per cent. of copper, whence $Cu = 63.605$. Three

---
* Fresenius' Zeitschrift, 13, 352.
† Proc. Amer. Acad., 26, 276. 1891.

other indirect results, similarly corrected, gave 79.900 per cent. Cu in CuO, or Cu = 63.603. If we assign all five experiments equal weight, and judge their value by the two detailed above, the mean percentage becomes 79.900, ± .0038. This figure need not be combined with the data given by previous observers, so far as practical purposes are concerned; but as this work is, in part at least, a study of the compensation of errors, it may not be wasted time to effect the combination, as follows:

Berzelius	79.823, ± .0020
Erdmann and Marchand	79.8645, ± .0038
Millon and Commaille	79.7787, ± .0043
Hampe	79.8347, ± .0013
Richards	79.900, ± .0038
General mean	79.8355, ± .0010

This result is practically identical with that of Hampe, whose work receives excessive weight, as does also that of Berzelius. The oxide of copper is evidently of doubtful value in the measurement of this atomic weight.

The composition of the sulphate has been studied, not only by Hampe, but also by Baubigny* and by Richards.† Baubigny merely ignited the anhydrous salt, weighing both it and the residual oxide, as follows:

4.022 grm. $CuSO_4$ gave	2.0035 CuO.	49.813 per cent.
2.596 "	1.293 "	49.807 "

Mean, 49.810, ± .002

The same ratio, in reverse—that is, the synthesis of the sulphate from the oxide—was investigated by Richards (p. 275), who shows that the results obtained are vitiated by the same errors which affect the copper oxide experiments previously cited. The weights given are reduced to vacuum standards. The percentage of oxide in the sulphate is stated in the third column of figures.

1.0084 grm. CuO gave	2.0235 grm. $CuSO_4$.	49.835 per cent.
2.7292 "	5.4770 "	49.830 "
1.0144 "	2.0350 "	49.848 "

Mean, 49.838, ± .0036

The two series combine thus:

Baubigny	49.810, ± .0020
Richards	49.838, ± .0036
General mean	49.816, ± .0017

Here, plainly, the rigorous discussion gives Baubigny's work weight in excess of its merits.

---

*Compt. Rend., 97, 906. 1883.
†Proc. Amer. Acad., 26, 240. 1891.

In the memoir by Richards now under consideration, his fourth upon copper, the greater part of his attention is devoted to the sulphate, Hampe being followed closely in order to ascertain what sources of error affected the work of the latter. Crystallized sulphate, $CuSO_4.5H_2O$ was purified with every precaution and made the basis of operations. Three series of experiments were carried out, the water being determined by loss of weight upon heating, and the copper being estimated electrolytically. In the first series the following data were found, the weights being reduced to a vacuum, as in all of Richards' determinations:

	$CuSO_4.5$ aq.	$CuSO_4$ at $250°$.	Cu.
1	2.8815	......	.7337
2	2.7152	......	.6911
3	3.4639	2.2184	.8817

Hence the subjoined percentages.

	Water at $250°$.	Cu in Cryst. Salt.	Cu in $CuSO_4$.
1	......	25.462	......
2	......	25.452	......
3	35.958	25.454	39.745
		Mean, 25.456	

In the second series of analyses, which are stated with much detail, several refinements were introduced, in order to estimate also the sulphuric acid. These will be considered later. The results, given below, are numbered consecutively with the former series.

	$CuSO_4.5$ aq.	$CuSO_4$ at $260°$.	$CuSO_4$ at $360°$.	Cu.
4	3.06006	1.9597	1.95637	.77886
5	2.81840	1.8048	......	.71740
6	7.50490	4.8064	4.79826	1.90973

Hence percentages as follows:

	Water, $260°$.	Water, $360°$.	Cu in Cryst. Salt.	Cu in $CuSO_4$, $260°$.	Ditto, $360°$.
4	35.959	36.068	25.452	39.744	39.811
5	35.964	......	25.454	39.750	......
6	35.957	36.065	25.446	39.733	39.799
Mean,	35.960	36.067	25.450	39.742	39.805

Hampe worked with a sulphate dried at $250°$, but these data show that a little water is retained at that temperature, and consequently that his results must have been too low. The third of Richards' series resembles the second, but extra precautions were taken to avoid conceivable errors.

	$CuSO_4.5$ aq.	$CuSO_4$ at $260°$.	$CuSO_4$ at $370°$.	Cu.
7	2.88307	......	......	.73380
8	3.62913	2.32373	......	.92344
9	5.81352	......	3.71680	1.47926

And the percentages are:

	Water at 260°.	At 370°.	Cu in Cryst. Salt.	Cu in CuSO$_4$.
7			25.452	
8		35.970	25.446	39.740 (260°)
9		36.067	25.445	39.799 (370°)
			25.448	

In this series the determinations of sulphuric acid gave essentially the same results for all three samples of sulphate, although one was not dehydrated, and the others were heated to 260° and 370° respectively. Hence the loss of weight in dehydration at either temperature represents water only, and does not involve partial decomposition of the sulphate. Between 360° and 400° copper sulphate is at essentially constant weight, but further experiments indicated that even at 400° it retained traces of water, and possibly as much as .042 per cent. The last trace is not expelled until the salt itself begins to decompose.

Richards also effected two syntheses of the sulphate directly from the metal by dissolving the latter in nitric acid, then evaporating to dryness with sulphuric acid, and heating to constant weight at 400°.

.67720 grm. Cu gave 1.7021 grm. CuSO$_4$.   39.786 per cent. Cu.
1.00613   "   2.5292   "   39.781   "

If we include these percentages in a series with the data from analyses 4, 6, and 9, which gave percentages of 39.811, 39.799, and 39.799 respectively of copper in sulphate dried at 360° and upwards, the mean becomes

$$CuSO_4 : Cu :: 100 : 39.795, \pm .0036$$

Since even this result is presumably too low, the other figures from sulphate dried at 250° must be rejected. Since Hampe's work on the sulphate is affected by the same sources of error, and apparently to a still greater extent, it need not be considered farther. As for Richards' nine determinations of Cu in CuSO$_4$.5H$_2$O, we may take them as one series giving a mean percentage of 25.451, $\pm$ .0011. This salt seems to retain occluded water, for the percentage of copper in it leads to a value for the atomic weight which is inconsistent with the best evidence, as will be seen later.

In the second and third series of Richards' experiments upon copper sulphate, the sulphuric acid was estimated by a method which gave valuable results. After the copper had been electrolytically precipitated, the acid which was set free was nearly neutralized by a weighed amount of pure sodium carbonate, and the slight excess remaining was determined by titration. Thus the weight of sodium carbonate equivalent to the copper was ascertained. The resulting solution of sodium sulphate was then evaporated to dryness, and a new ratio, connecting that salt with copper, was also determined. The cross ratio Na$_2$CO$_3$: Na$_2$SO$_4$ has already been utilized in a previous chapter. The results, ignoring the weights of hydrated copper sulphate, are as follows, with the experiments numbered as before:

	Cu.	Na_2CO_3.	Na_2SO_4
4	.77886	1.2993	1.7411
6	1.90973	3.1862	4.2679
7	.73380	1.22427	1.63994
8	.92344	1.54075	....
9	1.47926	....	3.30658

Hence,

$Cu : Na_2CO_3 :: 100 : x$.
166.824
166.840
166.840
166.849

Mean, 166.838, ± .0035

$Cu : Na_2SO_4 :: 100 : x$.
223.549
223.482
223.538
223.529

Mean, 223.525, ± .0098

In one more experiment the sulphuric acid was weighed as barium sulphate, the latter being corrected for occluded salts. 3.1902 grm. $CuSO_4.5H_2O$ gave 2.9761 $BaSO_4$; hence $CuSO_4.5H_2O : BaSO_4 :: 100 : 93.289$. The sulphate contained 25.448 per cent. of Cu; hence $BaSO_4 : Cu :: 93.289 : 25.448$. Still other ratios can be deduced from Richards' work on the sulphate, but in view of the uncertainties relative to the water in the salt they are hardly worth computing.

In his third paper upon the atomic weight of copper,* Richards studied the dibromide, $CuBr_2$. In preparing this salt he used hydrobromic acid made from pure materials, and further purified by ten distillations. This was saturated with copper oxide prepared from pure electrolytic copper, and the solution obtained was proved to be free from basic salts. As the crystallized compound was not easily obtained in a satisfactory condition, weighed quantities of the solution were taken for analysis, in which, after expulsion of bromine by nitric and sulphuric acids, the copper was determined by electrolysis. In other portions of solution the bromine was precipitated by silver nitrate, and weighed as silver bromide. The first preliminary series of experiments gave the subjoined results, with vacuum weights as usual:

In 25 Grammes of Solution.

Cu.	$AgBr$.
.4164	2.4599
.4164	2.4605
.4164	2.4605
.4165	2.4599

Hence 2 $AgBr : Cu :: 100 : 16.927, \pm .0013$.

---

* Proc. Amer. Acad., 25, 195. 1890.

The second, also preliminary series, was made with more dilute solutions, and came out as follows:

In 25 Grammes of Solution.

Cu.	AgBr.
.26190	1.5478
.26185	1.5477
	1.5479

Hence 2 AgBr : Cu :: 100 : 16.919, ± .0012.

In the third series, two distinct lots of crystallized bromide were dissolved, and the solutions examined in the same way.

Cu.	AgBr.	Ratio.
.2500	1.4771	16.925
.5473	3.2348	16.919

Mean, 16.922, ± .0020

In the final set of analyses, the materials used were purified even more scrupulously than before, and the process was distinctly modified, as regards the determination of the bromine. The solution of the bromide was added to a solution of pure silver in nitric acid, not quite sufficient for complete precipitation. The slight excess of bromine was then determined by titration with a solution containing one gramme of silver to the litre. Thus silver proportional to the copper in the bromide was determined, and the silver bromide was weighed in a Gooch crucible as before. The results are subjoined:

In 50 Grammes of Solution.

Cu.	Ag.	AgBr.
.54755	1.8586	3.2350
.54750	1.8579	3.2340
	1.8583	3.2348

Hence Cu : Ag_2 : : 100 : 339.392, ± .0108, and 2 AgBr : Cu : : 100 : 16.927, ± .0012.

The latter ratio, combined with the results of the three preceding series, gives a general mean of:

2 AgBr : Cu : : 100 : 16.924, ± .0007

In his two earlier papers * Richards determined the copper-silver ratio directly—that is, without the weighing of any compound of either metal. By placing pure copper in an *ice-cold* solution of silver nitrate, metallic silver is thrown down, and the weights of the two metals were in equiv-

---

* Proc. Amer. Acad., 22, 346, and 23, 177. 1886 and 1887.

alent proportions. In the first paper the following results were obtained. The third column gives the value of x in the ratio Cu : Ag_2 :: 100 : x.

Cu Taken.	Ag Found.	Ratio.
.53875	1.8192	339.527
.56190	1.9076	339.491
1.00220	3.4016	339.414
1.30135	4.4173	339.440
.99870	3.39035	339.477
1.02050	3.4646	339.500

Mean, 339.475, ± .0114

In the second paper Richards states that the silver of the fifth experiment, which had been dried at 150°, as were also the others, still retained water, to the extent of four-tenths milligramme in two grammes. If we assume this correction to be fairly uniform, as the concordance of the series indicates, and apply it throughout, the mean value for the ratio then becomes 339.408, ± .0114. This procedure, however, leaves the ratio in some uncertainty, and accordingly some new determinations were made, in which the silver, collected in a Gooch crucible, was heated to incipient redness before final weighing. Copper from two distinct sources was taken, and three experiments were carried out upon one sample to two with the other. Treating both sets as one series, the results were as follows:

Cu Taken.	Ag found.	Ratio.
.75760	2.5713	339.40
.95040	3.2256	339.39
.75993	2.5794	339.42
1.02060	3.4640	339.42
.90460	3.0701	339.39

Mean, 339.404, ± .0046

a value practically identical with the corrected mean of the previous determinations, and with that found in the later experiments upon copper bromide.

In various electrical investigations the same ratio, the electrochemical equivalent of copper, has been repeatedly measured, and the later results of Lord Rayleigh and Mrs. Sidgewick,* Gray,† Shaw,‡ and Vanni § may properly be included in this discussion. As the data are somewhat differently stated, I have reduced them all to the common standard adopted above. Gray gives two sets of measurements, one made with large and the other with small metallic plates:

* Phil. Trans., 175, 458.
† Phil. Mag. (5), 22, 389.
‡ British Assoc. Report, 1886. Abstract in Phil. Mag. (5), 23, 138.
§ Ann. der Phys. (Wiedemann's) (2), 44, 214.

COPPER. 99

Rayleigh and S.	Gray 1.	Gray 2.	Shaw.	Vanni.
340.483	341.297	340.252	339.68	340.483
340.832	341.413	339.674	340.05	340.600
340.367	340.815	340.020	339.84	340.367
———	340.252	339.905	339.71	340.252
340.561,	339.905	339.674	340.04	340.600
± .0935	341.064	339.328	339.94	340.136
	340.832	340.136	340.35	———
	341.297	340.136	339.82	340.406,
	341.064	340.136	340.09	± .0520
	341.413	340.020	339.84	
	———	340.020	339.90	
	340.935,	340.136	339.98	
	± .1072	———	340.14	
		339.953,	340.56	
		± .0521	339.82	
			———	
			339.983,	
			± .0411	

The lack of sharp concordance in these data and the consequently high probable errors seem to indicate a distinct superiority of the purely chemical method of determination over that adopted by the physicist. The eight distinct series now combine as follows:

Richards, first series corrected............ 339.408, ± .0114
Richards, second series.................. 339.404, ± .0046
Richards, CuBr$_2$ series.................. 339.392, ± .0108
Rayleigh and Sidgwick................. 340.561, ± .0935
Gray, with large plates.................. 340.935, ± .1072
Gray, with small plates.................. 339.953, ± .0521
Shaw........ .... 339.983, ± .0411
Vanni.................................. 340.406, ± .0520

General mean.................... 339.411, ± .0039

If we combine Richards' three series into a general mean separately, we get 339.402, ± .0040. Hence the other determinations, having high probable errors, practically vanish from the result, and it is a matter of indifference whether they are retained or rejected.

We now have the following ratios from which to compute the atomic weight of copper:

(1.) Percentage of Cu in CuO.......... 79.8355, ± .0010
(2.) " of Cu in CuSO$_4$........ 39.795, ± .0036
(3.) " of Cu in CuSO$_4$, 5H$_2$O.. 25.451, ± .0011
(4.) " of CuO in CuSO$_4$....... 49.816, ± .0017
(5.) Cu : Na$_2$CO$_3$: : 100 : 166.838, ± .0035
(6.) Cu : Na$_2$SO$_4$: : 100 : 223.525, ± .0098
(7.) BaSO$_4$: Cu : : 93.289 : 25.448.
(8.) 2AgBr : Cu : : 100 : 16.924, ± .0007
(9.) Cu : Ag$_2$: : 100 : 339.411, ± .0039

Reducing these ratios with the subjoined data:

O = 15.879, ± .0003
Ag = 107.108, ± .0031
S = 31.828, ± .0015
C = 11.920, ± .0004
Na = 22.881, ± .0046
Ba = 136.392, ± .0086
AgBr = 186.452, ± .0054

We have nine values for the atomic weight of copper. Since ratio (7) depends upon one experiment only, it is necessary to assign the value derived from it arbitrary weight. This will be taken as indicated by a probable error double that of the next highest, obtained from ratio (2). The values then are as follows:

From (1)............................ Cu = 62.869, ± .0034
From (2)............................ " = 63.022, ± .0070
From (3)............................ " = 63.070, ± .0030
From (4)............................ " = 63.003, ± .0042
From (5)............................ " = 63.127, ± .0051
From (6)............................ " = 63.128, ± .0050
From (7)............................ " = 63.215, ± .0140
From (8)............................ " = 63.110, ± .0032
From (9)............................ " = 63.114, ± .0020

General mean............... Cu = 63.070, ± .0012

If O = 16, Cu = 63.550. If we include Hampe's analyses of copper sulphate, which gave Cu = 62.839, ± .0035, the general mean becomes Cu = 63.046, ± .0011.

The foregoing means, however, are significant only as showing the effect and weight of the older data upon the newer determinations of Richards. The seventh of the individual values is also interesting, for the reason that the experiment upon which it depends was published by Richards previous to his investigation of the atomic weight of barium. With the old value for Ba, 137, it gives a value for copper in close agreement with Richards' other determinations. With the new value for barium it becomes discordant, although its weight is so low that it produces no appreciable effect upon the final mean.

Rejecting values 1 to 4, inclusive, the remaining five values give a general mean of

Cu = 63.119, ± .0015.

If O = 16, this becomes 63 600, and in the light of all the evidence these figures are to be preferred. If, again, we combine with this mean the results of Richards' work on the oxide and sulphate of copper, the final value becomes

Cu = 63.108, ± .0013,

and with O = 16, 63.589. This departs but little from the previous mean value, but it includes data which render it, in all probability, a trifle too low. The value Cu = 63.119 will be regarded as the best.

GOLD.

Among the early estimates of the atomic weight of gold the only ones worthy of consideration are those of Berzelius and Levol.

The earliest method adopted by Berzelius* was that of precipitating a solution of gold chloride by means of a weighed quantity of metallic mercury. The weight of gold thus thrown down gave the ratio between the atomic weights of the two metals. In the single experiment which Berzelius publishes, 142.9 parts of Hg precipitated 93.55 of Au. Hence if $Hg = 200$, $Au = 196.397$.

In a later investigation † Berzelius resorted to the analysis of potassio-auric chloride, $2KCl.AuCl_3$. Weighed quantities of this salt were ignited in hydrogen; the resulting gold and potassium chloride were separated by means of water, and both were collected and estimated. The loss of weight upon ignition was, of course, chlorine. As the salt could not be perfectly dried without loss of chlorine, the atomic weight under investigation must be determined by the ratio between the KCl and the Au. If we reduce to a common standard, and compare with 100 parts of KCl, the equivalent amounts of gold will be those which I give in the last of the subjoined columns:

4.1445 grm. K_2AuCl_5 gave	.8185 grm. KCl and	2.159 grm. Au.	263.775
2.2495	"	.44425	"	1.172	"	263.815
5.1300	"	1.01375	"	2.67225	"	263.600
3.4130	"	.674	"	1.77725	"	263.687
4.19975	"	.8295	"	2.188	"	263.773

Mean, 263.730, ± .026

Still a third series of experiments by Berzelius ‡ may be included here. In order to establish the atomic weight of phosphorus he employed that substance to precipitate gold from a solution of gold chloride in excess. Between the weight of phosphorus taken and the weight of gold obtained it was easy to fix a ratio. Since the atomic weight of phosphorus has been better established by other methods, we may properly reverse this ratio and apply it to our discussion of gold. 100 parts of P precipitate the quantities of Au given in the third column:

.829 grm. P precipitated	8.714 grm. Au.	1051.15
.754	"	7.930	"	1051.73

Mean, 1051.44, ± .196

Hence if $P = 31$, $Au = 195.568$.

---

* Poggend. Annalen, 8, 177.
† Lehrbuch, 5 Aufl., 3, 1212.
‡ Lehrbuch, 5 Aufl., 3, 1188.

Levol's* estimation of the atomic weight under consideration can hardly have much value. A weighed quantity of gold was converted in a flask into $AuCl_3$. This was reduced by a stream of sulphur dioxide, and the resulting sulphuric acid was determined as $BaSO_4$. One gramme of gold gave 1.782 grm. $BaSO_4$. Hence $Au = 195.06$.

All these values may be neglected as worthless, except that derived from Berzelius' K_2AuCl_5 series.

In 1886 Krüss† published the first of the recent determinations of the atomic weight under consideration, several distinct methods being recorded. First, in a solution of pure auric chloride the gold was precipitated by means of aqueous sulphurous acid. In the filtrate from the gold the chlorine was thrown down as silver chloride, and thus the ratio $Au : 3\,AgCl$ was measured. I subjoin Krüss' weights, together with a third column giving the gold equivalent to 100 parts of silver chloride:

Au.	AgCl.	Ratio.
7.72076	16.84737	45.828
5.68290	12.40425	45.814
3.24773	7.08667	45.828
4.49167	9.80475	45.811
3.47949	7.59300	45.825
3.26836	7.13132	45.832
5.16181	11.26524	45.821
4.86044	10.60431	45.834

Mean, $45.824, \pm .0020$

The remainder of Krüss' determinations were made with potassium auribromide, $KAuBr_4$, and with this salt several ratios were measured. The salt was prepared from pure materials, repeatedly recrystallized under precautions to exclude access of atmospheric dust, and dried over phosphorus pentoxide. First, its percentage of gold was determined, sometimes by reduction with sulphurous acid, sometimes by heating in a stream of hydrogen. For this ratio, the weights and percentages are as follows, the experiments being numbered for further reference, and the reducing agent being indicated.

	$KAuBr_4$.	Au.	Per cent.
1. SO_2	10.64821	3.77753	35.476
2. SO_2	4.71974	1.67330	35.453
3. H	7.05762	2.50122	35.440
4. H	4.49558	1.59434	35.465
5. SO_2	8.72302	3.09448	35.475
6. SO_2	7.66932	2.71860	35.448
7. SO_2	7.15498	2.53695	35.457
8. H	12.26334	4.34997	35.471
9. H	7.10342	2.51919	35.465

Mean, $35.461, \pm .0028$

---

* Ann. Chim. Phys. (3), 30, 355. 1850.
† Untersuchungen über das Atomgewicht des Goldes. München, 1886. 112 pp., 8vo.

In five of the foregoing experiments the reductions were effected with sulphurous acid; and in these, after filtering off the gold, the bromine was thrown down and weighed as silver bromide. This, in comparison with the gold, gives the ratio Au : 4AgBr : : 100 : x.

	Au.	4AgBr.	Ratio.
1	3.77753	14.39542	381.080
2	1.67330	6.37952	381.254
5	3.09448	11.78993	380.999
6	2.71860	10.35902	381.042
7	2.53695	9.66117	380.731

Mean, 381.021, ± .057

Hence Au : AgBr : : 100 : 95.255, ± .0142.

In the remaining experiments, Nos. 3, 4, 8, and 9, the $KAuBr_4$ was reduced in a stream of hydrogen, the loss of weight, Br_3, being noted. In the residue the gold was determined, as noted above, and the KBr was also collected and weighed. The weights were as follows:

	Au.	Loss, Br_3.	KBr.
3	2.50122	3.04422	1.51090
4	1.59434	1.93937	.96243
8	4.34997	5.29316	2.62700
9	2.51919	3.06534	1.52153

From these data we obtain two more ratios, viz., Au : Br_3 : : 100 : x, and Au : KBr : : 100 : x, thus:

	Au : Br_3.	Au : KBr.
3	121.710	60.405
4	121.641	60.365
8	121.683	60.391
9	121.680	60.398

Mean, 121.678, ± .0100   Mean, 60.390, ± .0059

From all the ratios, taken together, Krüss deduces a final value of Au = 197.13, if O = 16. It is obviously possible to derive still other ratios from the results given, but to do so would be to depart unnecessarily from the author's methods as stated by himself.

Thorpe and Laurie,[*] whose work appeared shortly after that of Krüss, also made use of the salt $KAuBr_4$, but, on account of difficulty in drying it without change, they did not weigh it directly. After proving the constancy in it of the ratio Au : KBr, even after repeated crystallizations, they adopted the following method: The unweighed salt was heated with gradual increase of temperature, up to about 160°, for several hours, and afterwards more strongly over a small Bunsen flame. This was done in a porcelain crucible, tared by another in weighing, which latter was treated in precisely the same way. The residue, KBr + Au, was weighed, the KBr dissolved out, and the gold then weighed separately. The

---

[*] Journ. Chem. Soc., 51, 565. 1887.

weight of KBr was taken by difference. The ratio Au : KBr :: 100 : x appears in a third column.

Au.	KBr.	Ratio.
6.19001	3.73440	60.329
4.76957	2.87715	60.323
4.14050	2.49822	60.336
3.60344	2.17440	60.342
3.67963	2.21978	60.326
4.57757	2.76195	60.337
5.36659	3.23821	60.326
5.16406	3.11533	60.327

Mean, 60.331, ± .0016

This mean combines with Krüss' thus:

Krüss.................................... 60.390, ± .0059
Thorpe and Laurie...................... 60.331, ± .0016

General mean...................... 60.338, ± .0015

The potassium bromide of the previous experiments was next titrated with a solution of pure silver by Stas' method, the operation being performed in red light. Thus we get the following data for the ratio Ag : Au :: 100 : x, using the weights of gold already obtained:

Ag.	Au.	Ratio.
3.38451	6.19001	182.893
2.60896	4.76957	182.813
2.28830	4.18266	182.786
2.26415	4.14050	182.868
1.97147	3.60344	182.775
2.01292	3.67963	182.801
2.50334	4.57757	182.863
2.93608	5.36659	182.780
2.82401	5.16406	182.865

Mean, 182.827, ± .0101

Finally, in eight of these experiments, the silver bromide formed during titration was collected and weighed, giving values for the ratio Au : AgBr :: 100 : x, as follows:

Au.	AgBr.	Ratio.
6.19001	5.89199	95.186
4.76957	4.54261	95.242
4.18266	3.98288	95.224
4.14050	3.94309	95.232
3.60344	3.43015	95.191
3.67963	3.50207	95.175
4.57757	4.35736	95.189
5.36659	5.11045	95.227

Mean, 95.208, ± .0061
Krüss found, 95.255, ± .0142

General mean, 95.222, ± .0056

From the second and third of the ratios measured by Thorpe and Laurie an independent value for the ratio Ag : Br may be computed. It becomes 100 : 74.072, which agrees closely with the determinations made by Stas and Marignac. Similarly, the ratios Ag : KBr and AgBr : KBr may be calculated, giving additional checks upon the accuracy of the manipulation, though not upon the purity of the original material studied.

Thorpe and Laurie suggest objections to the work done by Krüss, on the ground that the salt $KAuBr_4$ cannot be completely dried without loss of bromine. This suggestion led to a controversy between them and Krüss, which in effect was briefly as follows:

First, Krüss* urges that the potassium auribromide ordinarily contains traces of free gold, not belonging to the salt, produced by the reducing action of dust particles taken up from the air. He applies a correction for this supposed free gold to the determinations made by Thorpe and Laurie, and thus brings their results into harmony with his own. To this argument Thorpe and Laurie † reply, somewhat in detail, stating that the error indicated was guarded against by them, and that they had dissolved quantities of from eight to nineteen grammes of the auribromide without a trace of free gold becoming visible. A final note in defense of his own work was published by Krüss a little later.‡

In 1889 an elaborate set of determinations of this constant was published by Mallet,§ whose experiments are classified into seven distinct series. First, a neutral solution of auric chloride was prepared, which was weighed off in two approximately equal portions. In one of these the gold was precipitated by pure sulphurous acid, collected, washed, dried, ignited in a Sprengel vacuum, and weighed. To the second portion a solution containing a known weight of pure silver was added. After filtering, with all due precautions, the silver remaining in the filtrate was determined by titration with a weighed solution of pure hydrobromic acid. We have thus a weight of gold, and the weight of silver needed to precipitate the three atoms of chlorine combined with it; in other words, the ratio $Ag_3 : Au : : 100 : x$. All weights in this and the subsequent series are reduced to vacuum standards, and all weighings were made against corresponding tares.

$Au.$	$Ag_3.$	$Ratio.$
7.6075	12.4875	60.921
8.4212	13.8280	60.900
6.9407	11.3973	60.898
3.3682	5.5286	60.923
2.8244	4.6371	60.909

Mean, 60.910, ± .0034

Hence Ag : Au : : 100 : 182.730, ± .0102.

* Ber. Deutsch. Chem. Gesell., 20, 2365. 1887.
† Berichte, 20, 3036, and Journ. Chem. Soc., 51, 866. 1887.
‡ Berichte, 21, 126. 1888.
§ Philosophical Transactions, 180, 395. 1889.

The second series of determinations was essentially like the first, except that auric bromide was taken instead of the chloride. The ratio measured, $Ag_3 : Au$, is precisely the same as before. Results as follows:

$Au.$	$Ag_3.$	Ratio.
8.2345	13.5149	60.929
7.6901	12.6251	60.911
10.5233	17.2666	60.945
2.7498	4.5141	60.916
3.5620	5.8471	60.919
3.9081	6.4129	60.941

Mean, 60.927, ± .0038

Hence Ag : Au :: 100 : 182.781, ± .0114.

In the third series of experiments the salt $KAuBr_4$ was taken, purified by five recrystallizations. The solution of this was weighed out into nearly equal parts, the gold being measured as in the two preceding series in one portion, and the bromine thrown down by a standard silver solution as before. This gives the ratio $Ag_4 : Au :: 100 : x$.

$Au.$	$Ag.$	Ratio.
5.7048	12.4851	45.693
7.9612	17.4193	45.693
2.4455	5.3513	45.699
4.1632	9.1153	45.673

Mean, 45.689, ± .0040.

Hence Ag : Au :: 100 : 182.756, ± .0160.

The fifth series of determinations, which for present purposes naturally precedes the fourth, was electrolytic in character, gold and silver being simultaneously precipitated by the same current. The gold was in solution as potassium auro-cyanide, and the silver in the form of potassium silver cyanide. The equivalent weights of the two metals, thrown down in the same time, were as follows, giving directly the ratio Ag : Au :: 100 : x.

$Au.$	$Ag.$	Ratio.
5.2721	2.8849	182.748
6.3088	3.4487	182.933
4.2770	2.3393	182.832
3.5123	1.9223	182.713
3.6804	2.0132	182.814

Mean, 182.808, ± .0256

This mean may be combined with the preceding means, and also with the determination of the same ratio by Thorpe and Laurie, thus:

Thorpe and Laurie	182.827,	± .0101
Mallet, chloride series	182.730,	± .0102
Mallet, bromide series	182.781,	± .0114
Mallet, $KAuBr_4$ series	182.756,	± .0160
Mallet, electrolytic	182.808,	± .0256
General mean	182.778,	± .0055

In Mallet's fourth series a radically new method was employed. Trimethyl-ammonium aurichloride, $N(CH_3)_3HAuCl_4$, was decomposed by heat, and the residual gold was determined. In order to avoid loss by spattering, the salt was heated in a crucible under a layer of fine siliceous sand of known weight. Several crops of crystals of the salt were studied, as a check against impurities, but all gave concordant values.

Salt.	Residual Au.	Per cent. Au.
14.9072	7.3754	49.475
15.5263	7.6831	49.484
10.4523	5.1712	49.474
6.5912	3.2603	49.464
5.5744	2.7579	49.474

Mean, 49.474, ± .0021

In his sixth and seventh series Mallet seeks to establish, by direct measurement, the ratio between hydrogen and gold. In their experimental details his methods are somewhat elaborate, and only the processes, in the most general way, can be indicated here. First, gold was precipitated electrolytically from a solution of potassium aurocyanide, and its weight was compared with that of the amount of hydrogen simultaneously liberated in a voltameter by the same current in the same time. The hydrogen was measured, and its weight was then computed from its density. The volumes are given, of course, at 0° and 760 mm.

Wt. Au.	Vol. H, cc.	Wt. H.
4.0472	228.64	.0205483
4.0226	227.03	.0204046
4.0955	231.55	.0208103

These data, with the weight of one litre of hydrogen taken as 0.89872 gramme, give the subjoined values in the ratio $H : Au :: 1 : x$.

196.960
197.151
196.805

Mean, 196.972, ± .0675

In the last series of experiments a known quantity of metallic zinc was dissolved in dilute sulphuric acid, and the amount of hydrogen evolved was measured. Then a solution of pure auric chloride or bromide was treated with a definite weight of the same zinc, and the quantity of gold thrown down was determined. The zinc itself was purified by practical distillation in a Sprengel vacuum. From these data the ratio $H_3 : Au$ was computed by direct comparison of the weight of gold and that of the liberated hydrogen. The results were as follows:

THE ATOMIC WEIGHTS.

$Wt.\ Au.$	$Vol.\ H,\ cc.$	$Wt.\ H.$
10.3512	1756.10	.157824
8.2525	1400.38	.125857
8.1004	1374.87	.123565
3.2913	558.64	.050206
3.4835	590.93	.053109
3.6421	618.11	.055551

Hence for the ratio $H_2 : Au :: 1 : x$ we have:

65.587
65.571
65.557
65.556
65.593
65.563

Mean, $65.571, \pm .00436$

And $H : Au :: 1 : 196.713, \pm .0131$. This, combined with the value found in the preceding series, gives a general mean of $196.722, \pm .0129$. The ratios available for gold are now as follows:

(1.) $2KCl : Au :: 100 : 263.730, \pm .026$
(2.) $3AgCl : Au :: 100 : 45.824, \pm .0020$
(3.) $KAuBr_4 : Au :: 100 : 35.461, \pm .0028$
(4.) $Au : AgBr :: 100 : 95.222, \pm .0056$
(5.) $Au : Br_3 :: 100 : 121.678, \pm .0100$
(6.) $Au : KBr :: 100 : 60.338, \pm .0015$
(7.) $Ag : Au :: 100 : 182.778, \pm .0055$
(8.) $NC_3H_{10}AuCl_4 : Au :: 100 : 49.474, \pm .0021$
(9.) $H : Au :: 1 : 196.722, \pm .0129$

For the reduction of these ratios the antecedent data are:

$Ag = 107.108, \pm .0031$		$C = 11.920, \pm .0004$
$Cl = 35.179, \pm .0048$		$AgCl = 142.287, \pm .0037$
$Br = 79.344, \pm .0062$		$AgBr = 186.452, \pm .0054$
$K = 38.817, \pm .0051$		$KCl = 74.025, \pm .0019$
$N = 13.935, \pm .0021$		$KBr = 118.200, \pm .0073$

Hence for the atomic weight of gold we have nine values:

From (1) .... $Au = 195.226, \pm .0193$
From (2) .... " $= 195.605, \pm .0099$
From (3) .... " $= 195.711, \pm .0224$
From (4) .... " $= 195.808, \pm .0126$
From (5) .... " $= 195.624, \pm .0222$
From (6) .... " $= 195.896, \pm .0131$
From (7) .... " $= 195.770, \pm .0082$
From (8) .... " $= 196.238, \pm .0224$
From (9) .... " $= 196.722, \pm .0129$

General mean .... $Au = 195.850, \pm .0044$

If $O = 16$, this becomes $Au = 197.342$.

Of the foregoing values the first one, which is derived from Berzelius' work, should certainly be rejected. So also, apparently, should the eighth and ninth values. Excluding these, values 2 to 7, inclusive, give a general mean of Au = 195.743, ± .0049. With O = 16, this becomes Au = 197.235. Probably these values are more nearly correct than those which include all the determinations.

The ninth value in the list given above represents Mallet's comparisons of gold directly with hydrogen, and is peculiarly instructive. In Mallet's paper the other determinations are discussed upon the basis of O = 15.96, which brings them more nearly into harmony with the hydrogen series. The great divergence shown in this recalculation is due to the new value for oxygen, 15.879, and its effect upon the atomic weights of silver, bromine, etc. The former agreement between the several series of gold values was therefore only apparent, and we are now able to see that concordance among determinations may be only coincidence, and no proof of accuracy. It is probable, furthermore, that direct comparisons of metals with hydrogen cannot give good measurements of atomic weights, for several reasons. First, it is not possible to be certain that every trace of hydrogen has been collected and measured, and any loss tends to raise the apparent atomic weight of the metal studied; secondly, the weight of the hydrogen is computed from its volume, and a slight change in the factors used in reduction of the observations may make a considerable difference in the final result. These uncertainties exist in all determinations of atomic weights hitherto made by the hydrogen method.

CALCIUM.

For determining the atomic weight of calcium we have sets of experiments by Berzelius, Erdmann and Marchand, and Dumas. Salvétat* also has published an estimation, but without the details necessary to enable us to make use of his results. I also find a reference† to some work of Marignac, which, however, seems to have been of but little importance. The earlier work of Berzelius was very inexact as regards calcium, and it is not until we come down to the year 1824 that we find any material of decided value.

The most important factor in our present discussion is the composition of calcium carbonate, as worked out by Dumas and by Erdmann and Marchand.

In 1842 Dumas ‡ made three ignitions of Iceland spar, and determined the percentages of carbon dioxide driven off and of lime remaining. The impurities of the material were also determined, the correction for them applied, and the weighings reduced to a vacuum standard. The percentage of lime came out as follows:

$$56.12$$
$$56.04$$
$$56.06$$

Mean, $56.073, \pm .016$

About this same time Erdmann and Marchand § began their researches upon the same subject. Two ignitions of spar, containing .04 per cent. of impurity, gave respectively 56.09 and 56.18 per cent. of residue; but these results are not exact enough for us to consider further. Four other results obtained with artificial calcium carbonate are more noteworthy. The carbonate was precipitated from a solution of pure calcium chloride by ammonium carbonate, was washed thoroughly with hot water, and dried at a temperature of 180°. With this preparation the following residues of lime were obtained:

$$56.03$$
$$55.98$$
$$56.00$$
$$55.99$$

Mean, $56.00, \pm .007$

It was subsequently shown by Berzelius that calcium carbonate prepared by this method retains traces of water even at 200°, and that

---

* Compt. Rend., 17, 318. 1843.
† See Oudeman's monograph, p. 51.
‡ Compt. Rend., 14, 537. 1842.
§ Journ. für Prakt. Chem., 26, 472. 1842.

CALCIUM.

minute quantities of chloride are also held by it. These sources of error are, however, in opposite directions, since one would tend to diminish and the other to increase the weight of residue.

In the same paper there are also two direct estimations of carbonic acid in pure Iceland spar, which correspond to the following percentages of lime:

$$56.00$$
$$56.02$$

Mean, $56.01, \pm .007$

In a still later paper* the same investigators give another series of results based upon the ignition of Iceland spar. The impurities were carefully estimated, and the percentages of lime are suitably corrected:

4.2134 grm. $CaCO_3$ gave	2.3594 grm. CaO.	55.997 per cent.
15.1385 "	8.4810 "	56.022 "
23.5503 "	13.1958 "	56.031 "
23.6390 "	13.2456 "	56.032 "
42.0295 "	23.5533 "	56.044 "
49.7007 "	27.8536 "	56.042 "

Mean, $56.028, \pm .0047$

Six years later Erdmann and Marchand † published one more result upon the ignition of calcium carbonate. They found that the compound began giving off carbon dioxide below the temperature at which their previous samples had been dried, or about 200°, and that, on the other hand, traces of the dioxide were retained by the lime after ignition. These two errors do not compensate each other, since both tend to raise the percentage of lime. In the one experiment now under consideration these errors were accurately estimated, and the needful corrections were applied to the final result. The percentage of residual lime in this case came out 55.998. This agrees tolerably well with the figures found in the direct estimation of carbonic acid, and, if combined with those two, gives a mean for all three of $56.006, \pm .0043$.

Combining all these series, we get the following result:

Dumas	56.073,	± .016
Erdmann and Marchand	56.006,	± .007
Erdmann and Marchand	56.028,	± .0047
Erdmann and Marchand	56.006,	± .0043
General mean	56.0198,	± .0029

For reasons given above, this mean is probably vitiated by a slight constant error, which makes the figure a trifle too high.

---

* Journ. für Prakt. Chem., 31, 269. 1844.
† Journ. für Prakt. Chem., 50, 237. 1850.

In the earliest of the three papers by Erdmann and Marchand there is also given a series of determinations of the ratio between calcium carbonate and sulphate. Pure Iceland spar was carefully converted into calcium sulphate, and the gain in weight noted. One hundred parts of spar gave of sulphate:

$$136.07$$
$$136.06$$
$$136.02$$
$$136.06$$

Mean, $136.0525, \pm .0071$

In 1843 the atomic weight of calcium was redetermined by Berzelius,* who investigated the ratio between lime and calcium sulphate. The calcium was first precipitated from a pure solution of nitrate by means of ammonium carbonate, and the thoroughly washed precipitate was dried and strongly ignited in order to obtain lime wholly free from extraneous matter. This lime was then, with suitable precautions, treated with sulphuric acid, and the resulting sulphate was weighed. Correction was applied for the trace of solid impurity contained in the acid, but not for the weighing in air. The figures in the last column represent the percentage of weight gained by the lime upon conversion into sulphate:

1.80425 grm. CaO gained	2.56735 grm.		142.295
2.50400	"	3.57050 "	142.592
3.90000	"	5.55140 "	142.343
3.04250	"	4.32650 "	142.202
3.45900	"	4.93140 "	142.567

Mean, $142.3998, \pm .0518$

Last of all we have the ratio between calcium chloride and silver, as determined by Dumas.† Pure calcium chloride was first ignited in a stream of dry hydrochloric acid, and the solution of this salt was afterwards titrated with a silver solution in the usual way. The $CaCl_2$ proportional to 100 parts of Ag is given in a third column:

2.738 grm. $CaCl_2$ =	5.309 grm. Ag.		51.573
2.436	"	4.731 "	51.490
1.859	"	3.617 "	51.396
2.771	"	5.3885 "	51.424
2.240	"	4.3585 "	51.394

Mean, $51.4554, \pm .0230$

We have now four ratios to compute from, as follows:

(1.) Percentage CaO in $CaCO_3$, $56.0198, \pm .0029$
(2.) CaO : SO_3 :: 100 : $142.3998, \pm .0518$
(3.) $CaCO_3$: $CaSO_4$ :: 100 : $136.0525, \pm .0071$
(4.) Ag_2 : $CaCl_2$ :: 100 : $51.4554, \pm .0230$

---

* Journ. für Prakt. Chem., 31, 263. Ann. Chem. Pharm., 46, 241.
† Ann. Chim. Phys. (3), 55, 129. 1859. Ann. Chem. Pharm., 113, 34.

The antecedent values are—

$O = 15.879, \pm .0003$ $C = 11.920, \pm .0004$
$Ag = 107.108, \pm .0031$ $S = 31.828, \pm .0015$
$Cl = 35.179, \pm .0048$

Hence the subjoined values for the atomic weight of calcium:

From (1) .... $Ca = 39.757, \pm .0048$
From (2) .... " $= 39.925, \pm .0203$
From (3) .... " $= 39.706, \pm .0204$
From (4) .... " $= 39.868, \pm .0503$

Mean .... $Ca = 39.764, \pm .0045$

If $O = 16$, $Ca = 40.067$.

STRONTIUM.

The ratios which fix the atomic weight of strontium resemble in general terms those relating to barium, only they are fewer in number and represent a smaller amount of work. The early experiments of Stromeyer,* who measured the volume of CO_2 evolved from a known weight of strontium carbonate, are hardly available for the present discussion. So also we may exclude the determination by Salvétat,† who neglected to publish sufficient details.

Taking the ratio between strontium chloride and silver first in order, we have series of figures by Pelouze and by Dumas. Pelouze ‡ employed the volumetric method to be described under barium, and in two experiments obtained the subjoined results. In another column I append the ratio between $SrCl_2$ and 100 parts of silver:

1.480 grm. $SrCl_2 = 2.014$ grm. Ag.   73.486
2.210   "   3.008   "   73.471

Mean, 73.4781, $\pm .0050$

Dumas, § by the same general method, made sets of experiments with three samples of chloride which had previously been fused in a current of dry hydrochloric acid. His results, expressed in the usual way, are as follows:

---
* Schweigg. Journ., 19. 228. 1816.
† Compt. Rend., 17, 318 1843.
‡ Compt. Rend., 20, 1047. 1845.
₰ Ann. Chim. Phys. (3), 55, 29. 1859. Ann Chem. Pharm., 113, 34.

Series A.

3.137 grm. $SrCl_2$ =	4.280 grm. Ag.	Ratio,	73.2944
1.982 "	2.705 "	"	73.2717
3.041 "	4.142 "	"	73.4186
3.099 "	4.219 "	"	73.4534
		Mean,	73.3595

Series B.

3.356 grm. $SrCl_2$ =	4.574 grm. Ag.	Ratio,	73.3713
6.3645 "	8.667 "	"	73.4327
7.131 "	9.712 "	"	73.4246
		Mean,	73.4095

Series C.

7.213 grm. $SrCl_2$ =	9.811 grm. Ag.	Ratio,	73.5195
2.206 "	3.006 "	"	73.3866
4.268 "	5.816 "	"	73.5529
4.018 "	5.477 "	"	73.3613
		Mean,	73.4551

Mean of all as one series, 73.4079, ± .0170

Combining these data we have:

Pelouze . . . . 73.4781, ± .0050
Marignac . . . . 73.4079, ± .0170

General mean . . . . 73.4725, ± .0048

The foregoing figures apply to anhydrous strontium chloride. The ratio between silver and the crystallized salt, $SrCl_2.6H_2O$, has also been determined in two series of experiments by Marignac.* Five grammes of salt were used in each estimation, and, in the second series, the percentage of water was first determined. The quantities of the salt corresponding to 100 parts of silver are given in the last column:

Series A.

5 grm. $SrCl_2.6H_2O$ =	4.0515 grm. Ag.	123.411
" "	4.0495 "	123.472
" "	4.0505 "	123.442
	Mean,	123.442

Series B.

5 grm. $SrCl_2.6H_2O$ =	4.0490 grm. Ag.	123.487
" "	4.0500 "	123.457
" "	4.0490 "	123.487
	Mean,	123.477

Mean of all as one series, 123.460, ± .0082

---

* Journ. für Prakt. Chem., 74, 216. 1858.

In the same paper Marignac gives two sets of determinations of the percentage of water in crystallized strontium chloride. The first set, corresponding to "B" above, is as follows:

$$40.556$$
$$40.568$$
$$40.566$$
$$\text{Mean, } 40.563$$

In the second set ten grammes of salt were taken at a time, and the following percentages were found:

$$40.58$$
$$40.59$$
$$40.58$$
$$\text{Mean, } 40.583$$
$$\text{Mean of all as one series, } 40.573, \pm .0033$$

The chloride used in the series of estimations last given was subsequently employed for ascertaining the ratio between it and the sulphate. Converted directly into sulphate, 100 parts of chloride yield the quantities given in the third column:

5.942 grm. $SrCl_2$ gave	6.887 grm. $SrSO_4$.		115.932
5.941 "	6.8855 "		115.949
5.942 "	6.884 "		115.927
		Mean,	115.936, $\pm .004$

Richards,[*] in his study of strontium bromide, followed pretty much the lines laid down in his work on barium. The properties of the bromide itself were carefully investigated, and its purity established beyond reasonable doubt, and then the two usual ratios were determined. First, the ratio $Ag_2 : SrBr_2 :: 100 : x$, by titration with standard solutions of silver. For this ratio there are three series of measurements, by varied processes, concerning which full details are given. The data obtained, with weights reduced to a vacuum, are as follows:

First Series.

Wt. Ag.	Wt. $SrBr_2$.	Ratio.
1.30755	1.49962	114.689
2.10351	2.41225	114.677
2.23357	2.56153	114.683
5.3684	6.15663	114.683
	Mean,	114.683

---

[*] Proc. Amer. Acad. of Sciences, 1894, p. 369.

THE ATOMIC WEIGHTS.

Second Series.

Wt. Ag.	Wt. SrBr$_2$.	Ratio.
1.30762	1.49962	114.683
2.10322	2.41225	114.693
4.57502	5.24727	114.694
5.3680	6.15663	114.691

Mean, 114.690

Third Series.

2.5434	2.9172	114.697
3.3957	3.8946	114.692
3.9607	4.5426	114.692
4.5750	5.2473	114.695

Mean, 114.694

Mean of all as one series, 114.689, ± .0012

For the ratio, measured gravimetrically, 2AgBr : SrBr$_2$: : 100 : x, two series of determinations are given:

First Series.

Wt. AgBr.	Wt. SrBr$_2$.	Ratio.
2.4415	1.6086	65.886
2.8561	1.8817	65.884
6.9337	4.5681	65.883

Mean, 65.884

Second Series.

2.27625	1.49962	65.881
3.66140	2.41225	65.883
3.88776	2.56153	65.887
9.34497	6.15663	65.882

Mean, 65.883

Mean of all as one series, 65.884, ± .0006

For the atomic weight of strontium we now have the subjoined ratios:

(1.) Ag$_2$: SrCl$_2$: : 100 : 73.4725, ± .0048
(2.) Ag$_2$: SrCl$_2$.6H$_2$O : : 100 : 123.460, ± .0082
(3.) Per cent. H$_2$O in SrCl$_2$.6H$_2$O, 40.573, ± .0033
(4.) SrCl$_2$: SrSO$_4$: : 100 : 115.936, ± .0040
(5.) Ag$_2$: SrBr$_2$: : 100 : 114.689, ± .0012
(6.) 2AgBr : SrBr$_2$: : 100 : 65.884, ± .0006

The antecedent values are—

O. = 15.879, ± .0003 Br = 79.344, ± .0062
Ag = 107.108, ± .0031 S = 31.828, ± .0015
Cl = 35.179, ± .0048 AgBr = 186.452, ± .0054

STRONTIUM.

For the molecular weight of $SrCl_2$ three estimates are available:

From (1)............................	$SrCl_2 = 157.390, \pm .0112$
From (2)............................	" $= 157.197, \pm .0192$
From (3)............................	" $= 157.123, \pm .0157$
General mean............	$SrCl_2 = 157.281, \pm .0083$

For $SrBr_2$ there are two values:

From (5)............................	$SrBr_2 = 245.682, \pm .0076$
From (6)............................	" $= 245.684, \pm .0075$
General mean............	$SrBr_2 = 245.683, \pm .0053$

Finally, with these intermediate data we obtain three independent measures of the atomic weight of strontium, as follows:

From molecular weight $SrCl_2$.........	$Sr = 86.923, \pm .0127$
From molecular weight $SrBr_2$.........	" $= 86.995, \pm .0135$
From ratio (4)....................	" $= 86.434, \pm .0811$
General mean................	$Sr = 86.948, \pm .0092$

If $O = 16$, $Sr = 87.610$. Rejection of the third value, which is worthless, raises these means by 0.01 only. The second value, 86.995, which represents Richards' work, is undoubtedly the best of the three.

BARIUM.

For the atomic weight of barium we have a series of eight ratios, established by the labors of Berzelius, Turner, Struve, Marignac, Dumas, and Richards. Andrews* and Salvétat,† in their papers upon this subject, gave no details nor weighings, and therefore their work may be properly disregarded. First in order, we may consider the ratio between silver and barium chloride, as determined by Pelouze, Marignac, Dumas, and Richards.

Pelouze, ‡ in 1845, made the three subjoined estimations of this ratio, using his well known volumetric method. A quantity of pure silver was dissolved in nitric acid, and the amount of barium chloride needed to precipitate it was carefully ascertained. In the last column I give the quantity of barium chloride proportional to 100 parts of silver:

3.860 grm.	$BaCl_2$ ppt.	4.002 grm.	Ag.	96.452
5.790	"	6.003	"	96.452
2.895	"	3.001	"	96.468

Mean, 96.4573, ± .0036

Essentially the same method was adopted by Marignac § in 1848. His experiments were made upon four samples of barium chloride, as follows. A, commercial barium chloride, purified by recrystallization from water. B, the same salt, calcined, redissolved in water, the solution saturated with carbonic acid, filtered, and allowed to crystallize. C, the preceding salt, washed with alcohol, and again recrystallized. D, the same, again washed with alcohol. For 100 parts of silver the following quantities of chloride were required, as given in the third column:

	Ag.	$BaCl_2$.	Ratio.	
A.	3.4445	3.3190	96.356	
	3.7480	3.6110	96.345	Mean, 96.354
	6.3446	6.1140	96.362	
B.	4.3660	4.1780	96.356	
	4.8390	4.6625	96.352	Mean, 96.354
C.	6.9200	6.6680	96.358	
	5.6230	5.4185	96.363	Mean, 96.360
D.	5.8435	5.6300	96.346	
	8.5750	8.2650	96.384	
	4.8225	4.6470	96.361	Mean, 96.367
	6.8460	6.5980	96.377	

Mean, 96.360, ± .0024

---
* Chemical Gazette, October, 1852.
† Compt. Rend., 17, 318.
‡ Compt. Rend., 20, 1047. Journ. für Prakt. Chem., 35, 73.
§ Arch. d. Sci. Phys. et Nat., 8, 271.

Dumas* employed barium chloride prepared from pure barium nitrate, and took the extra precaution of fusing the salt at a red heat in a current of dry hydrochloric acid gas. Three series of experiments upon three samples of chloride gave the following results:

	$Ag.$	$BaCl_2.$	Ratio.	
A.	1.8260	1.7585	96.303	
	3.9980	3.8420	96.339	Mean, 96.333
	2.2405	2.1585	96.340	
	4.1680	4.0162	96.358	
B.	1.7270	1.6625	96.265	
	2.5946	2.4987	96.304	
	3.5790	3.4468	96.306	
	4.2395	4.0822	96.290	Mean, 96.290
	4.3683	4.2062	96.289	
	4.6290	4.4564	96.271	
	9.0310	8.6975	96.307	
C.	2.3835	2.2957	96.316	
	4.2930	4.1372	96.371	
	4.4300	4.2662	96.303	Mean, 96.338
	4.6470	4.4764	96.329	
	5.8520	5.6397	96.372	

Mean, 96.316, ± .0055

The work done by Richards† was of a much more elaborate kind, for it involved some collateral investigations as to the effect of heat upon barium chloride, etc. Every precaution was taken to secure the spectroscopic purity of the material, which was prepared from several sources, and similar care was taken with regard to the silver. For details upon these points the original paper must be consulted. As for the titrations, three methods were adopted, and a special study was made with reference to the accurate determination of the end point; in which particular the investigations of Pelouze, Marignac, and Dumas were at fault. In the first series of determinations, silver was added in excess, and the latter was measured with a standard solution of hydrochloric acid. The end point was ascertained by titrating backward and forward with silver solution and acid, and was taken as the mean between the two apparent end points thus observed. The results of this series, with weights reduced to vacuum standards, were as follows:

$Ag.$	$BaCl_2.$	Ratio.
6.1872	5.9717	96.517
5.6580	5.4597	96.495
3.5988	3.4728	96.499
9.4010	9.0726	96.507
.7199	.6950	96.541

Mean, 96.512, ± .0055

---

*Ann. Chem. Pharm., 113, 22. 1860. Ann. Chim. Phys. (3), 55, 129.
† Proc. Amer. Acad., 29, 55. 1893.

In the second series of experiments a small excess of silver was added as before, and the precipitate of silver chloride was removed by filtration. The filtrate and wash waters were concentrated to small bulk whereupon a trace of silver chloride was obtained and taken into account. The excess of silver remaining was then thrown down as silver bromide, and from the weight of the latter the silver was calculated, and subtracted from the original amount.

$Ag.$	$BaCl_2.$	$Ratio.$
6.59993	6.36974	96.512
5.55229	5.36010	96.539
4.06380	3.92244	96.522

Mean, $96.524, \pm .0054$

The third series involved mixing solutions of barium chloride and silver in as nearly as possible equivalent amounts, and then determining the actual quantities of silver and chlorine left unprecipitated. The filtrate and wash waters were divided into two portions, one-half being evaporated with hydrobromic acid and the other with silver nitrate. The small amounts of silver bromide and chloride thus obtained were determined by reduction and the use of Volhard's method:

$Ag.$	$BaCl_2.$	$Ratio.$
4.4355	4.2815	96.528
2.7440	2.6488	96.531
6.1865	5.9712	96.520
3.4023	3.2841	96.526

Mean, $96.526, \pm .0035$

Two final experiments were carried out by Stas' method, somewhat as in the first series, with variations and greater refinement in the observation of the end point. The results were as follows:

$Ag.$	$BaCl_2.$	$Ratio.$
6.7342	6.50022	96.525
10.6023	10.23365	96.523

Mean, $96.524, \pm .0007$

A careful study of Richards' paper will show that, although the last two experiments are probably the best, they are not entitled to such preponderance of weight as the "probable error" here computed would give them. I therefore treat Richards' work as I have already done that of Marignac and Dumas, regarding all of his series as one, which gives for the value of the ratio $96.520, \pm .0025$. This combines with the previous series thus:

Pelouze .... 96.457, ± .0036
Marignac .... 96.360, ± .0024
Dumas .... 96.316, ± .0055
Richards .... 96.520, ± .0025

General mean .... 96.434, ± .0015

The ratio between silver and crystallized barium chloride has also been fixed by Marignac.* The usual method was employed, and two series of experiments were made, in the second of which the water of crystallization was determined previous to the estimation. Five grammes of chloride were taken in each determination. The following quantities of $BaCl_2.2H_2O$ correspond to 100 parts of silver:

A. $\begin{cases} 113.109 \\ 113.135 \\ 113.097 \end{cases}$ Mean, 113.114

B. $\begin{cases} 113.135 \\ 113.122 \\ 113.060 \end{cases}$ Mean, 113.106

Mean, 113.110, ± .0079

The direct ratio between the chlorides of silver and barium has been measured by Berzelius, Turner, and Richards. Berzelius† found of barium chloride proportional to 100 parts of silver chloride—

72.432
72.422

Mean, 72.427

Turner‡ made five experiments, with the following results:

72.754
72.406
72.622
72.664
72.653

Mean, 72.680, ± .0154

Of these, Turner regards the fourth and fifth as the best; but for present purposes it is not desirable to so discriminate.

Richards' determinations § fall into three series, and all are characterized by their taking into account chloride of silver recovered from the wash waters. In the first series the barium chloride was ignited at low redness in air or nitrogen; in the second series it was fused in a stream of pure hydrochloric acid; and in the third series it was not ignited at all. In the last series it was weighed in the crystallized state, and the

---

* Journ. für Prakt. Chem., 74, 212. 1858.
† Poggend. Annalen, 8, 177.
‡ Phil. Trans., 1829. 291.
§ Proc. Amer. Acad., 29. 55. 1893.

amount of anhydrous chloride was computed from the data so obtained. The data, corrected to vacuum standards, are as follows:

	AgCl.	BaCl$_2$.	Ratio.	
A.	8.7673	6.3697	72.653	
	5.1979	3.7765	72.654	
	4.9342	3.5846	72.648	Mean, 72.649
	2.0765	1.5085	72.646	
	4.4271	3.2163	72.650	
B.	2.09750	1.52384	72.650	
	7.37610	5.36010	72.669	Mean, 72.6563
	5.39906	3.92244	72.650	
C.	8.2189	5.97123	72.6524	Mean, 72.6555
	4.5199	3.28410	72.6587	

Mean, 72.653, ± .0014

If we assign Berzelius' work equal weight with that of Turner, the three series representing the ratio 2AgCl : BaCl$_2$ combine as follows:

Berzelius .... 72.427, ± .0154
Turner................................. 72.680, ± .0154
Richards............................... 72.653, ± .0014

General mean.................... 72.650, ± .0014

Incidentally to some of his other work, Marignac* determined the percentage of water in crystallized barium chloride. Two sets of three experiments each were made, the first upon five grammes and the second upon ten grammes of salt. The following are the percentages obtained:

A.	14.790	
	14.796	Mean, 14.795
	14.800	
B.	14.80	
	14.81	Mean, 14.803
	14.80	

Mean, 14.799, ± .0018

The ratio between barium nitrate and barium sulphate has been determined only by Turner.† According to his experiments 100 parts of sulphate correspond to the following quantities of nitrate:

112.060
111.990
112.035

Mean, 112.028, ± .014

For the similar ratio between barium chloride and barium sulphate, there are available determinations by Turner, Berzelius, Struve, Marignac, and Richards.

---

* Journ. für Prakt. Chem., 74, 312. 1858.
† Phil. Trans., 1833. 538.

Turner* found that 100 parts of chloride ignited with sulphuric acid gave 112.19 parts of sulphate. By the common method of precipitation and filtration a lower figure was obtained, because of the slight solubility of the sulphate. This point bears directly upon many other atomic weight determinations.

Berzelius,† treating barium chloride with sulphuric acid, obtained the following results in $BaSO_4$ for 100 parts of $BaCl_2$:

112.17
112.18
———
Mean, 112.175

Struve,‡ in two experiments, found:

112.0912
112.0964
———
Mean, 112.0938

Marignac's § three results are as follows:

8.520 grm. $BaCl_2$ gave 9.543 $BaSO_4$.   Ratio, 112.007
8.519   "   9.544   "   "   112.032
8.520   "   9.542   "   "   111.995
                                    ———————
                              Mean, 112.011, ± .0071

Richards, in his work on this ratio, regards the results as of slight value, because of the occlusion of the chloride by the sulphate. This source of error he was never able to avoid entirely. Another error in the opposite direction is found in the retention of sulphuric acid by the precipitated sulphate. Eight experiments were made in two series, one set by adding sulphuric acid to a strong solution of barium chloride in a platinum crucible, the other by precipitation in the usual way. Richards gives in his published paper only the end results and the mean of his determinations; the details cited below I owe to his personal kindness. The weights are reduced to vacuum standards:

	$BaCl_2$.	$BaSO_4$.	Ratio.
First.	1.78934	2.0056	112.086
	2.07670	2.3274	112.072
	1.58311	1.7741	112.064
	3.27563	3.6712	112.076
	3.02489	3.3903	112.080
	3.87091	4.3385	112.080
Second.	3.02489	3.9726	112.076
	3.87091	3.4880	112.085

Mean, 112.077, ± .0017

---

* Phil. Trans., 1829, 291.
† Poggend. Annalen, 8, 177.
‡ Ann. Chem. Pharm., 80, 204. 1851.
§ Journ. für Prakt. Chem., 74, 212. 1858.

This mean is subject to a small correction due to loss of chlorine on drying the chloride, which reduces it to 112.073. Omitting Turner's single determination as unimportant, and assigning to the work of Berzelius and of Struve equal weight with that of Marignac, the measurements of this ratio combine thus:

Berzelius	112.175, ± .0071
Struve	112.094, ± .0071
Marignac	112.011, ± .0071
Richards	112.073, ± .0017
General mean	112.075, ± .0016

In an earlier paper than the one previously cited, Richards* studied with great care the ratios connecting barium bromide with silver and silver bromide. The barium bromide was prepared by several distinct processes, its behavior upon dehydration and even upon fusion was studied, and its specific gravity was determined. The ratio with silver was measured by titration, a solution of hydrobromic acid being used for titrating back. The data are subjoined, with the $BaBr_2$ equivalent to 100 parts of silver stated:

$BaBr_2$.	Ag.	Ratio.
2.28760	1.66074	137.746
3.47120	2.52019	137.736
2.19940	1.59687	137.732
2.35971	1.71323	137.735
2.94207	2.13584	137.748
1.61191	1.17020	137.747
2.10633	1.52921	137.740
2.19682	2.11740	137.755
2.37290	1.72276	137.738
1.84822	1.34175	137.747
5.66647	4.11360	137.750
3.52670	2.56010	137.756
4.31690	3.13430	137.731
3.36635	2.44385	137.748
3.46347	2.51415	137.759
	Mean,	137.745, ± .0015

The silver bromide in most of these determinations, and in some others, was collected and weighed in a Gooch crucible with all necessary precautions. Vacuum standards were used throughout for both ratios. I give in a third column the $BaBr_2$ equivalent to 100 parts of AgBr:

---

*Proc. Amer. Acad., 28. 1893.

$BaBr_2$.	$AgBr$.	Ratio.
2.28760	2.89026	79.149
3.47120	4.38635	79.136
3.81086	4.81688	79.133
2.35971	2.98230	79.124
2.94207	3.71809	79.129
2.10633	2.66191	79.128
2.91682	3.68615	79.129
2.37290	2.99868	79.131
1.84822	2.33530	79.143
1.90460	2.40733	79.116
5.66647	7.16120	79.127
3.52670	4.45670	79.133
2.87743	3.63644	79.127
3.46347	4.37669	79.135

Mean, 79.132, ± .0015

The ratios for barium now sum up as follows:

(1.) $Ag_2 : BaCl_2 :: 100 : 96.434$, ± .0015
(2.) $Ag_2 : BaCl_2.2H_2O :: 100 : 113.110$, ± .0079
(3.) $2AgCl : BaCl_2 :: 100 : 72.650$, ± .0014
(4.) Per cent. of H_2O in $BaCl_2.2H_2O$, 14.799, ± .0018
(5.) $BaSO_4 : BaN_2O_6 :: 100 : 112.028$, ± .014
(6.) $BaCl_2 : BaSO_4 :: 100 : 112.075$, ± .0016
(7.) $Ag_2 : BaBr_2 :: 100 : 137.745$, ± .0015
(8.) $2AgBr : BaBr_2 :: 100 : 79.132$, ± .0015

The reduction of these ratios depends upon the subjoined antecedent values:

$Ag = 107.108$, ± .0031
$Cl = 35.179$, ± .0048
$Br = 79.344$, ± .0062
$O = 15.879$, ± .0003
$N = 13.935$, ± .0021
$S = 31.828$, ± .0015
$AgCl = 142.287$, ± .0037
$AgBr = 186.452$, ± .0054

With these factors four estimates are obtainable for the molecular weight of barium chloride:

From (1)............... $BaCl_2 = 206.577$, ± .0068
From (2)............... " $= 206.542$, ± .0183
From (3)............... " $= 206.745$, ± .0067
From (4)............... " $= 205.866$, ± .0257

General mean.......... $BaCl_2 = 206.629$, ± .0045

For barium bromide we have:

From (7)............... $BaBr_2 = 295.070$, ± .0091
From (8)............... " $= 295.086$, ± .0102

General mean.......... $BaBr_2 = 295.078$, ± .0068

And for barium itself, four values are finally available, thus:

From molecular weight $BaCl_2$	$Ba = 136.271, \pm .0106$
From molecular weight $BaBr_2$	" $= 136.390, \pm .0141$
From ratio (5)	" $= 135.600, \pm .2711$
From ratio (6)	" $= 136.563, \pm .0946$
General mean	$Ba = 136.315, \pm .0085$

Or, if $O = 16$, $Ba = 137.354$.

In the foregoing computation all the data, good or bad, are included. Some of them, as shown by the weights, practically vanish; but others, as in the chloride series, carry an undue influence. A more trustworthy result can be deduced from Richards' experiments alone, which reduce as follows:

From $Ag_2 : BaCl_2$	$BaCl_2 = 206.761, \pm .0080$
From $2AgCl : BaCl_2$	" $= 206.754, \pm .0067$
General mean	$BaCl_2 = 206.755, \pm .0051$

From the bromide, as given above, $Ba = 136.390, \pm .0141$. From the value just found for the chloride, $Ba = 136.397, \pm .0109$. Combining the two values—

$$Ba = 136.392, \pm .0086.$$

Or, if $O = 16$, $Ba = 137.434$. This determination will be adopted in subsequent calculations as the most probable.

LEAD.

For the atomic weight of lead we have to consider experiments made upon the oxide, chloride, nitrate, and sulphate. The researches of Berzelius upon the carbonate and various organic salts need not now be considered, nor is it worth while to take into account any work of his done before the year 1818. The results obtained by Döbereiner* and by Longchamp † are also without special present value.

For the exact composition of lead oxide we have to depend upon the researches of Berzelius. His experiments were made at different times through quite a number of years; but were finally summed up in the last edition of his famous "Lehrbuch."‡ In general terms his method of experiment was very simple. Perfectly pure lead oxide was heated in a current of hydrogen, and the reduced metal weighed. From his weighings I have calculated the percentages of lead thus found and given them in a third column:

Earlier Results.

8.045 grm. PbO gave	7.4675 grm. Pb.	92.8217 per cent.
14.183 "	13.165 "	92.8224 "
10.8645 "	10.084 "	92.8160 "
13.1465 "	12.2045 "	92.8346 "
21.9425 "	20.3695 "	92.8313 "
11.159 "	10.359 "	92.8309 "

Latest.

6.6155 "	6.141 "	92.8275 "
14.487 "	13.448 "	92.8280 "
14.626 "	13.5775 "	92.8313 "

Mean, 92.8271, ± .0013

For the synthesis of lead sulphate we have data by Berzelius, Turner, and Stas. Berzelius,§ whose experiments were intended rather to fix the atomic weight of sulphur, dissolved in each estimation ten grammes of pure lead in nitric acid, then treated the resulting nitrate with sulphuric acid, brought the sulphate thus formed to dryness, and weighed. One hundred parts of metal yield of $PbSO_4$:

146.380
146.400
146.440
146.458

Mean, 146.419, ± .012

---

* Schweig. Journ., 17, 241. 1816.
† Ann. Chim. Phys., 34, 105. 1827.
‡ Bd. 3, s. 1218.
§ Lehrbuch, 5th ed., 3, 1187.

Turner,* in three similar experiments, found as follows:

$$146.430$$
$$146.398$$
$$146.375$$

Mean, $146.401, \pm .011$

In these results of Turner's, *absolute* weights are implied.

The results of Stas' syntheses,† effected after the same general method, but with variations in details, are as follows. Corrections for weighing in air were applied:

$$146.443$$
$$146.427$$
$$146.419$$
$$146.432$$
$$146.421$$
$$146.423$$

Mean, $146.4275, \pm .0024$

Combining, we get the subjoined result:

Berzelius .... $146.419, \pm .012$
Turner .... $146.401, \pm .011$
Stas .... $146.4275, \pm .0024$

General mean .... $146.4262, \pm .0023$

Turner, in the same paper, also gives a series of syntheses of lead sulphate, in which he starts from the oxide instead of from the metal. One hundred parts of PbO, upon conversion into $PbSO_4$, gained weight as follows:

$$35.84$$
$$35.71$$
$$35.84$$
$$35.75$$
$$35.79$$
$$35.78$$
$$35.92$$

Mean, $35.804, \pm .018$

These figures are not wholly reliable. Numbers one, two, and three represent lead oxide contaminated with traces of nitrate. The oxide of four, five, and six contained traces of minium. Number seven was free from these sources of error, and, therefore, deserves more consideration. The series as a whole undoubtedly gives too low a figure, and this error would tend to slightly raise the atomic weight of lead.

---

* Phil. Trans., 1833, 527-538.
† Aronstein's translation, 333.

LEAD.   129

Still a third series by Turner establishes the ratio between the nitrate and the sulphate, a known weight of the former being in each experiment converted into the latter. One hundred parts of sulphate represent of nitrate:

109.312
109.310
109.300

Mean, 109.307, ± .002

In all these experiments by Turner the necessary corrections were made for weighing in air.

In 1846 Marignac* published two sets of determinations of only moderate value. First, chlorine was conducted over weighed lead, and the amount of chloride so formed was determined. The lead chloride was fused before weighing. The ratio to 100 Pb is given in the last column:

20.506 grm. Pb gave 27.517 $PbCl_2$.   134.190
16.281   "   21.858   "   134.225
25.454   "   34.149   "   134.159

Mean, 134.191, ± .013

Secondly, lead chloride was precipitated by silver nitrate and the ratio between $PbCl_2$ and 2AgCl determined. The third column gives the AgCl formed by 100 parts of $PbCl_2$:

12.534 grm. $PbCl_2$ gave 12.911 AgCl.   103.01
14.052   "   14.506   "   103.23
25.533   "   26.399   "   103.39

Mean, 103.21, ± .0745

For the ratio between lead chloride and silver we have a series of results by Marignac and one experiment by Dumas. There are also unavailable data by Turner and by Berzelius.

Marignac,† applying the method used in his researches upon barium and strontium, and working with lead chloride which had been dried at 200°, obtained these results. The third column gives the ratio between $PbCl_2$ and 100 parts of Ag:

4.9975 grm. $PbCl_2$ = 3.8810 grm. Ag.   128.768
4.9980   "   3.8835   "   128.698
5.0000   "   3.8835   "   128.750
5.0000   "   3.8860   "   128.667

Mean, 128.721, ± .016

Dumas,‡ in his investigations, found that lead chloride retains traces

---
* Ann. Chem. Pharm., 59, 289, and 290. 1846.
† Journ. für Prakt. Chem., 74, 218. 1858.
‡ Ann. Chem. Pharm., 113, 35. 1860.

of water even at 250°, and is sometimes also contaminated with oxychloride. In one estimation 8.700 grammes $PbCl_2$ saturated 6.750 of Ag. The chloride contained .009 of impurity; hence, correcting, Ag : $PbCl_2$ :: 100 : 128.750. If we assign this figure equal weight with those of Marignac, we get as the mean of all 128.7266, ± .013. The sources of error indicated by Dumas, if they are really involved in this mean, would tend slightly to raise the atomic weight of lead.

The synthesis of lead nitrate, as carried out by Stas,* gives excellent results. Two series of experiments were made, with from 103 to 250 grammes of lead in each determination. The metal was dissolved in nitric acid, the solution evaporated to dryness with extreme care, and the nitrate weighed. All weighings were reduced to the vacuum standard. In series A the lead nitrate was dried in an air current at a temperature of about 155.° In series B the drying was effected in vacuo. 100 of lead yield of nitrate:

A.

159.973
159.975
159.982
159.975
159.968
159.973

Mean, 159.9743, ± .0012

B.

159.970
159.964
159.959
159.965

Mean, 159.9645, ± .0015
Mean from both series, 159.9704, ± .0010

There is still another set of experiments upon lead nitrate, originally intended to fix the atomic weight of nitrogen, which may properly be included here. It was carried out by Anderson† in Svanberg's laboratory, and has also appeared under Svanberg's name. Lead nitrate was carefully ignited, and the residual oxide weighed, with the following results:

5.19485 grm. PbN_2O_6 gave	3.5017 grm. PbO.	67.4071 per cent.	
9.7244 "	6.5546 "	67.4037 "	
9.2181 "	6.2134 "	67.4044 "	
9.6530 "	6.5057 "	67.3957 "	

Mean, 67.4027, ± .0016

---

* Aronstein's translation, 316.
† Ann. Chim. Phys. (3), 9. 254. 1843.

We have now nine ratios from which to compute:

(1.) Per cent. of Pb in PbO, 92.8271, ± .0013
(2.) Per cent of PbO in PbN_2O_6, 67.4027, ± .0016
(3.) Pb : $PbSO_4$ :: 100 : 146.4262, ± .0023
(4.) PbO : $PbSO_4$ :: 100 : 135.804, ± .0180
(5.) $PbSO_4$: PbN_2O_6 :: 100 : 109.307, ± .0020
(6.) Pb : PbN_2O_6 :: 100 : 159.9704, ± .0010
(7.) Pb : $PbCl_2$ :: 100 : 134.191, ± .013
(8.) $PbCl_2$: 2AgCl :: 100 : 103.21, ± .0745
(9.) Ag_2 : $PbCl_2$ :: 100 : 128.7266, ± .0130

To reduce these ratios we must use the following data:

O = 15.879, ± .0003   S = 31.828, ± .0015
Ag = 107.108, ± .0031   N = 13.935, ± .0021
Cl = 35.179, ± .0048   AgCl = 142.287, ± .0037

For the molecular weight of lead oxide we now get three estimates:

From (1) .... PbO = 221.375, ± .0403
From (2) .... "   = 221.796, ± .0132
From (4) .... "   = 221.944, ± .1116

General mean .... PbO = 221.757, ± .0125

For lead chloride we have—

From (8) .... $PbCl_2$ = 275.723, ± .1989
From (9) .... "   = 275.753, ± .0290

General mean .... $PbCl_2$ = 275.752, ± .0287

Including these results, six values are calculable for the atomic weight of lead:

From molecular weight of PbO ...... Pb = 205.878, ± .0126
From molecular weight of $PbCl_2$..... " = 205.394, ± .0302
From (3)......................... " = 205.367, ± .0051
From (5)......................... " = 203.352, ± .0479
From (6)......................... " = 205.341, ± .0068
From (7)......................... " = 205.779, ± .0831

General mean.............. Pb = 205.395, ± .0038

If O = 16, Pb = 206.960. If we reject the first, fourth, and sixth of these values, which are untrustworthy, the remaining second, third, and fifth give a general mean of Pb = 205.358, ± .0040. If O = 16, this becomes Pb = 206.923. From Stas' ratios alone Stas calculates Pb = 206.918 to 206.934; Ostwald finds 206.911; Van der Plaats (A), 206.9089, (B), 206.9308, and Thomsen 206.9042. The value adopted here represents mainly the work of Stas, and with H = 1 is

Pb = 205.358, ± .0040.

GLUCINUM.

Our knowledge of the atomic weight of glucinum is chiefly derived from experiments made upon the sulphate. Leaving out of account the single determination by Berzelius,[*] we have to consider the data furnished by Awdejew, Weeren, Klatzo, Debray, Nilson and Pettersson, and Krüss and Moraht.

Awdejew,[†] whose determination was the earliest of any value, analyzed the sulphate. The sulphuric acid was thrown down as barium sulphate; and in the filtrate, from which the excess of barium had been first removed, the glucina was precipitated by ammonia. The figures which Awdejew publishes represent the ratio between SO_3 and GlO, but not absolute weights. As, however, his calculations were made with $SO_3 = 501.165$, and Ba probably $= 855.29$, we may add a third column showing how much $BaSO_4$ is proportional to 100 parts of GlO:

SO_3.	GlO.	Ratio.
4457	1406	921.242
4531	1420	927.304
7816	2480	915.903
12880	4065	920.814

Mean, 921.316, ± 1.577

The same method was followed by Weeren and by Klatzo, except that Weeren used ammonium sulphide instead of ammonia for the precipitation of the glucina. Weeren[‡] gives the following weights of GlO and $BaSO_4$. The ratio is given in a third column, just as with the figures by Awdejew:

GlO.	$BaSO_4$.	Ratio.
.3163	2.9332	927.031
.2872	2.6377	918.419
.2954	2.7342	925.592
.5284	4.8823	902.946

Mean, 918.497, ± 3.624

Klatzo's [§] figures are as follows, with the third column added by the writer:

GlO.	$BaSO_4$.	Ratio.
.2339	2.1520	920.052
.1910	1.7556	919.162
.2673	2.4872	930.490
.3585	3.3115	923.710
.2800	2.5842	922.989

Mean, 923.281, ± 1.346

---

[*] Poggend. Annal., 8, 1.
[†] Poggend. Annal., 56, 106. 1842.
[‡] Poggend. Annal., 92, 124. 1854.
[§] Zeitschr. Anal. Chem., 8, 523. 1869.

Combining these series into a general mean, we get the subjoined result:

$$\begin{align}
\text{Awdejew} &\ldots\ldots\ldots\ldots\ 921.316, \pm 1.577 \\
\text{Weeren} &\ldots\ldots\ldots\ldots\ 918.497, \pm 3.624 \\
\text{Klatzo} &\ldots\ldots\ldots\ldots\ 923.281, \pm 1.346 \\
\text{General mean} &\ldots\ldots\ldots\ 922.164, \pm 0.985
\end{align}$$

Hence $GlO = 25.130, \pm .0269$.

Debray* analyzed a double oxalate of glucinum and ammonium, $Gl(NH_4)_2C_4O_8$. In this the glucina was estimated by calcination, after first converting the salt into nitrate. The following percentages were found:

$$\begin{align}
&11.5 \\
&11.2 \\
&11.6
\end{align}$$

Mean, $11.433, \pm .081$

The carbon was estimated by an organic combustion. I give the weights, and put in a third column the percentages of CO_2 thus obtained:

Salt.	CO_2.	Per cent. CO_2.
.600	.477	79.500
.603	.478	79.270
.600	.477	79.500

Mean, $79.423, \pm .052$

Calculating the ratio between CO_2 and GlO, we have for the molecular weight of the latter, $GlO = 25.151, \pm .1783$.

In 1880 the careful determinations of Nilson and Pettersson appeared.† These chemists first attempted to work with the sublimed chloride of glucinum, but abandoned the method upon finding the compound to be contaminated with traces of lime derived from a glass tube. They finally resorted to the crystallized sulphate as the most available salt for their purposes. This compound, upon strong ignition, yields pure glucina. The data are as follows:

$GlSO_4.4H_2O$.	GlO.	Per cent. GlO.
3.8014	.5387	14.171
2.6092	.3697	14.169
4.3072	.6099	14.160
3.0091	.4266	14.176

Mean, $14.169, \pm .0023$

Krüss and Moraht‡ in their work follow the general method adopted

---

* Ann. Chim. Phys. (3), 44, 37. 1855.
† Berichte d. Deutsch. Chem. Gesell., 13, 1451. 1880.
‡ Ann. d. Chem., 262, 38. 1891.

by Nilson and Pettersson, but with various added precautions and greater elaboration of detail. Their glucina was derived from three sources, namely, leucophane, beryl, and gadolinite, and the sulphate was repeatedly recrystallized. The results are subjoined:

$GlSO_4, 4H_2O$	GlO	Per cent. GlO
21.1916	3.0026	14.160
16.2056	2.3955	14.161
15.6826	2.2502	14.136
20.1056	2.8435	14.143
22.0265	3.1165	14.137
4.9609	.7009	14.126
16.3209	2.3502	14.145
21.3907	3.0053	14.143
20.1805	2.8755	14.135
20.0253	2.8316	14.146
18.9160	2.6832	14.134
17.2052	2.4073	14.155
22.5901	3.1805	14.133
20.8875	2.9565	14.154
19.0592	2.6995	14.130
17.8209	2.5206	14.153

Mean, 14.144 $\pm$ .0013

The first two determinations, which give the highest percentage, were made upon sulphate thrice crystallized. The others were made upon a salt four times crystallized, except in one instance, when there were five crystallizations. To the data derived from the four times crystallized compound Krüss and Moraht give preference, and so find a slightly lower value for the atomic weight of glucinum. Combining, we have for the mean percentages:

By Nilson and Pettersson .... 14.169, $\pm$ .0083
By Krüss and Moraht .... 14.144, $\pm$ .0017
                                        _____
General mean .... 14.153, $\pm$ .0014

Taking now all the data for glucinum, we have—

(1.) $GlO : BaSO_4 :: 100 : 922.164, \pm .465$
(2.) $4CO_2 : GlO :: 79.423, \pm .0032 : 11.433, \pm .061$
(3.) Percentage of GlO in $GlSO_4, 4H_2O$, 14.153, $\pm$ .0014

The antecedent atomic weights are—

$O = 15.879, \pm .0005$ $C = 11.920, \pm .0004$
$S = 31.828, \pm .0015$ $Ba = 136.392, \pm .0056$

Hence the subjoined values for glucina:

From (1)......................... GlO = 35.130, ± .0069
From (2)......................... " = 35.151, ± .1763
From (3)......................... " = 24.892, ± .0005

General mean............... GlO = 24.893, ± .0005
And Gl = 9.084, ± .0005

If O = 16, Gl = 9.083.

All the values but that derived from the third ratio might obviously be rejected. Their influence upon the final mean is altogether trivial.

MAGNESIUM.

There is perhaps no common metal of which the atomic weight has been subjected to closer scrutiny than that of magnesium. The value is low, and its determination should, therefore, be relatively free from many of the ordinary sources of error; it is extensively applied in chemical analysis, and ought consequently to be accurately ascertained. Strange discrepancies, however, exist between the results obtained by different investigators; so that the generally accepted figure cannot be regarded as absolutely free from doubt.

The early determinations made by Berzelius, Longchamp, and Gay-Lussac need not be considered here, as they have only antiquarian value. The investigations which demand attention are those of Scheerer, Svanberg and Nordenfeldt, Jacquelain, Macdonnell, Bahr, Marchand and Scheerer, Dumas, Marignac, Burton and Vorce, and Richards and Parker.

Scheerer's method of investigation was exceedingly simple.* He merely estimated the sulphuric acid in anhydrous magnesium sulphate, employing the usual process of precipitation as barium sulphate. He gives no weighings, but reports the percentages of SO_3 thus found. In his calculations, $O = 100$, $SO_3 = 500.75$, and $BaO = 955.29$. It is easy, therefore, to recalculate the figures which he gives, so as to establish what his method really represents, viz., the ratio between the sulphates of barium and magnesium.

Thus revised, his four analyses show that 100 parts of $MgSO_4$ yield the following quantities of $BaSO_4$:

	Per cent. SO_3
193.575	66.573
193.697	66.608
193.760	66.639
193.631	66.592

Mean, 193.6665, ± .0194

* Poggend. Annalen, 69, 535. 1846.

In a later note* Scheerer shows that the barium sulphate of these experiments carries down with it magnesium salts in such quantity as to make the atomic weight of magnesium 0.039 too low.

The work of Bahr, Jacquelain, Macdonnell, and Marignac, and in part that of Svanberg and Nordenfeldt, also relates to the composition of magnesium sulphate.

Jacquelain's experiments were as follows:† Dry magnesium sulphate was prepared by mixing the ordinary hydrous salt to a paste with sulphuric acid, and calcining the mass in a platinum crucible over a spirit lamp to constant weight and complete neutrality of reaction. This dry sulphate was weighed and intensely ignited three successive times. The weight of the residual MgO having been determined, it was moistened with sulphuric acid and recalcined over a spirit lamp, thus reproducing the original weight of $MgSO_4$. Jacquelain's weighings for these two experiments show that 100 parts of MgO correspond to the quantities of $MgSO_4$ given in the last column:

1.466 grm.	$MgSO_4$	gave	.492 grm.	MgO.		297.968
.492 "	MgO	"	1.466 "	$MgSO_4$.		297.968

Jacquelain also made one estimation of sulphuric acid in the foregoing sulphate as $BaSO_4$. His result (1.464 grm. $MgSO_4$ = 2.838 grm. $BaSO_4$), reduced to the standard adopted in dealing with Scheerer's experiments, gives for 100 parts of $MgSO_4$, 193.852 $BaSO_4$. If this figure be given equal weight with a single experiment in Scheerer's series, and combined with the latter, the mean will be 193.700, ± .0331. This again is subject to the correction pointed out by Scheerer for magnesium salts retained by the barium sulphate, but such a correction determined by Scheerer for a single experiment is only a rough approximation, and hardly worth applying.

The determinations published by Macdonnell‡ are of slight importance, and all depend upon magnesium sulphate. First, the crystallized salt, $MgSO_4.7H_2O$, was dried in vacuo over sulphuric acid and then dehydrated at a low red heat. The following percentages of water were found:

51.17
51.13
51.14
51.26
51.28
51.29

Mean, 51.21, ± .020

---

* Poggend. Annalen, 70, 407.
† Ann. Chim. Phys. (3), 32, 202.
‡ Proc. Royal Irish Acad., 5, 303. British Association Report, 1852, part 2, p. 36.

Secondly, anhydrous magnesium sulphate was precipitated with barium chloride. From the weight of the barium sulphate, with $SO_3 = 80$ and $Ba = 137$, Macdonnell computes the percentages of SO_3 given below. I calculate them back to the observed ratio in uniformity with Scheerer's work:

Per cent. SO_3.	Ratio, $MgSO_4 : BaSO_4$.
66.67	194.177
66.73	194.351
66.64	194.089
66.65	194.118
66.69	194.239

In another experiment 60.05 grains $MgSO_4$ gave 116.65 grains $BaSO_4$, a ratio of 100 : 194.254. Including this with the preceding figures, they give a mean of 194.205, ± .027. This, combined with the work of Scheerer and Jacquelain, 193.700, ± .033, gives a general mean of—

$$MgSO_4 : BaSO_4 :: 100 : 194.003, \pm .021.$$

In one final experiment Macdonnell found that 41.44 grains of pure magnesia gave 124.40 grains of $MgSO_4$, or 300.193 per cent.

Bahr's [*] work resembles in part that of Jacquelain. This chemist converted pure magnesium oxide into sulphate, and from the increase in weight determined the composition of the latter salt. From his weighings 100 parts of MgO equal the amounts of $MgSO_4$ given in the third column:

1.6938 grm. MgO gave	5.0157 grm. $MgSO_4$.	296.122
2.0459 "	6.0648 "	296.437
1.0784 "	3.1925 "	296.040

Mean, 296.200, ± .0815

About four years previous to the investigations of Bahr the paper of Svanberg and Nordenfeldt [†] appeared. These chemists started with the oxalate of magnesium, which was dried at a temperature of from 100° to 105° until it no longer lost weight. The salt then contained two molecules of water, and upon strong ignition it left a residue of MgO. The percentage of MgO in the oxalate comes out as follows:

7.2634 grm. oxalate gave	1.9872 grm. oxide.	27.359 per cent.
6.3795 "	1.7464 "	27.375 "
6.3653 "	1.7418 "	27.364 "
6.2216 "	1.7027 "	27.368 "

Mean, 27.3665, ± .0023

---

[*] Journ. für Prakt. Chem., 56, 310. 1852.
[†] Journ. für Prakt. Chem., 45, 473. 1848.

In three of these experiments the MgO was treated with H_2SO_4, and converted, as by Jacquelain and by Bahr in their later researches, into $MgSO_4$. One hundred parts of MgO gave of $MgSO_4$ as follows:

1.9872 grm. MgO gave	5.8995 grm.	$MgSO_4$.	296.875
1.7464 "	5.1783	"	296.513
1.7418 "	5.1666	"	296.624

Mean, 296.671, ± .072

In 1850 the elaborate investigations of Marchand and Scheerer * appeared. These chemists undertook to determine the composition of some natural magnesites, and, by applying corrections for impurities, to deduce from their results the sought-for atomic weight. The magnesite chosen for the investigation was, first, a yellow, transparent variety from Snarum; second, a white opaque mineral from the same locality; and, third, a very pure quality from Frankenstein. In each case the impurities were carefully determined; but only a part of the details need be cited here. Silica was of course easily corrected for by simple subtraction from the sum of all of the constituents; but iron and calcium, when found, having been present in the mineral as carbonates, required the assignment to them of a portion of the carbonic acid. In the atomic weight determinations the mineral was first dried at 300°. The loss in weight upon ignition was then carbon dioxide. It was found, however, that even here a correction was necessary. Magnesite, upon drying at 300°, loses a trace of CO_2, and still retains a little water; on the other hand, a minute quantity of CO_2 remains even after ignition. The CO_2 expelled at 300° amounted in one experiment to .054 per cent.; that retained after calcination to .055 per cent. Both errors tend in the same direction, and increase the apparent percentage of MgO in the magnesite. On the yellow mineral from Snarum the crude results are as follows, giving percentages of MgO, FeO, and CO_2 after eliminating silica:

CO_2.	MgO.	FeO.
51.8958	47.3278	.7764
51.8798	47.3393	.7809
51.8734	47.3154	.8112
51.8875	47.3372	.7753

Mean, 47.3299, ± .0037

After applying corrections for loss and retention of CO_2, as previously indicated, the mean results of the foregoing series become—

CO_2.	MgO.	FeO.
51.9931	47.2743	.7860

The ratio between the MgO and the CO_2, after correcting for the iron, will be considered further on.

---

* Journ. für Prakt. Chem., 50, 385.

MAGNESIUM.   139

Of the white magnesite from Snarum but a single analysis was made, which for present purposes may be ignored. Concerning the Frankenstein mineral three series of analyses were executed. In the first series the following results were obtained:

8.996 grm. CO_2 = 8.2245 grm. MgO.	47.760 per cent. MgO.
7.960 " 7.2775 "	47.761 "
9.3265 " 8.529 "	47.767 "
7.553 " 6.9095 "	47.775 "

Mean, 47.766, ± .0022

This mean, corrected for loss of CO_2 in drying, becomes 47.681. I give series second with corrections applied:

6.8195 grm. $MgCO_3$ gave 3.2500 grm. MgO.	47.658 per cent.
11.3061 " 5.3849 "	47.628 "
9.7375 " 4.635 "	47.599 "
12.3887 " 5.9033 "	47.650 "
32.4148 " 15.453 "	47.674 "
38.8912 " 18.5366 "	47.663 "
26.5223 " 12.6445 "	47.675 "

Mean, 47.650, ± .0069

The third series was made upon very pure material, so that the corrections, although applied, were less influential. The results were as follows:

4.2913 grm. $MgCO_3$ gave 2.0436 grm. MgO.	47.622 per cent.
27.8286 " 13.2539 "	47.627 "
14.6192 " 6.9692 "	47.672 "
18.3085 " 8.7237 "	47.648 "

Mean, 47.642, ± .0077

In a supplementary paper* by Scheerer, it was shown that an important correction to the foregoing data had been overlooked. Scheerer, reexamining the magnesites in question, discovered in them traces of lime, which had escaped notice in the original analyses. With this correction the two magnesites in question exhibit the following mean composition:

	Snarum.	*Frankenstein.*
CO_2	52.131	52.338
MgO	46.663	47.437
CaO	.430	.225
FeO	.776	......
	100.000	100.000

Correcting for lime and iron, by assigning each its share of CO_2, the Snarum magnesite gives as the true percentage of magnesia in pure

---

*Ann. d. Chem. und Pharm., 110, 240.

magnesium carbonate, the figure 47.624. To this, without serious mistake, we may assign the weight indicated by the probable error, ± .0037, the quantity previously deduced from the percentages of MgO given in the uncorrected analyses.

From the Frankenstein mineral, similarly corrected, the final mean percentage of MgO in $MgCO_3$ becomes 47.628. This, however, represents three series of analyses, whose combined probable errors may be properly assigned to it. The combination is as follows:

$$\pm .0022$$
$$\pm .0069$$
$$\pm .0077$$

Result, ± .0020, probable error of the general mean.

We may now combine the results obtained from both magnesites:

Snarum mineral............	Per cent. MgO, 47.624,	± .0037
Frankenstein mineral.......	" 47.628,	± .0020
General mean.......	Per cent. MgO, 47.627,	± .0018

The next investigation upon the atomic weight of magnesium which we have to consider is that of Dumas.* Pure magnesium chloride was placed in a boat of platinum, and ignited in a stream of dry hydrochloric acid gas. The excess of the latter having been expelled by a current of dry carbon dioxide, the platinum boat, still warm, was placed in a closed vessel and weighed therein. After weighing, the chloride was dissolved and titrated in the usual manner with a solution containing a known quantity of pure silver. The weighings which Dumas reports give, as proportional to 100 parts of silver, the quantities of $MgCl_2$ stated in the third column:

2.203 grm.	$MgCl_2$ =	4.964 grm. Ag.		44.380
2.5215	"	5.678	"	44.408
2.363	"	5.325	"	44.376
3.994	"	9.012	"	44.319
2.578	"	5.834	"	44.189
2.872	"	6.502	"	44.171
2.080	"	4.710	"	44.161
2.214	"	5.002	"	44.262
2.086	"	4.722	"	44.176
1.688	"	3.823	"	44.154
1.342	"	3.031	"	44.276

Mean, 44.261, ± .020

This determination gives a very high value to the atomic weight of magnesium, which is unquestionably wrong. The error, probably, is due to the presence of oxychloride in the magnesium chloride taken, an

* Ann. Chem. Pharm., 113, 33. 1860.

MAGNESIUM.   141

impurity tending to raise the apparent atomic weight of the metal. Richards' and Parker's revision of this ratio is more satisfactory.

Marignac,* in 1883, resorted to the old method of determination, depending upon the direct ratio between MgO and SO_3. This ratio he measured both synthetically and analytically. First, magnesia from various sources was converted into sulphate. The $MgSO_4$ from 100 parts of MgO is given in the third column:

	MgO.	$MgSO_4$.	Ratio.
1	1.5635	4.6620	298.17
2	1.4087	4.2025	298.32
3	1.5917	4.7480	298.30
4	1.4705	4.3855	298.23
5	1.4778	4.4060	298.15
6	1.6267	4.8530	298.33
7	1.3657	4.0740	298.37
8	1.9575	5.8390	298.29
9	1.6965	5.0600	298.26
10	1.8680	5.5715	298.26

Mean, 298.27, ± .0149

The magnesia for experiments 1 to 5 was prepared by calcination of the nitrate, that of 6 to 8 from the sulphate, and the remaining two from the carbonate. But Richards and Rogers† have shown that magnesia derived from the nitrate always contains occluded gaseous impurity, so that the experiments depending upon its use are somewhat questionable. The results tend to give an atomic weight for magnesium which is possibly too high. Whether the other samples of magnesia are subject to similar objections I cannot say.

Marignac's second series was obtained by the calcination of the sulphate, with results as follows:

$MgSO_4$.	MgO.	Ratio.
3.7705	1.2642	298.25
4.7396	1.5884	298.39
3.3830	1.1345	298.19
4.7154	1.5806	298.33
4.5662	1.5302	298.43
4.5640	1.5300	298.30
3.2733	1.0979	298.14
4.8856	1.6378	298.30
5.0092	1.6792	298.31
5.3396	1.7898	298.33
5.1775	1.7352	298.38
5.0126	1.6807	298.24
5.0398	1.6894	298.32

Mean, 298 30, ± .0150

---
* Arch. Sci. Phys. et Nat. (3), 10, 206.
† Am. Chem. Journ., 15, 567. 1893.

These data may now be combined with the work of previous investigators, giving Macdonnell's one result and Jacquelain's two, each equal weight with a single experiment in Bahr's series:

Macdonnell	300.193,	± .1413
Jacquelain	297.968,	± .0999
Bahr	296.200,	± .0815
Svanberg and Nordenfeldt	296.671,	± .0720
Marignac, synthetic	298.27,	± .0149
Marignac, calcination	298.30,	± .0150
General mean	298.210,	± .0103

Burton and Vorce,* who published their work on magnesium in 1890, started out with the metal itself, which had been purified by distillation in a Sprengel vacuum. This metal was dissolved in pure nitric acid, and the resulting nitrate was converted into oxide by calcination at a white heat. The oxide was carefully tested for oxides of nitrogen, which were proved to be absent, but occluded gases, the impurity pointed out by Richards and Rogers, were not suspected. This impurity must have been present, and it would tend to lower the apparent atomic weight of magnesium as calculated from the data obtained. The results were as follows, together with the percentage of Mg in MgO:

Mg Taken.	MgO Formed.	Per cent. Mg.
.33009	.54766	60.273
.34512	.57252	60.281
.26058	.43221	60.290
.28600	.47432	60.297
.30917	.51273	60.299
.27636	.45853	60.271
.36457	.60475	60.284
.32411	.53746	60.304
.32108	.53263	60.282
.28323	.46988	60.262

Mean, 60.2845, ± .0027

The latest determinations of all are those of Richards and Parker,† who studied magnesium chloride with all the precautions suggested by the most recent researches. The salt itself was not only free from oxychloride, but also spectroscopically pure as regards alkaline contaminations, and all weighings were reduced to a vacuum standard. The first series of experiments gives the ratio between silver chloride and magnesium chloride, and I have reduced the data to the form $2AgCl : MgCl_2 :: 100 : x$. The weighings and values for x are subjoined:

---

* Am. Chem. Journ., 12, 219. 1890.
† Zeitsch. Anorg. Chem., 13, 81. 1896.

MAGNESIUM. 143

$MgCl_2$.	$AgCl$.	Ratio.
1.33550	4.01952	33.225
1.51601	4.56369	33.219
1.32413	3.98528	33.226
1.40664	4.23297	33.231
1.25487	3.77670	33.227

Mean, 33.226, ± .0013

The remaining series of experiments, three in number, relate to the ratio $2Ag : MgCl_2$, which was earlier investigated by Dumas. For the elaborate details of manipulation the original memoir must be consulted. I can give little more than the weights found, and their reduction to the usual form of ratio, $Ag_2 : MgCl_2 :: 100 : x$:

Second Series.

$MgCl_2$.	Ag.	Ratio.
2.78284	6.30284	44.152
2.29360	5.19560	44.145
2.36579	5.35989	44.130

Mean, 44.142, ± .0043

This series gives slightly higher results than the others, and the authors, for reasons which they assign, discard it:

Third Series.

$MgCl_2$.	Ag.	Ratio.
1.99276	4.51554	44.131
1.78870	4.05256	44.138
2.12832	4.82174	44.140
2.51483	5.69714	44.141
2.40672	5.45294	44.136
1.95005	4.41747	44.144

Mean, 44.138, ± .0013

The fourth series, because of the experience gained in the conduct of the preceding determinations, is best of all, and the authors adopt its results in preference to the others:

Fourth Series.

$MgCl_2$.	Ag.	Ratio.
2.03402	4.60855	44.136
1.91048	4.32841	44.138
2.09932	4.75635	44.137
1.82041	4.12447	44.137
1.92065	4.35151	44.138
1.11172	2.51876	44.138

Mean, 44.137, ± .0003

These series combine with that of Dumas as follows:

Dumas........................... 44.261, ± .0200
Richards and Parker, second series..... 44.142, ± .0043
Richards and Parker, third series...... 44.138, ± .0013
Richards and Parker, fourth series..... 44.137, ± .0003

General mean................ 44.138, ± .0003

Here the first two values practically vanish, and the third and fourth series of Richards and Parker appear alone.

To sum up, we now have the subjoined ratios, bearing upon the atomic weight of magnesium:

(1.) $MgSO_4 : BaSO_4 :: 100 : 194.003$, ± .021
(2.) $MgO : MgSO_4 :: 100 : 298.210$, ± .0103
(3.) Per cent. of water in $MgSO_4, 7H_2O$, 51.21, ± .020
(4.) Per cent. of MgO in oxalate, 27.3665, ± .0023
(5.) Per cent. of MgO in carbonate, 47.627, ± .0018
(6.) Per cent. of Mg in MgO, 60.2845, ± .0027
(7.) $2Ag : MgCl_2 :: 100 : 44.138$, ± .0003
(8.) $2AgCl : MgCl_2 :: 100 : 33.226$, ± .0013

To reduce these ratios we have—

$O = 15.879$, ± .0003 $C = 11.920$, ± .0004
$Ag = 107.108$, ± .0031 $Ba = 136.392$, ± .0086
$Cl = 35.179$, ± .0048 $AgCl = 142.287$, ± .0037
$S = 31.828$, ± .0015

For the molecular weight of $MgSO_4$, two values are now calculable:

From (1)................... $MgSO_4 = 119.450$, ± .0137
From (3)................... " $= 119.239$, ± .0675

General mean.......... $MgSO_4 = 119.443$, ± .0135

Hence $Mg = 24.099$, ± .0136.
For MgO, three values are found:

From (2)................... $MgO = 40.091$, ± .0023
From (4)................... " $= 40.404$, ± .0037
From (5)................... " $= 39.721$, ± .0021

General mean............ $MgO = 39.974$, ± .0014

Hence $Mg = 24.095$, ± .0014.
For $MgCl_2$ there are two values:

From (7)................... $MgCl_2 = 94.551$, ± .0032
From (8)................... " $= 94.553$, ± .0044

General mean............ $MgCl_2 = 94.552$, ± .0026

Hence $Mg = 24.194$, ± .0099.

With the aid of these intermediate values, four estimates of the atomic weight of magnesium are available, as follows:

From molecular weight of $MgSO_4$....	$Mg = 24.099, \pm .0136$
From molecular weight of MgO......	" $= 24.095, \pm .0014$
From molecular weight of $MgCl_2$.....	" $= 24.194, \pm .0099$
From ratio (6).....................	" $= 24.103, \pm .0020$
General mean...............	$Mg = 24.100, \pm .0011$

If $O = 16$, this becomes $Mg = 24.283$.

On purely chemical grounds the third of the foregoing values, that derived from magnesium chloride, seems to be the best. I should unhesitatingly adopt it, rejecting the others, were it not for the fact that it rests upon one compound of magnesium alone, and therefore is not absolutely conclusive. It agrees admirably, however, with the sulphate determinations of Marignac, and it is highly probable that it may be fully confirmed later by evidence from other sources.

Marignac's data, taken alone, give $Mg = 24.197$. The fourth series of Richards and Parker, by itself, gives $Mg = 24.180$. The approximate mean of these, 24.19, may be preferred by many chemists to the general mean derived from all the observations.

ZINC.

The several determinations of the atomic weight of zinc are by no means closely concordant. The results obtained by Gay-Lussac* and Berzelius † were undoubtedly too low, and may be disregarded here. We need consider only the work done by later investigators.

In 1842 Jacquelain published the results of his investigations upon this important constant.‡ In two experiments a weighed quantity of zinc was converted into nitrate, and that by ignition in a *platinum* crucible was reduced to oxide. In two other experiments sulphuric acid took the place of nitric. As the zinc contained small quantities of lead and iron, these were estimated, and the necessary corrections applied. From the weights of metal and oxide given by Jacquelain the percentages have been calculated:

Nitric Series.

9.917 grm. Zn gave 12.3138 grm. ZnO.   80.536 per cent. Zn.
9.809   "   12.1800   "   80.534   "

Sulphuric Series.

2.398 grm. Zn gave 2.978 grm. ZnO.   80.524   "
3.197   "   3.968   "   80.570   "

Mean of all four, 80.541, ± .007

Hence Zn = 65.723.

The method adopted by Axel Erdmann § is essentially the same as that of Jacquelain, but varies from the latter in certain important details. First, pure zinc oxide was prepared, ignited in a covered crucible with sugar, and then, to complete the reduction, ignited in a porcelain tube in a current of hydrogen. The pure zinc thus obtained was converted into oxide by means of treatment with nitric acid and subsequent ignition in a *porcelain* crucible. Erdmann's figures give us the following percentages of metal in the oxide:

80.247
80.257
80.263
80.274

Mean, 80.260, ± .0037

Hence Zn = 64.562.

---

* Mémoire d'Arceuil, 2, 174.
† Gilb. Annal., 37, 460.
‡ Compt. Rend., 14, 636.
§ Poggend. Annal., 62, 611. Berz. Lehrb., 3, 1219.

Upon comparing Erdmann's results with those of Jacquelain two points are worth noticing: First, Erdmann worked with purer material than Jacquelain, although the latter applied corrections for the impurities which he knew were present; secondly, Erdmann calcined his zinc nitrate in a porcelain crucible, while Jacquelain used platinum. In the latter case it has been shown that portions of zinc may become reduced and alloy themselves with the platinum of the crucible; hence a lower weight of oxide from a given quantity of zinc, a higher percentage of metal, and an increased atomic weight. This source of constant error has undoubtedly affected Jacquelain's experiments, and vitiated his results. In Erdmann's work no such errors seem to be present.

Favre[*] employed two methods of investigation. First, zinc was dissolved in sulphuric acid, the hydrogen evolved was burned, and the weight of water thus formed was determined. To his weighings I append the ratio between metallic zinc and 100 parts of water:

25.389 grm. Zn gave	6.928 grm.	H_2O.	366.469
30.369 "	8.297	"	366.024
31.776 "	8.671	"	366.463
		Mean,	366.319, ± .088

Hence $Zn = 65.494$.

The second method adopted by Favre was to burn pure zinc oxalate, and to weigh the oxide and carbonic acid thus produced. From the ratio between these two sets of weights the atomic weight of zinc is easily deducible. From Favre's weighings, if $CO_2 = 100$, ZnO will be as given in the third column below:

7.796 grm. ZnO =	8.365 grm.	CO_2.	93.198
7.342 "	7.883	"	93.137
5.2065 "	5.588	"	93.173
		Mean,	93.169, ± .012

Hence $Zn = 65.521$.

Both of these determinations are open to objections. In the water series it was essential that the hydrogen should first be thoroughly dried before combustion, and then that every trace of water formed should be collected. A trivial loss of hydrogen or of water would tend to increase the apparent atomic weight of zinc.

In the combustion of the zinc oxalate equally great difficulties are encountered. Here a variety of errors are possible, such as are due, for example, to impurity of material, to imperfect drying of the carbon dioxide, and to incomplete collection of the latter. Indeed a fourth combustion is omitted from the series as given, having been rejected by Favre himself. In this case the oxide formed was contaminated by traces of sulphide.

---

[*]Ann. Chim. Phys. (3), 10, 163. 1844.

Baubigny,* in 1883, resorted to the well-known sulphate method. Zinc sulphate, elaborately purified, was dried at 440° to constant weight, and then calcined at a temperature equal to the fusing point of gold. These data were obtained:

$ZnSO_4$.	ZnO.	Per cent. ZnO.
6.699	3.377	50.410
8.776	4.4245	50.416

Mean, 50.413, ± .0020

Hence Zn = 64.909.

In Marignac's determinations of the atomic weight of zinc, published also in 1883,† there is a peculiar complication. After testing and criticising some other methods, he finally decided to study the double salt K_2ZnCl_4, which, however, is difficult to obtain in absolutely definite condition. Although the compound was purified by repeated crystallizations, it was found to deliquesce readily, and thereby to undergo partial dissociation, losing chloride of zinc, and leaving the porous layer on the crystalline surfaces richer in potassium. In order to evade this difficulty, Marignac placed a large quantity of the salt in a funnel, and collected the liquid product of deliquescence as it ran down. In this product he determined chlorine by volumetric titration with a standard solution of silver, and also estimated zinc by precipitation with sodium carbonate, and weighing as oxide. From the data thus obtained equations were formed, giving for each analysis an atomic weight of zinc which is independent of the proportion between $ZnCl_2$ and KCl in the substance analyzed. The data unfortunately are too bulky for reproduction here and the calculations are complex; but the results found for zinc, when Ag = 107.93, Cl = 35.457, and K = 39.137, are as follows:

1. One titration.................................. Zn = 65.22
2. Two titrations........................... 65.37
3. Two titrations........................... 65.31
4. Two titrations........................... 65.28
5. One titration........................... 65.26

Each of these values represents a distinct sample of the deliquesced material, and the number of chlorine determinations is indicated.

A second set of determinations was made by the same analytical method directly upon the recrystallized and carefully dried K_2ZnCl_4. The values for Zn are as follows:

6. Two titrations........................... Zn = 65.28
7. Two titrations........................... 65.39
8. One titration........................... 65.32

---

* Compt. Rend., 97, 906. 1883.
† Arch. Sci. Phys. et Nat. (3), 10, 194.

In order to adapt these data to the uniform scheme of calculation employed in this work, taking into account their probable error and the probable errors of the antecedent values for K, Cl, and Ag, it seems to be best to calculate them back with the atomic weights used by Marignac into the form of the ratio $Ag_4 : K_2ZnCl_4 :: 100 : x$. Doing this, and taking each value as many times as there are titrations represented in it—that is, giving the results of a double determination twice the weight of a single one—we have the following series of data for the ratio in question:

From 1...	66.090
From 2...	{ 66.124 66.124
From 3...	{ 66.110 66.110
From 4...	{ 66.104 66.104
From 5...	66.099
From 6...	{ 66.104 66.104
From 7...	{ 66.129 66.129
From 8...	66.113

Mean, 66.111, ± .0023

Hence, from Marignac's work, $Ag_4 : K_2ZnCl_4 :: 100 : 66.111, \pm .0023$, a ratio which can be discussed along with others at the close of this chapter.

During the years between 1883 and 1889, a number of determinations were made of the direct ratio between zinc and hydrogen—that is, weighed quantities of zinc were dissolved in acid, the hydrogen evolved was measured, and from its volume, with Regnault's data, the weight of H was computed. First in order are Van der Plaats' determinations,* whose results, as given by himself, are subjoined. The weights are reduced to a vacuum. Sulphuric acid was the solvent.

Zn, grms.	H, litres.	Zn =
6.6725	1.1424	65.21
9.1271	1.5643	65.14
13.8758	2.3767	65.18

Mean, 65.177, ± .0137

With the new value for the weight of hydrogen, .089872 gramme per litre, this becomes Zn = 64.980, ± .0137.

Reynolds and Ramsay made 29 déterminations of this ratio.† rejecting, however, all but 5. The weighings were reduced to vacuum, and in each experiment the volume of hydrogen was fixed by the mean of seven or eight readings. The values for Zn are as follows:

---

* Compt. Rend., 100, 52. 1885.
† Journ. Chem. Soc., 51, 854. 1887.

65.5060
65.4766
65.4450
65.5522
65.4141

Mean, 65.4787, ± .0161

These values were computed with Regnault's data for the weight of H. Corrected by the new value the mean becomes Zn = 65.280, ± .0161.

A few determinations by Mallet were made incidentally to his work on the atomic weight of gold, and appear in the same paper.* According to these experiments, one gramme of zinc gives—

341.85 cc. H., and Zn = 65.158
341.91   "   "   65.146
341.93   "   "   65.143
342.04   "   "   65.122

Mean, 65.142, ± .0039

In this case the Crafts-Regnault weight of H was taken, one litre = .08979 gramme. Corrected, the mean gives Zn = 65.082, ± .0039.

Two other series of determinations of questionable value remain to be noticed before leaving the consideration of the direct H : Zn ratio. They represent really the practice work of students, and are interesting as an illustration of the closeness with which such work can be done. The first series was made in the laboratory of the Johns Hopkins University, under the direction of Morse and Keiser,† and contains 51 determinations, as follows:

$Zn =$

64.68	65.74	65.40
65.26	64.72	64.80
65.32	65.26	65.20
65.20	64.74	64.40
65.60	64.72	65.00
64.60	65.10	64.40
65.00	64.76	65.24
65.68	64.90	64.60
65.38	64.92	64.80
65.06	64.64	65.14
64.84	65.24	64.84
64.88	64.72	64.82
65.00	65.20	64.80
65.08	65.12	64.40
65.06	66.40	64.60
64.74	64.60	64.80
65.12	65.60	64.74

Mean of all, Zn = 64.997, ± .0328

---

*Amer. Chem. Journ., 12, 205. 1890.
†Amer. Chem. Journ., 6, 347. 1884.

Corrected for the difference between Regnault's value for H and the new value, this becomes Zn = 64.800, ± .0328.

The second student series was published by Torrey,* who gives 15 determinations, as follows:

$Zn =$

65.36	64.96
65.30	64.70
64.92	65.00
64.72	64.78
65.04	64.44
64.80	65.24
65.20	64.92
64.90	

Mean, 64.952, ± .0436

Corrected as in the other series, this gives Zn = 64.755, ± .0436.

The five corrected means for the ratio H : Zn may now be combined, thus:

Van der Plaats	64.980, ± .0137
Reynolds and Ramsay	65.280, ± .0161
Mallet	65.082, ± .0039
Morse and Keiser	64.800, ± .0328
Torrey	64.755, ± .0436
General mean	65.079, ± .0036

Morse and Burton,† in their determinations of the atomic weight of zinc, returned essentially to the old method adopted by Erdmann and by Jacquelain. Their zinc was obtained spectroscopically pure by distillation in a vacuum, and was oxidized by nitric acid which left absolutely no residue upon evaporation. The conversion to oxide was effected in a porcelain crucible, which was enclosed in a larger one, and the ignition of the nitrate was carried out in a muffle. In weighing, the crucible was tared by one of nearly equal weight. Results as follows:

Wt. Zn.	Wt. ZnO.	Per cent. Zn in ZnO.
1.11616	1.38972	80.320
1.03423	1.28782	80.308
1.11628	1.38987	80.315
1.05760	1.31681	80.316
1.04801	1.30492	80.313
1.02957	1.28193	80.318
1.09181	1.35944	80.315
1.16413	1.44955	80.305
1.07814	1.34248	80.305
1.12754	1.40400	80.306
.91112	1.13446	80.310

---

* Amer. Chem. Journ., 10, 74. 1888.
† Amer. Chem. Journ., 10, 311. 1888.

1.10011	1.36981	80.311
1.17038	1.45726	80.313
1.03148	1.28436	80.310
1.05505	1.31365	80.308

Mean, 80.3115, ± .00084.

Combining this mean with the means found by the earlier investigators, we have—

Jacquelain	80.541,	± .0070
Erdmann	80.260,	± .0037
Morse and Burton	80.3115,	± .00084
General mean	80.317,	± .0008

Morse and Burton verified by experiment the stability of oxide of zinc at the temperatures of ignition, and found that it did not dissociate. They also proved the absence of oxides of nitrogen from the zinc oxide. The investigations of Richards and Rogers,* however, have shown that zinc oxide prepared by ignition of the nitrate always carries gaseous occlusions, so that the atomic weight of zinc computed from the data of Morse and Burton is probably too low. But for that objection, their work would leave little to be desired on the score of accuracy.

The determinations made by Gladstone and Hibbard † represent still another process for measuring the atomic weight of zinc. Zinc was dissolved in a voltameter, and the same current was used to precipitate metallic silver or copper in equivalent amount. The weight of zinc dissolved, compared with the weight of the other metal thrown down, gives the atomic weight sought for. Two voltameters were used in the experiments, giving duplicate estimates for zinc with reference to each weighing of silver or copper. The silver series is as follows, with the ratio $Ag_2 : Zn :: 100 : x$ in the third column:

Zn.	Ag.	Ratio.
.7767	2.5589	30.353
.7758	2.5589	30.318
.5927	1.9551	30.316
.5924	1.9551	30.300
.2277	.7517	30.291
.2281	.7517	30.345
.7452	2.4588	30.307
.7475	2.4588	30.401
.8770	2.9000	30.241
.8784	2.9000	30.290
.9341	3.0809	30.319
.9347	3.0809	30.339

Mean, 30.318, ± .0077

---

* Proc. Amer. Acad., 1893, 200.
† Journ. Chem. Soc., 55, 443. 1889.

ZINC.   153

To the copper series I add the ratio Cu : Zn : : 100 : z.

Zn.	Cu.	Ratio.
.7767	.7526	103.13
.7758	.7526	103.08
.5927	.5737	103.31
.5924	.5737	103.26
.2277	.2209	103.08
.2281	.2209	103.26
.8770	.8510	103.05
.8784	.8510	103.22
.9341	.9038	103.36
.9347	.9038	103.42

Mean, 103.22, ± .0261

Richards and Rogers,* in their investigation of the atomic weight of zinc, studied the anhydrous bromide. This was prepared by solution of zinc oxide in hydrobromic acid, evaporation to dryness, and subsequent distillation in an atmosphere of carbon dioxide. In some experiments, however, the bromide was heated in an atmosphere of nitrogen, mingled with gaseous hydrobromic acid. All water can thus be removed, without formation of oxybromides.

The zinc bromide so obtained was dissolved in water, and precipitated with a solution containing a known amount of silver in the form of nitrate. The silver bromide was weighed on a Gooch crucible, and the ratio $2AgBr : ZnBr_2$ thus found. An excess of silver was always used, and in one series of experiments it was estimated by precipitation with hydrobromic acid. Deducting the excess thus found from the original quantity of silver, the amount of the latter proportional to the zinc bromide was found; hence the ratio $Ag_2 : ZnBr_2$. The results, with vacuum weights, are as follows:

Series A.

$ZnBr_2$.	$AgBr$.	Ratio.
1.69616	2.82805	59.976
1.98198	3.30450	59.978
1.70920	2 84949	59.984
2.35079	3.91941	59.978
2.66078	4.43751	59.961

Mean, 59.975, ± .0034

Series B.

$ZnBr_2$.	Ag.	$AgBr$.	Ag Ratio.	$AgBr$ Ratio.
2.33882	2.24063	3.90067	104.382	59.959
1.97142	1.88837	3.28742	104.398	59.969
2.14985	2.05971	3.58539	104.376	59.961
2.00966	1.92476	3.35074	104 411	59.977

Mean, 104.392, ± .0054   Mean, 59.967, ± .0027

---

* Zeitsch. Anorg. Chem., 10, 1. 1895.

At the end of the same paper, Richards alone gives two more series of determinations made upon zinc bromide prepared by the action of pure bromine upon pure electrolytic zinc. The bromide so obtained was further refined by sublimation or distillation, and dried by heating in a stream of carbon dioxide and gaseous hydrobromic acid. Thus was ensured the absence of basic salts and of water. The weights and results found in the two series were as follows:

Series C.

$ZnBr_2$.	Ag.	Ratio.
6.23833	5.9766	104.379
5.26449	5.0436	104.380
9.36283	8.9702	104.377

Mean, 104.379, ± .0007

Series D.

$ZnBr_2$.	AgBr.	Ratio.
2.65847	4.43358	59.962
2.30939	3.85149	59.961
5.26449	8.77992	59.961

Mean, 59.961, ± .0004

In some details of manipulation these series differ from those given by Richards and Rogers jointly, but their minutiæ are not essential to the present discussion.

Combining these several series, we have—

For $Ag_2 : ZnBr_2 :: 100 : x$.

Series B	104.392, ± .0054
Series C	104.379, ± .0007
General mean	104.380, ± .0007

For $2AgBr : ZnBr_2 :: 100 : x$.

Series A	59.975, ± .0034
Series B	59.967, ± .0027
Series D	59.961, ± .0004
General mean	59.962, ± .0004

From the Ag ratio	$ZnBr_2$ = 223.599, ± .0066
From the AgBr ratio	" = 223.601, ± .0066
General mean	$ZnBr_2$ = 223.600, ± .0047
	And Zn = 64.912, ± .0133

For computing the atomic weight of zinc we now have these ratios:

(1.) Per cent. Zn in ZnO, 80.317, ± .0008
(2.) Per cent. ZnO in $ZnSO_4$, 50.413, ± .0020
(3.) H_2O : Zn : : 100 : 366.319, ± .088
(4.) $2CO_2$: Zn : : 100 : 93.169, ± .012
(5.) H : Zn : : 1 : 65.079, ± .0036
(6.) Ag_4 : K_2ZnCl_4 : : 100 : 66.111, ± .0023
(7.) Ag_2 : Zn : : 100 : 30.318, ± .0077
(8.) Cu : Zn : : 100 : 103.22, ± .0261
(9.) Ag_2 : $ZnBr_2$: : 100 : 104.38, ± .0007
(10.) 2AgBr : $ZnBr_2$: : 100 : 59.962, ± .0004

The antecedent atomic weights, with H = 1, are—

O = 15.879, ± .0003   C = 11.920, ± .0004
Cl = 35.179, ± .0048   S = 31.828, ± .0015
Br = 79.344, ± .0062   Cu = 63.119, ± .0015
Ag = 107.108, ± .0031   AgBr = 186.452, ± .0054
K = 38.817, ± .0051

With these data, combining ratios 9 and 10 into one (see preceding paragraphs), we have nine independent values for the atomic weight of zinc, as follows:

From (1) .... Zn = 64.795, ± .0030
From (2) .... " = 64.909, ± .0073
From (3) .... " = 65.494, ± .0019
From (4) .... " = 65.521, ± .0115
From (5) .... " = 65.079, ± .0036
From (6) .... " = 64.891, ± .0253
From (7) .... " = 64.947, ± .0166
From (8) .... " = 65.151, ± .0166
From (9) and (10) .... " = 64.912, ± .0133

General mean of all.......... Zn = 65.152, ± .0014
With O = 16................. Zn = 65.650

Of these values, Nos. 3 and 4, representing Favre's work, are unquestionably far wrong. Rejecting them, the general mean of the remaining seven values becomes—

$$Zn = 64.912, ± .0021.$$

If O = 16, this gives Zn = 65.407. These figures are identical, except as regards the lower probable error, with the result deduced from Richards and Rogers' determinations alone, and they may be taken as satisfactory.

CADMIUM.

The earliest determination of the atomic weight of this metal was by Stromeyer, who found that 100 parts of cadmium united with 14.352 of oxygen.* With our value for the atomic weight of oxygen, these figures make Cd = 110.64. This result has now only a historical interest.

The more modern estimates of the atomic weight of cadmium begin with the work of v. Hauer.† He heated pure anhydrous cadmium sulphate in a stream of dry hydrogen sulphide, and weighed the cadmium sulphide thus obtained. His results were as follows, with the percentage of CdS in $CdSO_4$ therefrom deduced:

7.7650 grm. $CdSO_4$ gave	5.3741 grm. CdS.	69.209 per cent.
6.6086 "	4.5746 "	69.222 "
7.3821 "	5.1117 "	69.245 "
6.8377 "	4.7336 "	69.228 "
8.1956 "	5.6736 "	69.227 "
7.6039 "	5.2634 "	69.220 "
7.1415 "	4.9431 "	69.217 "
5.8245 "	4.0335 "	69.251 "
6.8462 "	4.7415 "	69.257 "

Mean, 69.231, ± .0042

Lenssen‡ worked upon pure cadmium oxalate, handling, however, only small quantities of material. This salt, upon ignition, leaves the following percentages of oxide:

.5128 grm. oxalate gave	.3281 grm. CdO.	63.982 per cent.
.6552 "	.4193 "	63.996 "
.4017 "	.2573 "	64.053 "

Mean, 64.010, ± .014

Dumas∥ dissolved pure cadmium in hydrochloric acid, evaporated the solution to dryness, and fused the residue in hydrochloric acid gas. The cadmium chloride thus obtained was dissolved in water and titrated with a solution of silver after the usual manner. From Dumas' weighings I calculate the ratio between $CdCl_2$ and 100 parts of silver:

2.369 grm. $CdCl_2$ =	2.791 grm. Ag.	84.880
4.540 "	5.348 "	84.892
6.177 "	7.260 "	85.083
2.404 "	2.841 "	84.618
3.5325 "	4.166 "	84.794
4.042 "	4.767 "	84.791

Mean, 84.843, ± .026

---

* See Berz. Lehrbuch, 5th Aufl., 3, 1219.
† Journ. für Prakt. Chem., 72, 350. 1857.
‡ Journ. für Prakt. Chem., 79, 281. 1860.
∥ Ann. Chem. Pharm., 113, 27. 1860.

Next in order comes Huntington's* work, carried out in the laboratory of J. P. Cooke. Bromide of cadmium was prepared by dissolving the carbonate in hydrobromic acid, and the product, dried at 200°, was purified by sublimation in a porcelain tube. Upon the compound thus obtained two series of experiments were made.

In one series the bromide was dissolved in water, and a quantity of silver not quite sufficient for complete precipitation of the bromine was then added in nitric acid solution. After the precipitate had settled, the supernatant liquid was titrated with a standard solution of silver containing one gramme to the litre. The precipitate was washed by decantation, collected by reverse filtration, and weighed. To the weighings I append the ratio between $CdBr_2$ and 100 parts of silver bromide:

1.5592 grm. $CdBr_2$	gave	2.1529 grm. AgBr.	Ratio,	72.423
* 3.7456	"	5.1724	"	72.415
2.4267	"	3.3511	"	72.415
* 3.6645	"	5.0590	"	72.435
* 3.7679	"	5.2016	"	72.437
2.7938	"	3.8583	"	72.410
* 1.9225	"	2.6552	"	72.405
3.4473	"	4.7593	"	72.433

Mean, 72.4216, ± .0028

The second series was like the first, except that the weight of silver needed to effect precipitation was noted, instead of the weight of silver bromide formed. In the experiments marked with an asterisk, both the amount of silver required and the amount of silver bromide thrown down were determined in one set of weighings. The third column gives the $CdBr_2$ proportional to 100 parts of silver:

* 3.7456 grm. $CdBr_2$ =	2.9715 grm. Ag.		126.051
5.0270	"	3.9874 "	126.072
* 3.6645	"	2.9073 "	126.045
* 3.7679	"	2.9888 "	126.067
* 1.9225	"	1.5248 "	126.082
2.9101	"	2.3079 "	126.093
3.6510	"	2.8951 "	126.110
3.9782	"	3.1551 "	126.088

Mean, 126.076, ± .0052

According to Huntington's own calculations, these experiments fix the ratio between silver, bromine, and cadmium as Ag : Br : Cd :: 108 : 80 : 112.31.

In 1890, Partridge† published determinations of the atomic weight of cadmium, made by three methods, the weighings being reduced to

---

* Proc. Amer. Acad., 1881.
† Amer. Journ. Sci. (3), 40, 377. 1890.

vacuum standards throughout. First, Lenssen's method was followed, viz., the ignition of the oxalate, with the subjoined results:

CdC_2O_4.	CdO.	Per cent. CdO.
1.09898	.70299	63.966
1.21548	.77746	63.962
1.10711	.70807	63.957
1.17948	.75440	63.959
1.16066	.74327	63.959
1.17995	.75471	63.964
1.34227	.85864	63.968
1.43154	.91573	63.970
1.53510	.98197	63.968
1.41311	.90397	63.971

Mean, 63.964, ± .0010

Secondly, v. Hauer's experiments were repeated, cadmium sulphate being reduced to sulphide by heating in a stream of H_2S. The following data were obtained:

$CdSO_4$.	CdS.	Per cent. CdS.
1.60514	1.11076	69.204
1.55831	1.07834	69.197
1.67190	1.15669	69.185
1.66976	1.15554	69.200
1.40821	.97450	69.202
1.56290	1.08156	69.205
1.63278	1.12985	69.194
1.58270	1.09524	69.198
1.53873	1.06481	69.201
1.70462	1.17962	69.201

Mean, 69.199, ± .0012
v. Hauer found, 69.231, ± .0042

General mean, 69.202, ± .0012

In the third set of determinations cadmium oxalate was transformed to sulphide by heating in H_2S, giving the ratio $CdC_2O_4 : CdS :: 100 : x$.

CdC_2O_4	CdS.	Per cent CdS.
1.57092	1.13065	71.972
1.73654	1.24979	71.973
2.19276	1.57825	71.974
1.24337	.89492	71.974
1.18743	.85463	71.975
1.54038	1.10858	71.968
1.38905	.99974	71.976
2.03562	1.46517	71.979
2.03781	1.46658	71.970
1.91840	1.38075	71.971

Mean, 71.973, ± .0007

CADMIUM.   159

This work of Partridge was presently discussed by Clarke,* with reference to the concordance of the data, and it was shown that the three ratios determined could be discussed algebraically, giving values for the atomic weights of Cd, S, and C, when O = 16. These values are—

$$Cd = 111.7850$$
$$C = 11.9958$$
$$S = 32.0002,$$

and are independent of all antecedent values except that assumed for the standard, oxygen.

Morse and Jones,† starting out from cadmium purified by fractional distillation in vacuo, adopted two methods for their determinations. First, they effected the synthesis of the oxide from known weights of metal by dissolving the latter in nitric acid, evaporating to dryness, and subsequent ignition of the product. The oxide thus obtained was found to be completely free from oxides of nitrogen. The weighings, which are given below, were made in tared crucibles. The third column gives the percentage of Cd in CdO.

Cd Taken.	CdO Found.	Per cent. Cd.
1.77891	2.03288	87.507
1.82492	2.08544	87.508
1.74688	1.99626	87.507
1.57000	1.79418	87.505
1.98481	2.26820	87.506
2.27297	2.59751	87.504
1.75695	2.00775	87.508
1.70028	1.94305	87.505
1.92237	2.19679	87.508
1.92081	2.19502	87.508

Mean, 87.5066, ± .00032

The second method employed by Morse and Jones was that of Lenssen with cadmium oxalate. This salt they find to be somewhat hygroscopic, a property against which the operator must be on his guard. The data found are as follows:

CdC_2O_4.	CdO.	Per cent. CdO.
1.53937	.98526	64.004
1.77483	1.13582	63 996
1.70211	1.08949	64.008
1.70238	1.08967	64.004
1.74447	1.11651	64.003

Mean, 64.003, ± .0042

Lorimer and Smith, like Morse and Jones, determined the atomic weight of cadmium by means of the oxide, but by analysis instead of

---
*Am. Chem. Journ., 13, 34. 1891.
†Am. Chem. Journ., 14, 261. 1892.

synthesis. Weighed quantities of oxide were dissolved in potassium cyanide solution, from which metallic cadmium was thrown down electrolytically. The weights are reduced to vacuum standards.

CdO Taken.	Cd Found.	Per cent. Cd.
.34767	.30418	87.491
.41538	.36352	87.515
1.04698	.91618	87.507
1.04066	.91500	87.493
1.26447	1.10649	87.506
.78493	.68675	87.492
.86707	.75884	87.518
.67175	.58785	87.510
1.44362	1.26329	87.508

Mean, 87.5044, ± .0023

Mr. Bucher's dissertation * upon the atomic weight of cadmium does not claim to give any final measurements, but rather to discuss the various methods by which that constant has been determined. Nevertheless, it gives many data which seem to have positive value, and which are certainly fit for discussion along with those which have preceded this paragraph. Bucher begins with cadmium purified by distillation nine times in vacuo, and from this his various compounds were prepared. His first series of determinations was made by reducing cadmium oxalate to oxide, the oxalate having been dried fifty hours at 150°. The reduction was effected by heating in jacketed porcelain crucibles, with various precautions, and the results obtained, reduced to vacuum standards, are as follows:

Oxalate.	Oxide.	Per cent. Oxide.
1.97674	1.26414	63.951
1.94912	1.24682	63.968
1.96786	1.25886	63.971
1.87099	1.19675	63.958
1.37550	.87994	63.972
1.33313	.85308	63.991
1.94450	1.24452	64.002
2.01846	1.29210	64.014

Mean, 63.978, ± .0052

Combining this with the means found by previous experimenters, we have for the percentage of oxide in oxalate—

Lenssen	64.010, ± .0140
Partridge	63.962, ± .0010
Morse and Jones	64.003, ± .0042
Bucher	63.978, ± .0052
General mean	63.966, ± .0010

* "An examination of some methods employed in determining the atomic weight of cadmium." Johns Hopkins University doctoral dissertation. By John E. Bucher. Baltimore, 1895.

CADMIUM. 161

Bucher's next series of determinations was by Partridge's method—the conversion of cadmium oxalate into cadmium sulphide by heating in a stream of sulphuretted hydrogen. The sulphide was finally cooled in a current of dry nitrogen. The vacuum weights and ratios are subjoined:

Oxalate.	Sulphide.	Percentage.
2.56319	1.84716	72.065
2.18364	1.57341	72.055
2.11643	1.52462	72.037
3.13105	2.25582	72.047

Mean, 72.051, ± .0127
Partridge found, 71.973, ± .0007

General mean, 71.974, ± .0007

Here Bucher's mean practically vanishes.

The third method employed by Bucher was that of weighing cadmium chloride, dissolving in water, precipitating with silver nitrate, and weighing the silver chloride found. The cadmium chloride was prepared, partly by solution of cadmium in hydrochloric acid, evaporation to dryness, and sublimation in vacuo; and partly by the direct union of the metal with chlorine. The silver chloride was weighed in a Gooch crucible, with platinum sponge in place of the asbestos. To the vacuum weights I append the ratio $2AgCl : CdCl_2 :: 100 : x$.

$CdCl_2$.	$AgCl$.	Ratio.
3.09183	4.83856	63.900
2.26100	3.53854	63.896
1.35729	2.12431	63.893
2.05582	3.21727	63.899
1.89774	2.97041	63.886
3.50367	5.48473	63.880
2.70292	4.23087	63.886
4.24276	6.63598	63.936
3.40200	5.32314	63.910
4.60659	7.20386	63.946
2.40832	3.76715	63.930
2.19144	3.42724	63.942
2.84628	4.45477	63.893
2.56748	4.01651	63 923
2.31003	3.61370	63.924
1.25008	1.95652	63.893
1.96015	3.06541	63.944
2.29787	3.59391	63.938
1.94227	3.03811	63.915
1.10976	1.73547	63.946
1.63080	2.55016	63.949

Mean, 63.916, ± .0032

Bucher gives a rather full discussion of the presumable errors in this method, which, however, he regards as somewhat compensatory. The series is followed by a similar one with cadmium bromide, the latter having been sublimed in vacuo. Results as follows:

$CdBr_2$.	$AgBr$.	Ratio.
4.39941	6.07204	72.454
3.18030	4.38831	72.472
3.60336	4.97150	72.480
4.04240	5.58062	72.453
3.60505	4.97519	72.461

Mean, 72.464, ± .0035
Huntington found, 72.4216, ± .0028

General mean, 72.438, ± .0022

In order to fix a minimum value for the atomic weight of cadmium, Bucher effected the synthesis of the sulphate from the metal. 1.15781 grammes of cadmium gave 2.14776 of sulphate.

Hence Cd = 111.511.

The sulphate produced was dried at 400°, and afterwards examined for free sulphuric acid, giving a correction which was applied to the weighings. The corrected weight is given above. Any impurity in the sulphate would tend to lower the apparent atomic weight of cadmium, and therefore the result is believed by the author to be a minimum.

Finally, Bucher examined the oxide method followed by Morse and Jones. The syntheses of oxide were effected in double crucibles, first with both crucibles porcelain, and afterwards with the small inner crucible of platinum. Two experiments were made by the first method, three by the last. Weights and percentages (Cd in CdO) as follows:

Cd.	CdO.	Percentage.
1.26142	1.44144	87.511
.99785	1.14035	87.504

Mean, 87.508

1.11321	1.27247	87.484
1.02412	1.17054	87.491
2.80966	3.21152	87.487

Mean, 87.487
Mean of all as one series, 87.495, ± .0035

The two means given above, representing work done with porcelain and with platinum crucibles, correspond to a difference of about 0.2 in the atomic weight of cadmium. Experiments were made with pure oxide of cadmium by converting it into nitrate and then back to oxide, exactly as in the foregoing syntheses. In each case the oxide obtained at the end of the operation represented an increase in weight, but the increase was greater in platinum than in porcelain. Hence the weighings of cadmium oxide in the foregoing determinations probably are subject to constant errors, and cannot be trusted to fix the atomic weight of cadmium. Their mean, taken in one series, has really no significance; but as the computations in this work involve a study of compensation of errors, the data may be combined with their predecessors, as follows:

Morse and Jones	87.5066, ± .00032
Lorimer and Smith	87.5044, ± .0023
Bucher	87.495, ± .0035
General mean	87.5064, ± .0003

This is equivalent to the absolute rejection of Bucher's data, and is therefore not wholly fair to them. His work throws doubt upon the validity of the ratio, as determined, altogether.

The latest determinations relative to the atomic weight of cadmium are those of Hardin,* who effected the electrolysis of the chloride and bromide, and also made a direct comparison between cadmium and silver. The aqueous solutions of the salts, mixed with potassium cyanide, were electrolyzed in platinum dishes. The cadmium which served as the starting point for the investigation was purified by distillation in hydrogen. All weights are reduced to a vacuum. The data for the chloride series are as follows, with a column added for the percentage of Cd in $CdCl_2$:

Weight $CdCl_2$.	Weight Cd.	Percentage Cd.
.43140	.26422	61.247
.49165	.30112	61.247
.71752	.43942	61.241
.72188	.44208	61.241
.77264	.47319	61.245
.81224	.49742	61.240
.90022	.55135	61.246
1.02072	.62505	61.236
1.26322	.77365	61.244
1.52344	.93314	61.252

Mean, 61.244, ± .0010.

The results for the bromide, similarly stated, are these:

Weight $CdBr_2$.	Weight Cd.	Percentage Cd.
.57745	.23790	41.198
.76412	.31484	41.203
.91835	.37842	41.207
1.01460	.41808	41.206
1.15074	.47414	41.203
1.24751	.51392	41.196
1.25951	.51905	41.210
1.51805	.62556	41.208
1.63543	.67378	41.199
2.15342	.88722	41.200

Mean, 41.203, ± .0010.

---

* Journ. Amer. Chem. Soc., 18, 1016. 1896.

The direct comparison of cadmium and silver was effected by the simultaneous electrolysis, in the same current, of double cyanide solutions. Silver was thrown down in one platinum dish, and cadmium in another. The process was not altogether satisfactory, and gave divergent results, those which are cited below having been selected by Hardin from the mass of data obtained. I have added in a third column the cadmium proportional to 100 parts of silver:

Weight Cd.	Weight Ag.	Ratio.
.12624	.24335	51.876
.11032	.21262	51.886
.12720	.24515	51.887
.12616	.24331	51.852
.22058	.42520	51.877

Mean, 51.876, ± .0041

For cadmium we now have the following ratios:

(1.) Per cent. of Cd in CdO, 87.5064, ± .0003
(2.) Per cent. of CdO in CdC_2O_4, 63.966, ± .0010
(3.) Per cent. of CdS from CdC_2O_4, 71.974, ± .0007
(4.) Per cent. of CdS from $CdSO_4$, 69.202, ± .0012
(5.) Ag_2 : $CdCl_2$: : 100 : 84.843, ± .0260
(6.) $2AgCl$: $CdCl_2$: : 100 : 63.916, ± .0032
(7.) Ag_2 : $CdBr_2$: : 100 : 126.076, ± .0052
(8.) $2AgBr$: $CdBr_2$: : 100 : 72.438, ± .0022
(9.) Per cent. of Cd in $CdCl_2$, 61.244, ± .0010
(10.) Per cent of Cd in $CdBr_2$, 41.203, ± .0010
(11.) $2Ag$: Cd : : 100 : 51.876, ± .0041

Bucher's single experiment upon the synthesis of the sulphate, although important and interesting, cannot carry weight enough to warrant its consideration in connection with the other ratios, and is therefore not included.

The antecedent values, for use in computation are—

O = 15.879, ± .0003   S   = 31.828, ± .0015
Ag = 107.108, ± .0031   C   = 11.920, ± .0004
Cl = 35.179, ± .0048   AgCl = 142.287, ± .0037
Br = 79.344, ± .0062   AgBr = 186.452, ± .0054

For the molecular weight of cadmium chloride, two values are now deducible:

From (5)..................... $CdCl_2$ = 181.739, ± .0560
From (6)..................... "   = 181.888, ± .0103
General mean........... $CdCl_2$ = 181.883, ± .0138

Hence Cd = 111.525, ± .0138.

CADMIUM.   165

For cadmium bromide we have—

From (7)..................... $CdBr_2 = 270.073, \pm .0136$
From (8).....................   " $= 270.124, \pm .0113$

General mean........... $CdBr_2 = 270.105, \pm .0087$

Hence Cd = 111.417, ± .0151.
For cadmium there are nine independent values, as follows:

From (3)..................... $Cd = 110.793, \pm .0081$
From (4)..................... " $= 110.890, \pm .0069$
From (2)..................... " $= 111.004, \pm .0047$
From (11).................... " $= 111.127, \pm .0095$
From (9)..................... " $= 111.183, \pm .0155$
From (10).................... " $= 111.202, \pm .0093$
From (1)..................... " $= 111.227, \pm .0034$
From molecular weight $CdBr_2$....... " $= 111.417, \pm .0151$
From molecular weight $CdCl_2$....... " $= 111.525, \pm .0138$

General mean............... $Cd = 111.100, \pm .0022$

If O = 16, Cd = 111.947.
This result is obviously uncertain. The data are far from being conclusive, however, and I am therefore inclined to trust the mean rather than any one of the values taken separately. It is quite possible that the highest of all the figures may be nearest the truth, as Bucher's experiments seem to indicate; but until new evidence is obtained it would hardly be wise to make any selection. The mean obtained agrees well with the data of Morse and Jones, Lorimer and Smith, and Hardin.

MERCURY.

In dealing with the atomic weight of mercury we may reject the early determinations by Sefström * and a large part of the work done by Turner.† The latter chemist, in addition to the data which will be cited below, gives figures to represent the percentage composition of both the chlorides of mercury; but these results are neither reliable nor in proper shape to be used.

First in order we may consider the percentage composition of mercuric oxide, as established by Turner and by Erdmann and Marchand. In both investigations the oxide was decomposed by heat, and the mercury was accurately weighed. Gold leaf served to collect the last traces of mercurial vapor.

Turner gives four estimations. Two represent oxide obtained by the ignition of the nitrate, and two are from commercial oxide. In the first two the oxide still contained traces of nitrate, but hardly in weighable proportions. A comparison of the figures from this source with the others is sufficiently conclusive on this point. The third column represents the percentage of mercury in HgO:

144.805 grains Hg =	11.54 grains O.	92.619 per cent.
125.980 "	10.08 "	92.592 "
173.561 "	13.82 "	92.625 "
114.294 "	9.101 "	92.670 "

Mean, 92.614, ± .0050

In the experiments of Erdmann and Marchand ‡ every precaution was taken to ensure accuracy. Their weighings, reduced to a vacuum standard, give the subjoined percentages:

82.0079 grm. HgO gave	75.9347 grm. Hg.	92.594 per cent.
51.0320 "	47.2538 "	92.597 "
84.4996 "	78.2501 "	92.604 "
44.6283 "	41.3285 "	92.606 "
118.4066 "	109.6408 "	92.597 "

Mean, 92.5996, ± .0015

Hardin's determination of the same ratio, being different in character, will be considered later.

With a view to establishing the atomic weight of sulphur, Erdmann and Marchand also made a series of analyses of pure mercuric sulphide. These data are now best available for discussion under mercury. The

---

* Sefström. Berz. Lehrb., 5th ed., 3, 1215. Work done in 1812.
† Phil. Trans., 1833, 531-535.
‡ Journ. für Prakt. Chem., 31, 395. 1844.

sulphide was mixed with pure copper and ignited, mercury distilling over and copper sulphide remaining behind. Gold leaf was used to retain traces of mercurial vapor, and the weighings were reduced to vacuum:

34.3568 grm. HgS gave	29.6207 grm. Hg.	86.215 per cent. Hg.	
24.8278 "	21.40295 "	86.206 "	
37.2177 "	32.08416 "	86.207 "	
80.7641 "	69.6372 "	86.223 "	

Mean, 86.2127, ± .0027

For the percentage of mercury in mercuric chloride we have data by Turner, Millon, Svanberg, and Hardin. Turner,* in addition to some precipitations of mercuric chloride by silver nitrate, gives two experiments in which the compound was decomposed by pure stannous chloride, and the mercury thus set free was collected and weighed. The results were as follows:

44.782 grains Hg =	15.90 grains Cl.	73.798 per cent.
73.09 "	25.97 "	73.784 "

Mean, 73.791, ± .005

Millon † purified mercuric chloride by solution in ether and sublimation, and then subjected it to distillation with lime. The mercury was collected as in Erdmann and Marchand's experiments. Percentages of metal as follows:

73.87
73.81
73.83
73.87

Mean, 73.845, ± .010

Svanberg, ‡ following the general method of Erdmann and Marchand, made three distillations of mercuric chloride with lime, and got the following results:

12.048 grm. $HgCl_2$ gave	8.889 grm. Hg.	73.780 per cent.
12.529 "	9.2456 "	73.794 "
12.6491 "	9.3363 "	73.810 "

Mean, 73.795, ± .006

The most recent determinations of the atomic weight of mercury are due to Hardin,§ whose methods were entirely electrolytic. First, pure mercuric oxide was dissolved in dilute, aqueous potassium cyanide, and

---

* Phil. Trans., 1833, 531-535.
† Ann. Chim. Phys. (3), 18, 345. 1846.
‡ Journ. für Prakt. Chem., 45, 472. 1848.
§ Journ. Amer. Chem. Soc., 18, 1003. 1896.

electrolyzed in a platinum dish. Six determinations are published, out of a larger number, but without reduction of the weights to a vacuum. The data, with a percentage column added, are as follows:

Weight HgO.	Weight Hg.	Per cent. Hg.
.26223	.24281	92.594
.23830	.22065	92.593
.23200	.21482	92.595
.14148	.13100	92.593
.29799	.27592	92.594
.19631	.18177	92.593

Mean, 92.594, ± .0003.

Various sources of error were detected in these experiments, and the series is therefore rejected by Hardin. It combines with previous series as follows:

Turner	92.614, ± .0050
Erdmann and Marchand	92.5996, ± .0015
Hardin	92.594, ± .0003
General mean	92.595, ± .0003

Hardin also studied mercuric chloride, bromide, and cyanide, and the direct ratio between mercury and silver, with reduction of weights to a vacuum. Electrolysis was conducted in a platinum dish, as usual. With the chloride and bromide, the solutions were mixed with dilute potassium cyanide. The data for the chloride are as follows, the percentage column being added by myself:

Weight $HgCl_2$.	Weight Hg.	Per cent. Hg.
.45932	.33912	73.831
.54735	.40415	73.838
.56002	.41348	73.833
.63586	.46941	73.823
.64365	.47521	73.831
.73281	.54101	73.827
.86467	.63840	73.832
1.06776	.78825	73.823
1.07945	.79685	73.820
1.51402	1.11780	73.830

Mean, 73.829, ± .0012

Combining this with the earlier determinations, we have—

Turner	73.791, ± .0050
Millon	73.845, ± .0100
Svanberg	73.795, ± .0060
Hardin	73.829, ± .0012
General mean	73.826, ± .0011

For the bromide Hardin's data are—

Weight $HgBr_2$.	Weight Hg.	Per cent. Hg.
.70002	.38892	55.558
.56430	.31350	55.555
.57142	.31750	55.563
.77285	.42932	55.550
.80930	.44955	55.548
.85342	.47416	55.560
1.11076	.61708	55.555
1 17270	.65145	55.551
1.26186	.70107	55.559
1.40142	.77870	55.565

Mean, 55.556, ± .0012

And for the cyanide—

Weight HgC_2N_2.	Weight Hg.	Per cent. Hg.
.55776	.44252	79.337
.63290	.50215	79.341
.70652	.56053	79.337
.80241	.63663	79.340
.65706	.52130	79.338
.81678	.64805	79.342
1.07628	.85392	79.340
1.22615	.97282	79.339
1.66225	1.31880	79.338
2.11170	1.67541	79.339

Mean, 79.339, ± .0004

In the last series cited no potassium cyanide was used, but the solution of mercuric cyanide, with the addition of one drop of sulphuric acid, was electrolyzed directly.

The direct ratio between silver and mercury was determined by throwing down the two metals, simultaneously, in the same electric current. Both metals were taken in double cyanide solution. With Hardin's equivalent weights I give a third column, showing the quantity of mercury corresponding to 100 parts of silver. Many experiments were rejected, and only the following seven are published by the author:

Weight Hg.	Weight Ag.	Ratio.
.06126	.06610	92.678
.06190	.06680	92.665
.07814	.08432	92.671
.10361	.11181	92.666
.15201	.16402	92.678
.26806	.28940	92.626
.82808	.89388	92.639

Mean, 92.660, ± .0051

We now have six ratios involving the atomic weight of mercury, as follows:

(1.) Per cent. of Hg in HgO, 92.595, ± .0003
(2.) Per cent. of Hg in HgS, 86.2127, ± .0027
(3.) Per cent. of Hg in $HgCl_2$, 73.826, ± .0011
(4.) Per cent. of Hg in $HgBr_2$, 55.556, ± .0012
(5.) Per cent. of Hg in HgC_2N_2, 79.339, ± .0004
(6.) 2Ag : Hg :: 100 : 92.660, ± .0051

The calculations involve the following values:

O = 15.879, ± .0003
Ag = 107.108, ± .0031
Cl = 35.179, ± .0048
Br = 79.344, ± .0062
S = 31.828, ± .0015
C = 11.920, ± .0004
N = 13.935, ± .0021

Hence the values for mercury are—

From (1)..................... Hg = 198.557, ± .0084
From (2)..................... " = 199.027, ± .0406
From (3)..................... " = 198.482, ± .0285
From (4)..................... " = 198.364, ± .0170
From (5)..................... " = 198.568, ± .0170
From (6)..................... " = 198.493, ± .0124

General mean.............. Hg = 198.532, ± .0059

If O = 16, Hg = 200.045.

But according to Hardin the value derived from the analyses of mercuric oxide is untrustworthy. Rejecting this, and also the abnormally high result from the sulphide series, the general mean of the four remaining values is—

Hg = 198.491, ± .0083,

or, with O = 16, Hg = 200.004. These figures seem to be the best for the atomic weight of mercury.

BORON.

In the former edition of this work the data relative to boron were few and unimportant. There was a little work on record by Berzelius and by Laurent, and this was eked out by a discussion of Deville's analyses of boron chloride and bromide. As the latter were not intended for atomic weight determinations they will be omitted from the present recalculation, which includes the later researches of Hoskyns-Abrahall, Ramsay and Aston, and Rimbach.

Berzelius* based his determination upon three concordant estimations of the percentage of water in borax. Laurent† made use of two similar estimations, and all five may be properly put in one series, thus:

$$\left.\begin{array}{l}47.10\\47.10\\47.10\end{array}\right\} \text{Berzelius.}$$
$$\left.\begin{array}{l}47.15\\47.20\end{array}\right\} \text{Laurent.}$$

Mean, $47.13, \pm .013$

In 1892 the posthumous notes of the late Hoskyns-Abrahall were edited and published by Ewan and Hartog.‡ This chemist especially studied the ratio between boron bromide and silver, and also redetermined the percentage of water in crystallized borax. The latter work, which was purely preliminary, although carried out with great care, gave the following results, reduced to vacuum standards:

$Na_2B_4O_7.10H_2O.$	$Na_2B_4O_7.$	Per cent. $H_2O.$
7.00667	3.69587	47.2069
12.95936	6.82560	47.3308
4.65812	2.45248	47.3504
4.47208	3.93956	47.2763
4.94504	2.60759	47.2686

Mean, $47.2866, \pm .0171$

Two sets of determinations were made with the bromide, which was prepared from boron and bromine directly, freed from excess of the latter by standing over mercury, and finally collected, after distillation, in small, weighed, glass bulbs. It was titrated with a solution of silver after all the usual precautions. The first series of experiments was as follows, with BBr_3 proportional to 100 parts of silver stated as the ratio:

---

* Poggend. Annalen, 8, 1. 1826.
† Journ. für Prakt. Chem., 47, 415. 1849.
‡ Journ. Chem. Soc., 61, 650. August, 1892.

THE ATOMIC WEIGHTS.

RBr	Ag	Ratio.
4.3?305	1.69006	77-449
4.38904	5.69829	77-426
5.09702	6.5?8?0	77-444
6.5?30?	8.58919	77-435
7.75303	10.00055	77-430

Mean, 77-440, ± .0053

This series of data is regarded by the editors as preliminary, and not entitled to much consideration. The second series, which follows, was the final one; both represent vacuum standards:

RBr	Ag	Ratio.
4.06?35	5.27?1?0?	77-415
8.0?3?51	10.?0?0?6	77-4?4
1.655111	2.13?305	77-4?9
8.09?35?	10.3?0?0?	77-4?6
4.09?3?3	5.?63?0?	77-4?7
2.38?3?5	3.0??6?1?	77-4?5
7.?0?0?04	9.?0?0?54	77-4?0

Mean, 77-4?2, ± .000?
First series, 77-4?0, ± .0053

General mean, 77-4?5, ± .004?

Ramsay and Aston,* in their paper upon the atomic weight of boron, suggest that Abrahall's bromide may have contained hydrobromic acid, which would fully account for the low result obtained. They themselves adopt two distinct methods, the first one being the time-honored determination of water in crystallized borax. The latter was prepared from pure boric acid and pure sodium hydroxide. Results as follows, reduced to a vacuum:

$Na_2B_4O_7 \cdot 10H_2O$	$Na_2B_4O_7$	Per cent. H_2O
10.?5?0?0?	5.46?0.357	47-1?99
5.30?0?0?0	2.80?46?7	47-1433
4.99?0?3?0	2.65?8?9?4	47-3?0?6
5.?00?0?56	3.00?1?27	47-1?1?
5.3?4?5.?5	2.80?6?5?4?6	47-1?8??
4.???7??4	2.6???0?6	47-1865
5.?3?8?4?1	2.?6?0?6??	47-1?24

Mean, 47-1?9?, ± .00?6.

Thus we may combine with the previous determinations, thus:

Rosales with Lonnet........ 47-?3, ± .01?0
Hodges-Marshall.......... 47-1?66, ± .0?71
Ramsay and Aston......... 47-1?9?, ± .00?6

General mean........... 47-17?6, ± .00?6

---

* Journ. Chem. Soc., 65, xxx. 1894.

The second method adopted by Ramsay and Aston was to distil anhydrous borax with hydrochloric acid and methyl alcohol, both scrupulously pure, thereby converting it into sodium chloride. The operation was conducted in a glass flask, and in the first series of determinations ordinary soft glass was used. This, however, was somewhat attacked, so that the sodium chloride contained silica; hence oxygen in the material of the flask had been replaced by chlorine, thereby increasing its weight, and lowering the apparent atomic weight of boron. In a second series flasks of hard combustion tubing were taken, and the error, though not absolutely avoided, was reduced to a very small amount. Both series are subjoined, together with the percentage of chloride formed; but the weights, given by the authors to seven decimal places, are only quoted to the nearest tenth milligramme. They are reduced to vacuum standards.

First Series.

$Na_4B_4O_7$	$NaCl$	Per cent. $NaCl$
4.7664	2.7798	57.877
5.2340	3.0578	57.928
3.9304	2.2925	57.829
4.0862	2.3713	58.032
3.0970	2.0060	57.953

Mean, 57.926, ± .0019

Second Series.

$Na_4B_4O_7$	$NaCl$	Per cent. $NaCl$
5.3018	3.0964	57.911
4.7806	2.7900	57.945
4.9995	2.8930	57.966
4.1131	2.3960	57.928
3.5138	1.9989	57.900

Mean, 57.930, ± .0012
First series, 57.926, ± .0019
General mean of both, 57.928, ± .0011

As a check upon the last series of results, the sodium chloride was dissolved in water, and precipitated with silver nitrate. The silver chloride was collected and weighed in a Gooch crucible, and its weight gives a new ratio with anhydrous borax. The atom ratio between the two chlorides, silver and sodium, has already been used in the discussion upon sodium. The new ratio I give in terms of $Na_4B_4O_7$ equivalent to 100 parts of $AgCl$.

$Na_2B_4O_7$.	$AgCl$.	Ratio.
5.3118	7.5259	70.580
4.7806	6.7794	70.517
4.9907	7.0801	70.489
4.7231	6.6960	70.536
3.3138	4.6931	70.610

Mean, 70.546, ± .0146

Rimbach* based his determination of the atomic weight of boron upon the fact that boric acid is neutral to methyl orange, and that therefore it is possible to titrate a solution of borax directly with hydrochloric acid. His borax was prepared from carefully purified boric acid and sodium carbonate, and his hydrochloric acid was standardized by a series of precipitations and weighings as silver chloride. It contained 1.84983 per cent. of actual HCl. The borax, dissolved in water, was titrated by means of a weight-burette. I give the weights found in the first and second columns of the following table, and in the third column, calculated by myself, the HCl proportional to 100 parts of crystallized borax. Rimbach himself computes the percentage of Na_2O and thence the atomic weight of boron, but the ratio $Na_2B_4O_7.10H_2O : 2HCl$ is the ratio actually determined.

$Na_2B_4O_7.10H_2O$.	HCl Solution.	Ratio.
10.00214	103.1951	19.0853
15.32772	158.1503	19.0864
15.08870	155.7271	19.0917
10.12930	104.5448	19.0922
5.25732	54.2571	19.0908
15.04324	155.2307	19.0883
15.04761	155.2959	19.0908
10.43409	107.6602	19.0868
5.04713	52.0897	19.0915

Mean, 19.0893, ± .0006

Obviously, this error should be increased by the probable errors involved in standardizing the acid, but they are too small to be worth considering.

The following ratios are now available for boron:

(1) Percentage of water in $Na_2B_4O_7.10H_2O$, 47.1756, ± .0066
(2) $3Ag : BBr_3 : : 100 : 77.425$, ± .0017
(3) $Na_2B_4O_7 : 2NaCl : : 100 : 57.933$, ± .0074
(4) $2AgCl : Na_2B_4O_7 : : 100 : 70.546$, ± .0146
(5) $Na_2B_4O_7.10H_2O : 2HCl : : 100 : 19.0893$, ± .0006

* Berichte Deutsch. Chem. Gesell., 26, 164. 1893.

For reduction we have the antecedent atomic and molecular weights—

$O = 15.879, \pm .0003$ $Na = 22.881, \pm .0046$
$Ag = 107.108, \pm .0031$ $NaCl = 58.060, \pm .0017$
$Cl = 35.179, \pm .0048$ $AgCl = 142.287, \pm .0037$
$Br = 79.344, \pm .0062$

For the molecular weight of $Na_2B_4O_7$ we now have—

From (1).................... $Na_2B_4O_7 = 200.198, \pm .0377$
From (3).................... " $= 200.439, \pm .0263$
From (4).................... " $= 200.756, \pm .0419$
From (5).................... " $= 200.260, \pm .0518$

General mean.......... $Na_2B_4O_7 = 200.421, \pm .0180$

Hence $B = 10.876, \pm .0051$.

From ratio (2), $B = 10.753, \pm .0207$. The two values combined give—

$B = 10.863, \pm .0050$.

Or, if $O = 16$, $B = 10.946$.

If we consider ratios (1), (3), (4), and (5) separately, they give the following values for B:

From (1)... $B = 10.821$
From (3)... " $= 10.881$
From (4)... " $= 10.960$
From (5)... " $= 10.836$

Of these, the second and third involve the data from which, in a previous section of this work, the ratio NaCl : AgCl was computed. In using that ratio for measuring the molecular weights of its component molecules, discordance was noted, which again appears here. The chief uncertainty in it seems to be connected with ratio (4), which is therefore entitled to comparatively little credence, although its rejection is not necessary at this point. In ratio (2), Abrahall's determination, the high probable error of B is due to the also high probable error of 3Br, and it is quite likely that the result is undervalued. The general mean, $B = 10.863, \pm .0050$, however, can hardly be much out of the way. It is certainly more probable than any one of the individual values.

ALUMINUM.

The atomic weight of aluminum has been determined by Berzelius, Mather, Tissier, Dumas, Isnard, Terreil, Mallet, and Baubigny. The early calculations of Davy and of Thomson we may properly disregard.

Berzelius'[*] determination rests upon a single experiment. He ignited 10 grammes of dry aluminum sulphate, $Al_2(SO_4)_3$, and obtained 2.9934 grammes of Al_2O_3 as residue.

Hence $Al = 27.103$.

In 1835[†] Mather published a single analysis of aluminum chloride, from which he sought to fix the atomic weight of the metal. 0.646 grm. of $AlCl_3$ gave him 2.056 of AgCl and 0.2975 of Al_2O_3. These figures give worthless values for Al, and are included here only for the sake of completeness. From the ratio between AgCl and $AlCl_3$, $Al = 28.584$.

Tissier's [‡] determination, also resting on a single experiment, appeared in 1858. Metallic aluminum, containing .135 per cent. of sodium, was dissolved in hydrochloric acid. The solution was evaporated with nitric acid to expel all chlorine, and the residue was strongly ignited until only alumina remained. 1.935 grm. of Al gave 3.645 grm. of Al_2O_3. If we correct for the trace of sodium in the aluminum, we have $Al = 26.930$.

Essentially the same method of determination was adopted by Isnard, [§] who, although not next in chronological order, may fittingly be mentioned here. He found that 9 grm. of aluminum gave 17 grm. of Al_2O_3. Hence $Al = 26.8$.

In 1858 Dumas,[||] in connection with his celebrated revision of the atomic weights, made seven experiments with aluminum chloride. The material was prepared in quantity, sublimed over iron filings, and finally resublimed from metallic aluminum. Each sample used was collected in a small glass tube, after sublimation from aluminum in a stream of dry hydrogen, and hermetically enclosed. Having been weighed in the tube, it was dissolved in water, and the quantity of silver necessary for precipitating the chlorine was determined. Reducing to a common standard, his weighings give the quantities of $AlCl_3$ stated in the third column, as proportional to 100 parts of silver:

1.8786 grm. $AlCl_3$	= 4.543 grm. Ag.		41.352
3.021	"	7.292 "	41.459—Bad.
2.399	"	5.802 "	41.348
1.922	"	4.6525 "	41.311
1.697	"	4.1015 "	41.375
4.3165	"	10.448 "	41.314
6.728	"	16.265 "	41.365

---

[*] Poggend. Annal., 8, 177.
[†] Silliman's Amer. Journ., 27, 241.
[‡] Compt. Rend., 46, 1105.
[§] Compt. Rend., 66, 508. 1868.
[||] Ann. Chim. Phys. (3), 55, 151. Ann. Chem. Pharm., 113, 26.

In the second experiment the AlCl₃ contained traces of iron. Rejecting this experiment, the remaining six give a mean of 41.344, ± .007. These data give a value for Al approximating to 27.5, and were for many years regarded as satisfactory. It now seems probable that the chloride contained traces of an oxy-compound, which would tend to raise the atomic weight.

In 1879 Terreil* published a new determination of the atomic weight under consideration, based upon a direct comparison of the metal with hydrogen. Metallic aluminum, contained in a tube of hard glass, was heated strongly in a current of dry hydrochloric acid. Hydrogen was set free, and was collected over a strong solution of caustic potash. 0.410 grm. of aluminum thus were found equivalent to 508.2 cc., or .045671 grm. of hydrogen. Hence Al = 26.932.

About a year after Terreil's determination appeared, the lower value for aluminum was thoroughly confirmed by J. W. Mallet.† After giving a full résumé of the work done by others, exclusive of Isnard, the author describes his own experiments, which may be summarized as follows:

Four methods of determination were employed, each one simple and direct, and at the same time independent of the others. First, pure ammonia alum was calcined, and the residue of aluminum oxide was estimated. Second, aluminum bromide was titrated with a standard solution of silver. Third, metallic aluminum was attacked by caustic soda, and the hydrogen evolved was measured. Fourth, hydrogen was set free by aluminum, and weighed as water. Every weight was carefully verified, the verification being based upon the direct comparison, by J. E. Hilgard, of a kilogramme weight with the standard kilogramme at Washington. The specific gravity of each piece was determined, and also of all materials and vessels used in the weighings. During each weighing both barometer and thermometer were observed, so that every result represents a real weight in vacuo.

The ammonium alum used in the first series of experiments was specially prepared, and was absolutely free from ascertainable impurities. The salt was found, however, to lose traces of water at ordinary temperatures—a circumstance which tended towards a slight elevation of the apparent atomic weight of aluminum as calculated from the weighings. Two sets of experiments were made with the alum; one upon a sample air-dried for two hours at 21°–25°, the other upon material dried for twenty-four hours at 19°–26°. These sets, marked A and B respectively, differ slightly, B being the less trustworthy of the two, judged from a chemical standpoint. Mathematically it is the better of the two. Calcination was effected with a great variety of precautions, concerning which the original memoir must be consulted. To Mallet's weighings I append the percentages of Al_2O_3 deduced from them:

---

* Bulletin de la Soc. Chimique, 31, 153.
† Phil. Trans., 1880, p. 1003.

Series A.

8.2144 grm. of the alum gave	.9258 grm. Al$_2$O$_3$.	11.270 per cent.
14.0378 "	1.5825 "	11.273 "
5.6201 "	.6337 "	11.275 "
11.2227 "	1.2657 "	11.278 "
10.8435 "	1.2216 "	11.266 "

Mean, 11.2724, ± .0014

Series B.

12.1023 grm. of the alum gave	1.3660 grm. Al$_2$O$_3$.	11.287 per cent.
10.4544 "	1.1796 "	11.283 "
6.7962 "	.7670 "	11.286 "
8.5601 "	.9654 "	11.278 "
4.8992 "	.5528 "	11.283 "

Mean, 11.2834, ± .0011

Combined, these series give a general mean of 11.2793, ±.0008. Hence Al = 26.952.

The aluminum bromide used in the second series of experiments was prepared by the direct action of bromine upon the metal. The product was repeatedly distilled, the earlier portions of each distillate being rejected, until a constant boiling point of 263.°3 at 747 mm. pressure was noted. The last distillation was effected in an atmosphere of pure nitrogen, in order to avoid the possible formation of oxide or oxy-bromide of aluminum; and the distillate was collected in three portions, which proved to be sensibly identical. The individual samples of bromide were collected in thin glass tubes, which were hermetically sealed after nearly filling. For the titration pure silver was prepared, and after fusion upon charcoal it was heated in a Sprengel vacuum in order to eliminate occluded gases. This silver was dissolved in specially purified nitric acid, the latter but very slightly in excess. The aluminum bromide, weighed in the sealed tube, was dissolved in water, precautions being taken to avoid any loss by splashing or fuming which might result from the violence of the action. To the solution thus obtained the silver solution was added, the silver being something less than a decigramme in deficiency. The remaining amount of silver needed to complete the precipitation of the bromine was added from a burette, in the form of a standard solution containing one milligramme of metal to each cubic centimetre. The final results were as follows, the figures in the third column representing the quantities of bromide proportional to 100 parts of silver. Series A is from the first portion of the last distillate of AlBr$_3$; series B from the second portion, and series C from the third portion:

Series A.

6.0024 grm. AlBr$_3$ =	7.2793 grm. Ag.	82.458
8.6492 "	10.4897 "	82.454
3.1808 "	3.8573 "	82.462

Series B.

6.9617 grm. AlBr$_3$ =	8.4429 grm. Ag.		82.456
11.2041 "	13.5897 "		82.445
3.7621 "	4.5624 "		82.459
5.2842 "	6.4085 "		82.456
9.7338 "	11.8047 "		82.457

Series C.

9.3515 grm. AlBr$_3$ =	11.3424 grm. Ag.		82.447
4.4426 "	5.3877 "		82.458
5.2750 "	6.3975 "		82.454

Mean, 82.455, ± .001

Hence Al = 26.916.

The experiments to determine the amount of hydrogen evolved by the action of caustic soda upon metallic aluminum were conducted with pure metal, specially prepared, and with caustic soda made from sodium. The soda solution was so strong as to scarcely lose a perceptible amount of water by the passage through it of a dry gas at ordinary temperature. As the details of the experiments are somewhat complex, the original memoir must be consulted for them. The following results were obtained, the weight of the hydrogen being calculated from the volume, reckoned at .089872 gramme per litre.

Wt. Al.	Vol. H.	Wt. H.	At. Wt.
.3697	458.8	.041234	26.898
.3769	467.9	.042051	26.889
.3620	449.1	.040362	26.907
.7579	941.5	.084614	26.872
.7314	907.9	.081595	26.891
.7541	936.4	.084156	26.882

Mean, 26.890, ± .0034

The closing series of experiments was made with larger quantities of aluminum than were used in the foregoing set. The hydrogen, evolved by the action of the caustic alkali, was dried by passing it through two drying tubes containing pumice stone and sulphuric acid, and two others containing asbestos and phosphorus pentoxide. Thence it passed through a combustion tube containing copper oxide heated to redness. A stream of dry nitrogen was employed to sweep the last traces of hydrogen into the combustion tube, and dry air was afterwards passed through the entire apparatus to reoxidize the surface of reduced copper, and to prevent the retention of occluded hydrogen. The water formed by the oxidation of the hydrogen was collected in three drying tubes.

The results obtained were as follows. The third column gives the amount of water formed from 10 grammes of aluminum.

2.1704 grm. Al gave	2.1661 grm. H$_2$O.		9.9802
2.9355	"	2.9292	"	9.9785
5.2632	"	5.2562	"	9.9867

Mean, 9.9818, ± .0017

Hence Al = 26.867.

From the last two series of experiments an independent value for the atomic weight of oxygen may be calculated. It becomes O = 15.895. The closeness of this figure to some of the best determinations affords a good indication of the accuracy of Mallet's work.

In connection with Mallet's work it is worth noting that Torrey* published a series of measurements of the H : Al ratio, representing determinations made under his direction by elementary students. These measurements are thirteen in number, and calculated with Regnault's old value for the weight of hydrogen, range from 26.661 to 27.360, or in mean, 27.049, ± .323. Corrected by the latest value for the weight of H, this mean becomes 26.967. The result, of course, has only confirmatory significance.

By Baubigny† we have only two determinations, based upon the calcination of anhydrous aluminum sulphate, Al$_2$(SO$_4$)$_3$.

3.6745 grm. salt gave	1.0965 Al$_2$O$_3$.	29.841 per cent.
2.539	"	.7572	"	29.823 "

Mean, 29.832, ± .0061

Hence Al = 26.858.

It is clear that the single determinations of Berzelius, Mather, Tissier, Isnard, and Terreil may now be safely left out of account, for the reason that none of them could affect appreciably the final value for Al. The ratios to consider are as follows:

(1.) 3Ag : AlCl$_3$ :: 100 : 41.344, ± .0070
(2.) Percentage of Al$_2$O$_3$ in ammonium alum, 11.2793, ± .0008
(3.) 3Ag : AlBr$_3$ :: 100 : 82.455, ± .0010
(4.) H : Al :: 1 : 26.890, ± .0034
(5.) Al$_2$: 3H$_2$O :: 10 : 9.9818, ± .0017
(6.) Percentage of Al$_2$O$_3$ in Al$_2$(SO$_4$)$_3$, 29.832, ± .0061

The antecedent values are—

O = 15.879, ± .0003   Br = 79.344, ± .0062
Ag = 107.108, ± .0031   N = 13.935, ± .0021
Cl = 35.179, ± .0048   S = 31.828, ± .0015

---

* Am. Chem. Journ., 10, 74. 1888.
† Compt. Rend., 97, 1369. 1883.

Hence for aluminum we have—

From (1)	$Al = 27.311$,	$\pm .0270$
From (2)	" $= 26.952$,	$\pm .0037$
From (3)	" $= 26.916$,	$\pm .0201$
From (4)	" $= 26.890$,	$\pm .0034$
From (5)	" $= 26.867$,	$\pm .0046$
From (6)	" $= 26.858$,	$\pm .0113$
General mean	$Al = 26.906$,	$\pm .0021$

With $O = 16$, $Al = 27.111$. The rejection of Dumas' data only lowers the result to 26.903.

GALLIUM.

Gallium has been so recently discovered, and obtained in such small quantities, that its atomic weight has not as yet been determined with much precision. The following data were fixed by the discoverer, Lecoq de Boisbaudran:*

3.1044 grammes gallium ammonium alum, upon ignition, left .5885 grm. Ga_2O_3.

Hence $Ga = 69.595$. If $O = 16$, $Ga = 70.125$.

.4481 grammes gallium, converted into nitrate and ignited, gave .6024 grm. Ga_2O_3.

Hence $Ga = 69.171$. If $O = 16$, $Ga = 69.698$.

These values, assigned equal weight, give these means:

With $H = 1$, $Ga = 69.383$.   With $O = 16$, $Ga = 69.912$.

* Journ. Chem. Soc., 1878, p. 646.

INDIUM.

Reich and Richter, the discoverers of indium, were also the first to determine its atomic weight.* They dissolved weighed quantities of the metal in nitric acid, precipitated the solution with ammonia, ignited the precipitate, and ascertained its weight. Two experiments were made, as follows:

.5135 grm. indium gave .6243 grm. In_2O_3.
.699 " .8515 "

Hence, in mean, In = 110.61, if $O = 16$; a value known now to be too low.

An unweighed quantity of fresh, moist indium sulphide was also dissolved in nitric acid, yielding, on precipitation,

.2105 grm. In_2O_3 and .542 grm. $BaSO_4$.

Hence, with $BaSO_4$ = 233.505, In = 112.03; also too low.

Soon after the publication of Reich and Richter's paper the subject was taken up by Winkler.† He dissolved indium in nitric acid, evaporated to dryness, ignited the residue, and weighed the oxide thus obtained.

.5574 grm. In gave .6817 grm. In_2O_3.
.6661 " .8144 "
.5011 " .6126 "

Hence, in mean, if $O = 16$, In = 107.76; a result even lower than the values already cited.

In a later paper by Winkler ‡ better results were obtained. Two methods were employed. First, metallic indium was placed in a solution of pure, neutral, sodio-auric chloride, and the amount of gold precipitated was weighed. I give the weighings and, in a third column, the amount of indium proportional to 100 parts of gold:

In.	Au.	Ratio.
.4471 grm.	.8205 grm.	57.782
.8445 "	1.4596 "	57.858
	Mean,	57.820, ± .026

Hence, if Au = 195.743, ± .0049, In = 113.179, ± .0517.

Winkler also repeated his earlier process, converting indium into oxide by solution in nitric acid and ignition of the residue. An ad-

---

* Journ. für Prakt. Chem., 92, 484.
† Journ. für Prakt. Chem., 94, 8.
‡ Journ. für Prakt. Chem., 102, 282.

ditional experiment, the third as given below, was made after the method of Reich and Richter. The third column gives the percentage of In in In_2O_3:

1.124 grm.	In gave	1.3616 grm.	In_2O_3.	Per cent.,	82.550
1.015	"	1.2291	"	"	82.581
.6376	"	.7725	"	"	82.537

These figures were confirmed by a single experiment of Bunsen's,[*] published simultaneously with the specific heat determinations which showed that the oxide of indium was In_2O_3, and not InO, as had been previously supposed:

1.0592 grm. In gave 1.2825 grm. In_2O_3. Per cent. In, 82.589

For convenience we may add this figure in with Winkler's series, which gives us a mean percentage of In in In_2O_3 of 82.564, ± .0082. Hence, if $O = 15.879$, ± .0003, $In = 112.787$, ± .0542.
Combining both values, we have—

From gold series.................. In = 113.179, ± .0517
From oxide series................ " = 112.787, ± .0542
General mean.............. In = 112.992, ± .0374

If $O = 16$, $In = 113.853$.

---

[*] Poggend. Annal., 141, 28.

THALLIUM.

The atomic weight of this interesting metal has been fixed by the researches of Lamy, Werther, Hebberling, Crookes, and Lepierre.

Lamy and Hebberling investigated the chloride and sulphate; Werther studied the iodide; Crookes' experiments involved the synthesis of the nitrate. Lepierre's work is still more recent, and is based upon several compounds.

Lamy * gives the results of one analysis of thallium sulphate and three of thallium chloride. 3.423 grammes Tl_2SO_4 gave 1.578 grm. $BaSO_4$; whence 100 parts of the latter are equivalent to 216.920 of the former. In the thallium chloride the chlorine was estimated as silver chloride. The following results were obtained. In the third column I give the amount of TlCl proportional to 100 parts of AgCl:

3.912 grm. TlCl gave	2.346 grm. AgCl.		166.752
3.000 "	1.8015 "		166.528
3.912 "	2.336 "		167.466

Mean, 166.915, ± .1905

Hebberling's † work resembles that of Lamy. Reducing his weighings to the standards adopted above, we have from his sulphate series, as equivalent to 100 parts of $BaSO_4$, the amounts of Tl_2SO_4 given in the third column:

1.4195 grm. Tl_2SO_4 gave	.6534 grm. $BaSO_4$.		217.248
1.1924 "	.5507 "		216.524
.8560 "	.3957 "		216.325

Mean, 216.699

Including Lamy's single result as of equal weight, we get a mean of 216.754, ± .1387.

From the chloride series we have these results, with the ratio stated as usual:

.2984 grm. TlCl gave	.1791 grm. AgCl.		166.611
.5452 "	.3278 "		166.321

Mean, 166,465, ± .097

Lamy's mean was 166.915, ± .1905. Both means combined give a general mean of 166.555, ± .0865.

Werther's ‡ determinations of iodine in thallium iodide were made by two methods. In the first series TlI was decomposed by zinc and potassium hydroxide, and in the filtrate the iodine was estimated as AgI.

---

* Zeit. Anal. Chem., 2, 211. 1863.
† Ann. Chem. Pharm., 134, 11. 1865.
‡ Journ. für Prakt. Chem., 92, 128. 1864.

One hundred parts of AgI correspond to the amounts of TlI given in the last column:

.720 grm. TlI gave .51 grm. AgI.   141.176
2.072   "   1.472   "   140.761
 .960   "   .679   "   141.384
 .385   "   .273   "   141.026
1.068   "   .759   "   140.711

Mean, 141.012, ± .085

In the second series the thallium iodide was decomposed by ammonia in presence of silver nitrate, and the resulting AgI was weighed. Expressed according to the foregoing standard, the results are as follows:

1.375 grm. TlI gave .978 grm. AgI.   Ratio, 140.593
1.540   "   1.095   "   "   140.639
1.380   "   .981   "   "   140.673

Mean, 140.635, ± .016

General mean of both series, 140.648, ± .016.

In 1873 Crookes,* the discoverer of thallium, published his final determination of its atomic weight. His method was to effect the synthesis of thallium nitrate from weighed quantities of absolutely pure thallium. No precaution necessary to ensure purity of materials was neglected; the balances were constructed especially for the research; the weights were accurately tested and all their errors ascertained; weighings were made partly in air and partly in vacuo, but all were reduced to *absolute* standards; and unusually large quantities of thallium were employed in each experiment. In short, no effort was spared to attain as nearly as possible absolute precision of results. The details of the investigation are too voluminous, however, to be cited here; the reader who wishes to become familiar with them must consult the original memoir. Suffice it to say that the research is a model which other chemists will do well to copy.

The results of ten experiments by Professor Crookes may be stated as follows. In a final column I give the quantity of nitrate producible from 100 parts of thallium. The weights given are in grains:

Thallium.	$TlNO_3$ + Glass.	Glass Vessel.	Ratio.
497.972995	1121.851852	472.557319	130.3875
293.193507	1111.387014	729.082713	130.3930
288.562777	971.214142	594.949719	130.3926
324.963740	1142.569408	718.849078	130.3900
183.790232	1005.779897	766.133831	130.3912
190.842532	997.334615	748.491271	130.3920
195.544324	1022.176679	767.203451	130.3915
201.816345	1013.480135	750.332401	130.3897
295.683523	1153.947672	768.403621	130.3908
299.203036	1159.870052	769.734201	130.3917

Mean, 130.3910, ± .00034

* Phil. Trans., 1873, p. 277.

Lepierre's* determinations were published in 1893, and represented several distinct methods. First, thallous sulphate was subjected to electrolysis in presence of an excess of ammonium oxalate, the reduced metal being dried and weighed in an atmosphere of hydrogen. The corrected weights, etc., are as follows:

1.8935 grm. Tl_2SO_4 gave	1.5327 Tl.	80.945 per cent.
2.7243 "	2.2055 "	80.957 "
2.8112 "	2.2759 "	80.958 "

Mean, 80.953, ± .0030

Secondly, weighed quantities of crystallized thallic oxide were converted into thallous sulphate by means of sulphurous acid, and the solution was then subjected to electrolysis, as in the preceding series.

3.2216 grm. Tl_2O_3 gave	2.8829 Tl.	89.487 per cent.
2.5417 "	2.2742 "	89.475 "

Mean, 89.481, ± .0040

In the third set of experiments a definite amount of thallous sulphate or nitrate was fused in a polished silver crucible with ten times its weight of absolutely pure caustic potash. Thallic oxide was thus formed, which, with various precautions, was washed with water and alcohol, and finally weighed in the original crucible. One experiment with the nitrate gave—

2.7591 grm. $TlNO_3$ yields 2.3649 Tl_2O_3.   85.713 per cent.

Two experiments were made with the sulphate, as follows:

3.1012 grm. Tl_2SO_4 gave	2.8056 Tl_2O_3.	90.468 per cent.
2.3478 "	2.1239 "	90.463 "

Mean, 90.465, ± .0020

Finally, crystallized thallic oxide was reduced by heat in a stream of hydrogen, and the water so formed was collected and weighed.

2.7873 grm. Tl_2O_3 gave	.3301 H_2O.	11.843 per cent.
3.9871 "	.4716 "	11.828 "
4.0213 "	.4761 "	11.839 "

Mean, 11.837, ± .0029

In a supplementary note† Lepierre states that his weights were all reduced to vacuum standards.

Some work by Wells and Penfield, ‡ incidentally involving a determination of atomic weight, but primarily intended for another purpose, may also be taken into account. Their question was as to the constancy of thallium itself. The nitrate was repeatedly crystallized, and the last crystallization, with the mother liquor representing the opposite end of

---

* Bull. Soc. Chim. (3), 9, 166.
† Bull. Soc. Chim. (3), 11, 423. 1894.
‡ Amer. Journ. Sci. (3), 47, 466. 1894.

the series, were both converted into chloride. In the latter the chlorine was estimated as silver chloride, which was weighed on a Gooch filter, with the results given below, which are sensibly identical. The TlCl equivalent to 100 parts of AgCl is stated in the last column.

	TlCl.	AgCl.	Ratio.
Crystals............	3.9146	2.3393	167.341
Mother liquor.......	3.3415	1.9968	167.343
		Mean,	167.342

The general mean of Lamy's and Hebberling's determinations of this ratio gave 166.555, ± .0865. If we arbitrarily assign Wells and Penfield's mean equal weight with that, we get a new general mean of 166.948, ± .0610.

The ratios to be considered are now as follows:

(1.) $BaSO_4 : Tl_2SO_4 :: 100 : 216.754, \pm .1387$
(2.) $AgCl : TlCl :: 100 : 166.948, \pm .0610$
(3.) $AgI : TlI :: 100 : 140.648, \pm .016$
(4.) $Tl : TlNO_3 :: 100 : 130.391, \pm .00034$
(5.) $Tl_2SO_4 : Tl_2 :: 100 : 80.953, \pm .0030$
(6.) $Tl_2O_3 : Tl_2 :: 100 : 89.481, \pm .0040$
(7.) $2TlNO_3 : Tl_2O_3 :: 100 : 85.713$
(8.) $Tl_2SO_4 : Tl_2O_3 :: 100 : 90.465, \pm .0020$
(9.) $Tl_2O_3 : 3H_2O :: 100 : 11.837, \pm .0029$

And the antecedent data are these:

$O = 15.879, \pm .0003$ $N = 13.935, \pm .0021$
$Ag = 107.108, \pm .0031$ $S = 31.828, \pm .0015$
$Cl = 35.179, \pm .0048$ $AgCl = 142.287, \pm .0037$
$I = 125.888, \pm .0069$ $AgI = 232.996, \pm .0062$

Ratio number seven rests upon a single experiment, and the atomic weight of thallium derived from it must therefore be arbitrarily weighted. It has been assumed, therefore, that its probable error is the same as that from number eight. Taking this much for granted, we have nine values for thallium, as given below:

From (1) .... $Tl = 203.478, \pm .1610$
From (2) .... " $= 202.366, \pm .0872$
From (3) .... " $= 201.816, \pm .0389$
From (4) .... " $= 202.595, \pm .0117$
From (5) .... " $= 202.614, \pm .0330$
From (6) .... " $= 202.620, \pm .0775$
From (7) .... " $= 202.679, \pm .0483$
From (8) .... " $= 202.496, \pm .0483$
From (9) .... " $= 202.746, \pm .0576$

General mean .... $Tl = 202.555, \pm .0098$

If $O = 16$, $Tl = 204.098$.

If we reject the first three values, retaining only those due to the experiments of Crookes and Lepierre, we have—

$$Tl = 202.605, \pm .0103$$

If $O = 16$, this becomes 204.149. This mean exceeds Crookes' determination only by 0.01, and may be regarded as fairly satisfactory. Crookes' ratio evidently outweighs all the others.

SILICON.

Although Berzelius * attempted to ascertain the atomic weight of silicon, first by converting pure Si into SiO_2, and later from the analysis of $BaSiF_6$, his results were not satisfactory. We need consider only the work of Pelouze, Schiel, Dumas, and Thorpe and Young.

Pelouze,† experimenting upon silicon tetrachloride, employed his usual method of titration with a solution containing a known weight of silver. One hundred parts of Ag gave the following equivalencies of $SiCl_4$:

$$39.4325$$
$$39.4570$$

Mean, $39.4447, \pm .0083$

Essentially the same method was adopted by Dumas.‡ Pure $SiCl_4$ was weighed in a sealed glass bulb, then decomposed by water, and titrated. The results for 100 Ag are given in the third column:

2.899 grm. $SiCl_4$ =	7.3558 grm. Ag.		39.411
1.242 "	3.154 "		39.379
3.221 "	8.1875 "		39.340

Mean, $39.377, \pm .014$

Dumas' and Pelouze's series combine as follows:

Pelouze .... $39.4447, \pm .0083$
Dumas .... $39.377, \pm .014$

General mean .... $39.4265, \pm .0071$

Schiel,§ also studying the chloride of silicon, decomposed it by ammonia. After warming and long standing it was filtered, and in the

---
* Lehrbuch, 5 Aufl., 3, 1200.
† Compt. Rend., 20, 1047. 1845.
‡ Ann. Chem. Pharm., 113, 31. 1860.
§ Ann. Chem. Pharm., 120, 94.

filtrate the chlorine was estimated as AgCl. One hundred parts of AgCl correspond to the quantities of $SiCl_4$ given in the last column:

.6738 grm. $SiCl_4$ gave 2.277 grm. AgCl.   29.592
1.3092   "   4.418   "   29.633

Mean, 29.6125, ± .0138

Thorpe and Young,* working with silicon bromide, seem to have obtained fairly good results. The bromide was perfectly clear and colorless, and boiled constantly at 153°. It was weighed, decomposed with water, and evaporated to dryness, the crucible containing it being finally ignited. The crucible was tared by one precisely similar, in which an equal volume of water was also evaporated. Results as follows, with weights at vacuum standards:

9.63007 grm.	$SiBr_4$ gave	1.67070	SiO_2.	17.349	per cent.
12.36099	"	2.14318	"	17.338	"
12.98336	"	2.25244	"	17.349	"
9.02269	"	1.56542	"	17.350	"
15.38426	"	2.66518	"	17.324	"
9.74550	"	1.69020	"	17.343	"
6.19159	"	1.07536	"	17.368	"
9.51204	"	1.65065	"	17.353	"
10.69317	"	1.85555	"	17.353	"

Mean, 17.347, ± .0027

The ratios now available are—

(1.) $4Ag : SiCl_4 :: 100 : 39.4265$, ± .0071
(2.) $4AgCl : SiCl_4 :: 100 : 29.6125$, ± .0138
(3.) $SiBr_4 : SiO_2 :: 100 : 17.347$, ± .0027

Reducing these ratios with—

O = 15.879, ± .0003   Br = 79.344, ± .0062
Ag = 107.108, ± .0031   AgCl = 142.287, ± .0037,
Cl = 35.179, ± .0048

we have the following values for the atomic weight of silicon:

From (1) .... Si = 28.200, ± .0363
From (2) .... " = 27.823, ± .0810
From (3) .... " = 28.187, ± .0122

General mean .... Si = 28.181, ± .0114

If O = 16, Si = 28.305.

---

* Journ. Chem. Soc., 51, 576. 1887.

TITANIUM.

The earliest determinations of the atomic weight of titanium are due to Heinrich Rose.* In his first investigation he studied the conversion of titanium sulphide into titanic acid, and obtained erroneous results; later, in 1829, he published his analyses of the chloride.† This compound was purified by repeated rectifications over mercury and over potassium, and was weighed in bulbs of thin glass. These were broken under water in tightly stoppered flasks; the titanic acid was precipitated by ammonia, and the chlorine was estimated as silver chloride. The following results were obtained. In a fourth column I give the TiO_2 in percentages referred to $TiCl_4$ as 100, and in a fifth column the quantity of $TiCl_4$ proportional to 100 parts of AgCl:

$TiCl_4$.	TiO_2.	AgCl.	Per cent. TiO_2.	AgCl Ratio.
.885 grm.	.379 grm.	2.661 grm.	42.825	33.258
2.6365 "	1.120 "	7.954 "	42.481	33.147
1.7157 "	.732 "	5.172 "	42.665	33.173
3.0455 "	1.322 "	9.198 "	43.423	33.100
2.4403 "	1.056 "	7.372 "	43.273	33.102
			Mean, 42.933, ± .121	33.156, ± .019

If we directly compare the AgCl with the TiO_2 we shall find 100 parts of the former proportional to the following quantities of the latter:

14.243
14.081
14.153
14.373
14.324

Mean, 14.235, ± .036

Shortly after the appearance of Rose's paper, Mosander‡ published some figures giving the percentage of oxygen in titanium dioxide, from which a value for the atomic weight of titanium was deduced. Although no details are furnished as to experimental methods, and no actual weighings are given, I cite his percentages for whatever they may be worth:

40.814
40.825
40.610
40.180
40.107
40.050
40.780
40.660
39.830

Mean, 40.428

---

* Gilbert's Annalen, 1823, 67 and 129.
† Poggend. Annalen, 15, 145. Berz. Lehrbuch, 3, 1210.
‡ Berz. Jahresbericht, 10, 108. 1831.

These figures, with $O = 15.879$, give values for Ti ranging from 46.03 to 47.98; or, in mean, $Ti = 46.80$. They are not, however, sufficiently explicit to deserve any farther consideration.

In 1847 Isidor Pierre made public a series of important determinations.* Titanium chloride, free from silicon and from iron, was prepared by the action of chlorine upon a mixture of carbon with pure, artificial titanic acid. This chloride was weighed in sealed tubes, these were broken under water, and the resulting hydrochloric acid was titrated with a standard solution of silver after the method of Pelouze. I subjoin Pierre's weighings, and add, in a third column, the ratio of $TiCl_4$ to 100 parts of silver:

$TiCl_4$.	Ag.	Ratio.
.8215 grm.	1.84523 grm.	44.520
.7740 "	1.73909 "	44.506
.7775 "	1.74613 "	44.527
.7160 "	1.61219 "	44.412
.8085 "	1.82344 "	44.339
.6325 "	1.42230 "	44.470
.8155 "	1.83705 "	44.392
.8165 "	1.83899 "	44.399
.8065 "	1.81965 "	44.322

Mean, $44.432, \pm .0173$

It will be seen that the first three of these results agree well with each other and are much higher than the remaining six. The last four experiments were made purposely with tubes which had been previously opened, in order to determine the cause of the discrepancy. According to Pierre, the opening of a tube of titanium chloride admits a trace of atmospheric moisture. This causes a deposit of titanic acid near the mouth of the tube, and liberates hydrochloric acid. The latter gas being heavy, a part of it falls back into the tube, so that the remaining chloride is richer in chlorine and poorer in titanium than it should be. Hence, upon titration, too low figures for the atomic weight of titanium are obtained. Pierre accordingly rejects all but the first three of the above estimations.

The memoir of Pierre upon the atomic weight of titanium was soon followed by a paper from Demoly,† who obtained much higher results. He also started out from titanic chloride, which was prepared from rutile. The latter substance was found to contain 1.8 per cent. of silica; whence Demoly inferred that the $TiCl_4$ investigated by Rose and by Pierre might have been contaminated with $SiCl_4$, an impurity which would lower the value deduced for the atomic weight under consideration. Accordingly, in order to eliminate all such possible impurities, this process was resorted

---

* Ann. Chim. Phys. (3), 20, 257.
† Ann. Chem. Pharm., 72, 214. 1849.

to: the chloride, after rectification over mercury and potassium, was acted upon by dry ammonia, whereupon the compound $TiCl_4.4NH_3$ was deposited as a white powder. This was ignited in dry ammonia gas, and the residue, by means of chlorine, was reconverted into titanic chloride, which was again repeatedly rectified over mercury, potassium, and potassium amalgam. The product boiled steadily at 135°. This chloride, after weighing in a glass bulb, was decomposed by water, the titanic acid was precipitated by ammonia, and the chlorine was estimated in the filtrate as silver chloride. Three analyses were performed, yielding the following results. I give the actual weighings:

1.470 grm. $TiCl_4$ gave	4.241 grm. AgCl and	.565 grm. TiO_2
2.330 "	6.752 "	.801 "
2.880 "	8.330 "	1.088 "

The ".801" in the last column is certainly a misprint for .901. Assuming this correction, the results may be given in three ratios, thus:

Per cent. TiO_2 from $TiCl_4$.	$TiCl_4$: $100\,AgCl$.	TiO_2 : $100\,AgCl$.
38.435	34.662	13.322
38.669	34.508	13.344
37.778	34.574	13.061
Mean, 38.294, ± .180	34.581, ± .030	13.242, ± .061

These three ratios give three widely divergent values for the atomic weight of titanium, ranging from about 36 to more than 56, the latter figure being derived from the ratio between AgCl and $TiCl_4$. This value, 56, is assumed by Demoly to be the best, the others being practically ignored.

Upon comparing Demoly's figures with those obtained by Rose, certain points of similarity are plainly to be noted. Both sets of results were reached by essentially the same method, and in both the discordance between the percentages of titanic acid and of silver chloride is glaring. This discordance can rationally be accounted for by assuming that the titanic chloride was in neither case absolutely what it purported to be; that, in brief, it must have contained impurities, such for example as hydrochloric acid, as shown in the experiments of Pierre, or possibly traces of oxychlorides. Considerations of this kind also throw doubt upon the results attained by Pierre, for he neglected the direct estimation of the titanic acid altogether, thus leaving us without means for correctly judging as to the character of his material.

In 1883[*] Thorpe published a series of experiments upon titanium tetrachloride, determining three distinct ratios and getting sharply concordant results. The first ratio, which was essentially like Pierre's, by

---

[*] Berichte Deutsch. Chem. Gesell., 16, 3014. 1883.

TITANIUM.

decomposition with water and titration with silver, was in detail as follows:

$TiCl_4$.	Ag.	$TiCl_4 : 100\,Ag$.
2.43275	5.52797	44.008
5.42332	12.32260	44.015
3.59601	8.17461	44.000
3.31222	7.52721	44.003
4.20093	9.54679	44.004
5.68888	12.92686	44.008
5.65346	12.85490	43.979
4.08247	9.28305	43.978

Mean, 43.999, ± .0032
Pierre found, 44.432, ± .0073

General mean, 44.017, ± .0031

The second ratio, which involved the weights of $TiCl_4$ taken in the last five determinations of the preceding series, included the weighing of the silver chloride formed. The $TiCl_4$ proportional to 100 parts of AgCl is given in a third column:

$TiCl_4$.	$AgCl$.	Ratio.
3.31222	10.00235	33.114
4.20093	12.68762	33.111
5.68888	17.17842	33.117
5.65346	17.06703	33.125
4.08247	12.32442	33.125

Mean, 33.118, ± .0019
Rose found, 33.156, ± .019
Demoly found, 34.581, ± .030

General mean, 33.123, ± .0019.

In the third series the chloride was decomposed by water, and after evaporation to dryness the resulting TiO_2 was strongly ignited.

$TiCl_4$.	TiO_2.	Per cent. TiO_2.
6.23398	2.62825	42.160
8.96938	3.78335	42.181
10.19853	4.30128	42.176
6.56894	2.77011	42.170
8.99981	3.79575	42.176
8.32885	3.51158	42.162

Mean, 42.171, ± .0022
Rose found, 42.933, ± .121
Demoly found, 38.294, ± .180

General mean, 42.171, ± .0022

In short, the work of Rose, Pierre, and Demoly practically vanishes. Furthermore, as will be seen later, the three ratios now give closely agreeing values for the atomic weight of titanium. The cross ratio, $4AgCl : TiO_2$ is not directly given by either of Thorpe's series; but the data furnished by Rose and Demoly combine into a general mean of $4AgCl : TiO_2 :: 100 : 13.980, \pm .0303$.

Some two years later Thorpe published his work more in detail,* and added a set of determinations, like those made upon the chloride, in which titanium tetrabromide was studied. Three ratios were measured, as was the case with the chloride. In the first, the bromide was decomposed by water and titrated with a silver solution.

$TiBr_4$.	Ag.	$TiBr_4 : 100\,Ag$.
2.854735	3.34927	85.235
3.120848	3.66522	85.241
4.731118	5.55097	85.230
6.969075	8.17645	85.234
6.678099	7.83493	85.234

Mean, $85.235, \pm .0027$

In the four last experiments of the preceding series, the silver bromide formed was weighed. The third column gives the $TiBr_4$ proportional to 100 parts of AgBr.

$TiBr_4$.	$AgBr$.	Ratio.
3.120848	6.375391	48.951
4.731118	9.663901	48.957
6.969075	14.227716	48.982
6.678099	13.639956	48.959

Mean, $48.962, \pm .0049$

For the third ratio the bromide was decomposed by water; and after evaporation with ammonia the residual titanic oxide was ignited and weighed:

$TiBr_4$.	TiO_2.	Per cent. TiO_2.
6.969730	1.518722	21.790
8.836783	1.923609	21.768
9.096309	1.979513	21.762

Mean, $21.773, \pm .0062$

Ignoring Mosander's work as unavailable, we have the following ratios to consider:

(1.) $4Ag : TiCl_4 :: 100 : 44.017, \pm .0031$
(2.) $4AgCl : TiCl_4 :: 100 : 33.123, \pm .0019$
(3.) $4AgCl : TiO_2 :: 100 : 13.980, \pm .0303$
(4.) $TiCl_4 : TiO_2 :: 100 : 42.171, \pm .0022$
(5.) $4Ag : TiBr_4 :: 100 : 85.235, \pm .0027$
(6.) $4AgBr : TiBr_4 :: 100 : 48.962, \pm .0049$
(7.) $TiBr_4 : TiO_2 :: 100 : 21.773, \pm .0062$

---

* Journ. Chem. Soc., Feb., 1885, p. 108, and March, p. 129.

These are to be computed with—

O = 15.879, ± .0003	Br = 79.344, ± .0062
Ag = 107.108, ± .0031	AgCl = 142.287, ± .0037
Cl = 35.179, ± .0048	AgBr = 186.454, ± .0054

For the molecular weight of titanium chloride they give two values:

From (1) .... $TiCl_4$ = 188.583, ± .0144
From (2) .... "   = 188.519, ± .0119

General mean............ $TiCl_4$ = 188.545, ± .0092

For $TiBr_4$ we have—

From (5) .... $TiBr_4$ = 365.174, ± .0157
From (6) ....   "   = 365.163, ± .0380

General mean............ $TiBr_4$ = 365.172, ± .0145

And for the atomic weight of titanium five values are calculable, as follows:

From molecular weight of $TiCl_4$ ...... Ti = 47.829, ± .0213
From molecular weight of $TiBr_4$...... " = 47.796, ± .0260
From (3).......................... " = 47.809, ± .1725
From (4).......................... " = 47.698, ± .0268
From (7)....... .... " = 47.738, ± .0787

General mean................. Ti = 47.786, ± .0138

If O = 16, this becomes Ti = 48.150.

GERMANIUM.

The data relative to the atomic weight of germanium are rather scanty, and are due entirely to the discoverer of the element, Winkler.* The pure tetrachloride was decomposed by sodium carbonate, mixed with a known excess of standard silver solution, and then titrated back with ammonium sulphocyanate. The data given are as follows:

$GeCl_4$.	Cl Found.	Per cent. Cl.
.1067	.076112	66.177
.1258	.083212	66.146
.2223	.147136	66.188
.2904	.192190	66.182
	Mean,	66.173

Hence, with Cl = 35.179, Ge = 71.933. If O = 16, Ge = 72.481.

* Journ. für Prakt. Chem. (2), 34, 177. 1886.

ZIRCONIUM.

The atomic weight of zirconium has been determined by Berzelius, Hermann, Marignac, Weibull, and Bailey. Berzelius* ignited the neutral sulphate, and thus ascertained the ratio in it between the ZrO_2 and the SO_3. Putting SO_3 at 100, he gives the following proportional quantities of ZrO_2:

75.84
75.92
75.80
75.74
75.97
75.85

Mean, 75.853, ± .023

This gives 43.134, ± .0142 as the percentage of zirconia in the sulphate.

Hermann's† estimate of the atomic weight of zirconium was based upon analyses of the chloride, concerning which he gives no details nor weighings. From sublimed zirconium chloride he finds Zr = 831.8, when O = 100; and from two lots of the basic chloride $2ZrOCl_2.9H_2O$, Zr = 835.65 and 851.40 respectively. The mean of all three is 839.62; whence, with modern formulæ and O = 15.879, Zr becomes = 88.882.

Marignac's results ‡ were obtained by analyzing the double fluoride of zirconium and potassium. His weights are as follows:

1.000 grm. gave .431 grm. ZrO_2 and .613 grm. K_2SO_4.
2.000   "   .864   "   1.232   "
.654   "   .282   "   .399   "
5.000   "   2.169   "   3.078   "

These figures give us three ratios. A, the ZrO_2 from 100 parts of salt; B, the K_2SO_4 from 100 parts of salt; and C, the ZrO_2 proportional to 100 parts of K_2SO_4:

A.	B.	C.
43.100	61.300	70.310
43.200	61.600	70.130
43.119	61.000	70.677
43.380	61.560	70.468

Mean, 43.200, ± .043   Mean, 61.365, ± .094   Mean, 70.396, ± .079.

Weibull,§ following Berzelius, ignited the sulphate, and also made a

---

*Poggend. Annal, 4, 126. 1825.
†Journ. für Prakt. Chem., 31, 77. Berz. Jahresb., 25, 147.
‡Ann. Chim. Phys. (3), 60, 270. 1860.
§Lund. Arsskrift, v. 18. 1881–'82.

similar set of experiments with the selenate of zirconium, obtaining results as follows:

Sulphate. $Zr(SO_4)_2$.

1.5499 grm. salt gave	.6684	ZrO_2.	43.126 per cent.
1.5445 "	.6665 "		43.153 "
2.1683 "	.9360 "		43.168 "
1.0840 "	.4670 "		43.081 "
.7913 "	.3422 "		43.321 "
.6251 "	.2695 "		43.113 "
.4704 "	.2027 "		43.091 "

Mean, 43.150, ± .0207

Selenate. $Zr(SeO_4)_2$.

1.0212 grm. salt gave	.3323	ZrO_2.	32.540 per cent.
.8418 "	.2744 "		32.597 "
.6035 "	.1964 "		32.544 "
.8793 "	.2870 "		32.640 "
.3089 "	.1003 "		32.470 "

Mean, 32.558, ± .0192

Bailey * also ignited the sulphate, after careful investigation of his material, and of the conditions needful to ensure success. He found that the salt was perfectly stable at 400°, while every trace of free sulphuric acid was expelled at 350°. The chief difficulty in the process arises from the fact that the zirconia produced by the ignition is very light, and easily carried off mechanically, so that the percentage found is likely to be too low. This difficulty was avoided by the use of a double crucible, the outer one retaining particles of zirconia which otherwise might be lost. The results, corrected for buoyancy of the air, are as follows:

2.02357 salt gave	.87785	ZrO_2.	43.381 per cent.
2.6185 "	1.1354 "		43.360 "
2.27709 "	.98713 "		43.350 "
2.21645 "	.96152 "		43.385 "
1.75358 "	.76107 "		43.402 "
1.64065 "	.7120 "		43.397 "
2.33255 "	1.01143 "		43.361 "
1.81105 "	.78485 "		43.337 "

Mean, 43.372, ± .0056

This, combined with previous determinations, gives—

Berzelius	43.134, ± .0142
Weibull	43.150, ± .0207
Bailey	43.372, ± .0056
General mean	43.317, ± .0051

---

* Proc. Roy. Soc., 46, 74. Chem. News, 60, 32.

For computing the atomic weight of zirconium we now have the subjoined ratios:

(1.) Percentage ZrO_2 in $Zr(SO_4)_2$, 43.317, $\pm$ .0051
(2.) Percentage ZrO_2 in $Zr(SeO_4)_2$, 32.558, $\pm$ .0192
(3.) Percentage ZrO_2 from K_2ZrF_6, 43.200, $\pm$ .043
(4.) Percentage K_2SO_4 from K_2ZrF_6, 61.365, $\pm$ .094
(5.) $K_2SO_4 : ZrO_2 :: 100 : 70.396$, $\pm$ .079

The antecedent atomic weights are—

$O = 15.879, \pm .0003$ $K = 38.817, \pm .0051$
$S = 31.828, \pm .0015$ $F = 18.912, \pm .0029$
$Se = 78.419, \pm .0042$

With these data we first get three values for the molecular weight of zirconia:

From (1) .... $ZrO_2 = 121.454, \pm .0182$
From (2) ....   " $= 121.708, \pm .0798$
From (5) ....   " $= 121.770, \pm .1370$

General mean .... $ZrO_2 = 121.471, \pm .0176$

Finally, there are three independent estimates for the atomic weight of zirconium:

From molecular weight ZrO_2. .... $Zr = 89.713, \pm .0177$
From ratio (3)......................   " $= 89.437, \pm .2390$
From ratio (4)......................   " $= 90.778, \pm .4326$

General mean .... $Zr = 89.716, \pm .0175$

If $O = 16$, $Zr = 90.400$.

Here the first value alone carries appreciable weight.

TIN.

The atomic weight of tin has been determined by means of the oxide, the chloride, the bromide, the sulphide, and the stannichlorides of potassium and ammonium.

The composition of stannic oxide has been fixed in two ways: by synthesis from the metal and by reduction in hydrogen. For the first method we may consider the work of Berzelius, Mulder and Vlaanderen, Dumas, Van der Plaats, and Bongartz and Classen.

Berzelius[*] oxidized 100 parts of tin by nitric acid, and found that 127.2 parts of SnO_2 were formed.

The work done by Mulder and Vlaanderen [†] was done in connection with a long investigation into the composition of Banca tin, which was found to be almost absolutely pure. For the atomic weight determinations, however, really pure tin was taken prepared from pure tin oxide. This metal was oxidized by nitric acid, with the following results. 100 parts of tin gave of SnO_2:

        127.56—Mulder.
        127.56—Vlaanderen.
        127.43—Vlaanderen.

      Mean, 127.517, $\pm$ .029

Dumas [‡] oxidized pure tin by nitric acid in a flask of glass. The resulting SnO_2 was strongly ignited, first in the flask and afterwards in platinum. His weighings, reduced to the foregoing standard, give for dioxide from 100 parts of tin the amounts stated in the third column:

  12.443 grm. Sn gave 15.820 grm. SnO_2.   127.14
  15.976   "   20.301   "   127.07

                Mean, 127.105, $\pm$ .024

In an investigation later than that previously cited, Vlaanderen § found that when tin was oxidized in glass or porcelain vessels, and the resulting oxide ignited in them, traces of nitric acid were retained. When, on the other hand, the oxide was strongly heated in platinum, the latter was perceptibly attacked, so much so as to render the results uncertain. He therefore, in order to fix the atomic weight of tin, reduced the oxide by heating it in a porcelain boat in a stream of hydrogen. Two experiments gave Sn = 118.08, and Sn = 118.24. These, when O = 16, become, if reduced to the above common standard,

---

      * Poggend. Annal., 8, 177.
      † Journ. für Prakt. Chem., 49, 35. 1849.
      ‡ Ann. Chem. Pharm., 113, 26.
      § Jahresbericht, 1858, 183.

127.100
127.064

Mean, 127.082, ± .012

Van der Plaats* prepared pure stannic oxide from East Indian tin (Banca), and upon the material obtained made two series of experiments; one by reduction and one by oxidation. The results, with vacuum weights, are as follows, the ratio between Sn and SnO_2 appearing in the third column:

Oxidation Series.

9.6756 grm. tin gave	12.2967	SnO_2.	127.091
12.7356 "	16.1885	"	127.114
23.4211 "	29.7667	"	127.093

Reduction Series.

5.5015 grm. SnO_2 gave	4.3280	tin.	127.114
4.9760 "	3.9145	"	127.117
3.8225 "	3.0078	"	127.086
2.9935 "	2.3553	"	127.096

Mean of both series as one, 127.102, ± .0033

The reductions were effected in a porcelain crucible.

Bongartz and Classen† purified tin by electrolysis, and oxidized the electrolytic metal by means of nitric acid. The oxide found was dried over a water-bath, then heated over a weak flame, and finally ignited for several hours in a gas-muffle. Some reduction experiments gave values which were too low. The oxidation series was as follows, with the usual ratio added by me in a third column:

Sn.	SnO_2.	Ratio.
2.5673	3.2570	126.865
3.8414	4.8729	126.852
7.3321	9.2994	126.831
5.4367	6.8962	126.845
7.3321	9.2994	126.831
9.8306	12.4785	126.935
11.2424	14.2665	126.896
5.5719	7.0685	126.860
9.8252	12.4713	126.932
4.3959	5.5795	126.925
6.3400	8.0440	126.877

Mean, 126.877, ± .0080

We now have six series of experiments showing the amount of SnO_2 formed from 100 parts of tin. To Berzelius' single determination may be assigned the weight of one experiment in Mulder and Vlaanderen's series:

---

* Compt. Rend., 100, 52. 1885.
† Berichte Deutsch. Chem. Gesell., 21, 2900. 1888.

TIN.

Berzelius,.................. .... 127.200, ± .041
Mulder and Vlaanderen,.......... ...... 127.517, ± .029
Dumas .... 127.105, ± .024
Vlaanderen .... 127.082, ± .012
Van der Plaats,....................... 127.102, ± .0033
Bongartz and Classen......... .... 126.877, ± .0080

General mean .... 127.076, ± .0026

Dumas, in the paper previously quoted, also gives the results of some experiments with stannic chloride, $SnCl_4$. This was titrated with a solution containing a known weight of silver. From the weighings given, 100 parts of silver correspond to the quantities of $SnCl_4$ named in the third column:

1.839 grm. $SnCl_4$ =	3.054 grm. Ag.	60.216
2.665 "	4.427 "	60.199

Mean, 60.207, ± .006

Tin tetrabromide and the stannichlorides of potassium and ammonium were all studied by Bongartz and Classen; who, in each compound, carefully purified, determined the tin electrolytically. The data given are as follows, the percentage columns being added by myself:

Tin Tetrabromide.

$SnBr_4$ Taken.	Sn Found.	Per cent. Sn.
8.5781	2.3270	27.127
9.5850	2.6000	27.126
9.9889	2.7115	27.145
10.4914	2.8445	27.113
16.8620	4.5735	27.123
16.6752	4.5236	27.119
11.1086	3.0125	27.116
10.6356	2.8840	27.113
11.0871	3.0060	27.123
19.5167	5.2935	27.128

Mean, 27.123, ± .0020

Potassium Stannichloride.

K_2SnCl_6.	Sn Found.	Per cent. Sn.
2.5718	.7472	29.054
2.2464	.6524	29.042
9.3353	2.7100	29.030
12.1525	3.5285	29.035
12.4223	3.6070	29.036
15.0870	4.3812	29.040
10.4465	3.0330	29.034
18.9377	5.5029	29.058
18.4743	5.3630	29.029
17.6432	5.1244	29.045

Mean, 29.040, ± .0021

Ammonium Stannichloride.

Am_2SnCl_6.	Sn Found.	Per cent. Sn.
1.6448	.5328	32.393
1.8984	.6141	32.347
2.0445	.6620	32.381
2.0654	.6690	32.391
2.0058	.6496	32.386
2.4389	.7895	32.371
4.0970	1.3254	32.351
3.4202	1.1078	32.390
3.6588	1.1836	32.349
1.5784	.5108	32.362
7.3248	2.3710	32.370
13.1460	4.2528	32.351
11.9483	3.8650	32.348
18.4747	5.9788	32.362
18.6635	6.0415	32.371
17.8894	5.7923	32.378

Mean, 32.369, ± .0088

One other method of determination for the atomic weight of tin was employed by Bongartz and Classen. Electrolytic tin was converted into sulphide, and the sulphur so taken up was oxidized by means of hydrogen peroxide, by Classen's method, and weighed as barium sulphate. The results, as given by the authors, are subjoined:

Sn Taken.	Per cent. of S Gained.
2.6285	53.91
.7495	53.87
1.4785	53.94
2.5690	53.94
2.1765	53.85
1.3245	53.88
.9897	53.83
2.7160	53.86

Mean, 53.885, ± .0098

This percentage of sulphur, however, was computed from weighings of barium sulphate. What values were assigned to the atomic weights of barium and sulphur is not stated, but as Meyer and Seubert's figures are used for other elements throughout this paper, we may assume that they apply here also. Putting $O = 15.96$, $S = 31.98$, and $Ba = 136.86$, the 53.885 per cent. of sulphur becomes 392.056, ± .0713 of $BaSO_4$, the compound actually weighed. This gives us the ratio—

$$Sn : 2BaSO_4 :: 100 : 392.056, \pm .0713$$

as the real result of the experiments, from which, with the later values for Ba, S, and O, the atomic weight of tin may be calculated.

We now have, for tin, the following available ratios:

(1.) $Sn : SnO_2 :: 100 : 127.076, \pm .0026$
(2.) $4Ag : SnCl_4 :: 100 : 60.207, \pm .0060$
(3.) Percentage of tin in $SnBr_4$, $27.123, \pm .0020$
(4.) Percentage of tin in K_2SnCl_6, $29.040, \pm .0021$.
(5.) Percentage of tin in Am_2SnCl_6, $32.369, \pm .0088$
(6.) $Sn : 2BaSO_4 :: 100 : 392.056, \pm .0713$

The antecedent values are—

$O = 15.879, \pm .0003$ $K = 38.817, \pm .0051$
$Ag = 107.108, \pm .0031$ $N = 13.935, \pm .0021$
$Cl = 35.179, \pm .0048$ $S = 31.828, \pm .0015$
$Br = 79.344, \pm .0062$ $Ba = 136.392, \pm .0086$

With these, six independent values for Sn are computable, as follows:

From (1)............................ $Sn = 117.292, \pm .0115$
From (2)............................ " $= 117.230, \pm .0331$
From (3)............................ " $= 118.120, \pm .0131$
From (4)............................ " $= 118.152, \pm .0155$
From (5)............................ " $= 118.190, \pm .0382$
From (6)............................ " $= 118.216, \pm .0220$

General mean................. $Sn = 117.805, \pm .0069$

If $O = 16$, $Sn = 118.701$.

If we reject the first two of these values, which include all of the older work, and take only the last four, which represent the concordant results of Bongartz and Classen, the general mean becomes—

$$Sn = 118.150, \pm .0089$$

Or, with $O = 16$, $Sn = 119.050$. This mean I regard as having higher probability than the other.

A single determination of the atomic weight of tin, made by Schmidt,[*] ought not to be overlooked, although it was only incidental to his research upon tin sulphide. In one experiment, 0.5243 grm. Sn gave 0.6659 SnO_2. Hence, with $O = 16$, $Sn = 118.49$. This lies about midway between the two sets of values already computed.

---

[*] Berichte, 27, 2743. 1894.

THORIUM.

The atomic weight of thorium has been determined from analyses of the sulphate, oxalate, formate, and acetate, with widely varying results. The earliest figures are due to Berzelius,* who worked with the sulphate, and with the double sulphate of potassium and thorium. The thoria was precipitated by ammonia, and the sulphuric acid was estimated as $BaSO_4$. The sulphate gave the following ratios in two experiments. The third column represents the weight of ThO_2 proportional to 100 parts of $BaSO_4$:

.6754 grm. ThO_2 = 1.159 grm. $BaSO_4$.   Ratio, 58.274
1.0515   "   1.832   "   "   57.396

The double potassium sulphate gave .265 grm. ThO_2, .156 grm. SO_3, and .3435 K_2SO_4. The SO_3, with the Berzelian atomic weights, represents .4537 grm. $BaSO_4$. Hence 100 $BaSO_4$ is equivalent to 58.408 ThO_2. This figure, combined with the two previous values for the same ratio, gives a mean of 58.026, ± .214.

From the ratio between the K_2SO_4 and the ThO_2 in the double sulphate, ThO_2 = 266.895.

In 1861 new determinations were published by Chydenius,† whose memoir is accessible to me only in an abstract‡ which gives results without details. Thoria is regarded as a monoxide, ThO, and the old equivalents (O = 8) are used. The following values are assigned for the molecular weight of ThO, as found from analyses of several salts:

From Sulphate.	From K. Th. Sulphate.
66.33	67.02
67.13	
67.75	
68.03	

Mean, 67.252, ± .201

From Acetate.	From Formate.	From Oxalate.	
67.31	68.06	65.87	Two results
66.59	67.89	65.95	by Berlin.
67.27	68.94	65.75	
67.06	—	65.13	
68.40	Mean, 68.297, ± .219	66.54	
		65.85	
Mean, 67.326, ± .201		Mean, 65.85, ± .123	

---

* Poggend. Annal., 16, 398. 1829. Lehrbuch, 3, 1224.
† Kemisk undersökning af Thorjord och Thorsalter. Helsingfors, 1861. An academic dissertation.
‡ Poggend. Annal., 119, 55. 1863.

We may fairly assume that these figures were calculated with $O = 8$, $C = 6$, and $S = 16$. Correcting by the values for these elements which have been found in previous chapters, ThO_2 becomes as follows:

From sulphate	$ThO_2 = 267.170, \pm .7950$
From acetate	" $= 267.488, \pm .7950$
From formate	" $= 271.239, \pm .8698$
From oxalate	" $= 261.478, \pm .4884$
General mean	$ThO_2 = 265.103, \pm .3394$

The single result from the double potassium sulphate is included with the column from the ordinary sulphate, and the influence of the atomic weight of potassium is ignored.

Chydenius was soon followed by Marc Delafontaine, whose researches appeared in 1863.* This chemist especially studied thorium sulphate; partly in its most hydrous form, partly as thrown down by boiling. In $Th(SO_4)_2.9H_2O$, the following percentages of ThO_2 were found:

$$45.08$$
$$44.90$$
$$45.06$$
$$45.21$$
$$45.06$$

Mean, $45.062, \pm .0332$

The lower hydrate, $2Th(SO_4)_2.9H_2O$, was more thoroughly investigated. The thoria was estimated in two ways: First (A), by precipitation as oxalate and subsequent ignition; second (B), by direct calcination. These percentages of ThO_2 were found:

$$52.83$$
$$52.52$$
$$52.72$$ A.
$$52.13$$

$$52.47$$
$$52.49$$
$$52.53$$
$$52.13$$
$$52.13$$ B.
$$52.43$$
$$52.60$$
$$52.40$$
$$52.96$$
$$52.82$$

Mean, $52.511, \pm .047$

In three experiments with this lower hydrate the sulphuric acid was also estimated, being thrown down as barium sulphate after removal of the thoria:

---

*Arch. Sci. Phys. et Nat. (2), 18, 343.

```
1.2425 grm. gave .400 SO₃.      (1.1656 grm. BaSO₄.)
1.138    "    .366 "            (1.0665   "       )
 .734    "    .2306 "           ( .6720   "       )
```

The figures in parentheses are reproduced by myself from Delafontaine's results, he having calculated his analyses with $O = 100$, $S = 200$, and $Ba = 857$. These data may be reduced to a common standard, so as to represent the quantity of $2Th(SO_4)_x.9H_2O$, equivalent to 100 parts of $BaSO_4$. We then have the following results:

$$106.597$$
$$106.704$$
$$109.226$$

Mean, $107.509, \pm .585$

Delafontaine was soon followed by Hermann,* who published a single analysis of the lower hydrated sulphate, as follows:

```
ThO₂............................  52.87
SO₃.............................  32.11
H₂O.............................  15.02
                                 ───────
                                 100.00
```

Hence, from the ratio between SO_3 and ThO_2, $ThO_2 = 262.286$. Probably the SO_3 percentage was loss upon calcination.

Both Hermann's results and those of Delafontaine are affected by one serious doubt, namely, as to the true composition of the lower hydrated sulphate. The latest and best evidence seems to establish the fact that it contains four molecules of water instead of four and a half,† a fact which tends to lower the resulting atomic weight of thorium considerably. In the final discussion of these data, therefore, the formula $Th(SO_4)_x.4H_2O$ will be adopted. As for Hermann's single analysis, his percentage of ThO_2, 52.87, may be included in one series with Delafontaine's, giving a mean of $52.535, \pm .0473$.

The next determinations to consider are those of Cleve,‡ whose results, obtained from both the sulphate and the oxalate of thorium, agree admirably. The anhydrous sulphate, calcined, gave the subjoined percentages of thoria:

$$62.442$$
$$62.477$$
$$62.430$$
$$62.470$$
$$62.357$$
$$62.366$$

Mean, $62.423, \pm .014$

---

* Journ. für Prakt. Chem., 93, 114.
† See Hillebrand, Bull. 90, U. S. Geol. Survey, p. 29.
‡ K. Svenska Vet. Akad. Handling., Bd. 2, No. 6, 1874.

The oxalate was subjected to a combustion analysis, whereby both thoria and carbonic acid could be estimated. From the direct percentages of these constituents no accurate value can be deduced, there having undoubtedly been moisture in the material studied. From the ratio between CO_2 and ThO_2, however, good results are attainable. This ratio I put in a fourth column, making the thoria proportional to 100 parts of carbon dioxide:

Oxalate.	ThO_2.	CO_2.	Ratio.
1.7135 grm.	1.0189 grm.	.6736 grm.	151.262
1.3800 "	.8210 "	.5433 "	151.114
1.1850 "	.7030 "	.4650 "	151.183
1.0755 "	.6398 "	.4240 "	150.896

Mean, 151.114, ± .053

In 1882, Nilson's determinations appeared.* This chemist studied both the anhydrous sulphate, and the salt with nine molecules of water, using the usual calcination method, but guarding especially against the hygroscopic character of the dry $Th(SO_4)_2$ and the calcined ThO_2. The hydrated sulphate gave results as follows:

$Th(SO_4)_2.9H_2O$.	ThO_2.	Per cent. ThO_2.
2.0549	.9267	45.097
2.1323	.9615	45.092
3.0017	1.3532	45.081
2.7137	1.2235	45.086
2.6280	1.1849	45.088
1.9479	.8785	45.099

Mean, 45.091, ± .0019
Delafontaine found, 45.062, ± .0332

General mean, 45.090, ± .0019

The anhydrous sulphate gave data as follows:

$Th(SO_4)_2$.	ThO_2.	Per cent. ThO_2.
1.4467	.9013	62.300
1.6970	1.0572	62.298
2.0896	1.3017	62.294
1.5710	.9787	62.298

Mean, 62.297, ± .0009

The last four determinations appear again in a paper published five years later by Krüss and Nilson,† who, however, give four more made

---
* Ber. Deutsch. Chem. Gesell., 15, 2519. 1882.
† Ber. Deutsch. Chem. Gesell., 20, 1665. 1887.

upon material obtained from a different source. The new data are subjoined:

$Th(SO_4)_2$	ThO_2	Per cent. ThO_2
1.1630	.7245	62.296
.8607	.5362	62.298
1.5417	.9605	62.301
1.5217	.9479	62.292

Mean, 62.297, ± .0013
Nilson's series, 62.297, ± .0009
Cleve found, 62.423, ± .0140

General mean, 62.298, ± .0007

From Chydenius' work we have four values for the molecular weight of thoria, which, combined as usual, give a general mean of $ThO_2 =$ 265.103, ± .3394. We also have the following ratios:

(1.) $2BaSO_4 : ThO_2 :: 100 : 58.026$, ± .214
(2.) $2BaSO_4 : Th(SO_4)_2.4H_2O :: 100 : 107.509$, ± .585
(3.) $4CO_2 : ThO_2 :: 100 : 151.114$, ± .053
(4.) Percentage of ThO_2 in $Th(SO_4)_2.9H_2O$, 45.090, ± .0019
(5.) Percentage of ThO_2 in $Th(SO_4)_2.4H_2O$, 52.535, ± .0473
(6.) Percentage of ThO_2 in $Th(SO_4)_2$, 62.298, ± .0007

Reducing with the following data, seven values for the atomic weight of thoria are calculable:

O = 15.879, ± .0003   C = 11.920, ± .0004
S = 31.828, ± .0015   Ba = 136.392, ± .0086

The values for ThO_2 are—

Chydenius' determinations	ThO_2 = 265.103, ± .3394
From (1)	" = 268.937, ± .9919
From (2)	" = 268.021, ± 2.7115
From (3)	" = 264.120, ± .0927
From (4)	" = 262.641, ± .0149
From (5)	" = 255.061, ± .3426
From (6)	" = 262.613, ± .0081
General mean	ThO_2 = 262.626, ± .0071

Hence Th = 230.868, ± .0071.
If O = 16, Th = 232.626.

PHOSPHORUS.

The material from which we are to calculate the atomic weight of phosphorus is by no means abundant. Berzelius, in his Lehrbuch,* adduces only his own experiments upon the precipitation of gold by phosphorus, and ignores all the earlier work relating to the composition of the phosphates. These experiments have been considered with reference to gold.

Pelouze,† in a single titration of phosphorus trichloride with a standard solution of silver, obtained a wholly erroneous result; and Jacquelain, ‡ in his similar experiments, did even worse. Schrötter's criticism upon Jacquelain sufficiently disposes of the latter. §

Only the determinations made by Schrötter, Dumas, and Van der Plaats remain to be considered.

Schrötter || burned pure amorphous phosphorus in dry oxygen, and weighed the pentoxide thus formed. One gramme of P yielded P_2O_5 in the following proportions:

2.28909
2.28783
2.29300
2.28831
2.29040
2.28788
2.28848
2.28856
2.28959
2.28872

Mean, 2.289186, ± .00033

Dumas ¶ prepared pure phosphorus trichloride by the action of dry chlorine upon red phosphorus. The portion used in his experiments boiled between 76° and 78°. This was titrated with a standard solution of silver in the usual manner. Dumas publishes weights, from which I calculate the figures given in the third column, representing the quantity of trichloride proportional to 100 parts of silver:

1.787 grm. PCl_3 =	4.208 grm. Ag.		42.4667
1.466 "	3.454 "		42.4435
2.056 "	4.844 "		42.4443
2.925 "	6.890 "		42.4528
3.220 "	7.582 "		42.4690

Mean, 42.4553, ± .0036

---

* 5th ed., 1188.
† Compt. Rend., 20, 1047.
‡ Compt. Rend., 33, 693.
§ Journ. für Prakt. Chem., 57, 315.
|| Journ. für Prakt. Chem., 53, 435. 1851.
¶ Ann. Chem. Pharm., 113, 29. 1860.

By Van der Plaats* three methods of determination were adopted, and all weights were reduced to vacuum standards. First, silver was precipitated from a solution of the sulphate by means of phosphorus. The latter had been twice distilled in a current of nitrogen. The silver, before weighing, was heated to redness. The phosphorus equivalent to 100 parts of silver is given in the third column.

.9096 grm. P gave 15.8865 Ag.   5.7256
.5832   "   10.1622 "   5.7389

Mean, 5.7322, ± .0045

The second method consisted in the analysis of silver phosphate; but the process is not given. Van der Plaats states that it is difficult to be sure of the purity of this salt.

6.6300 grm. Ag_3PO_4 gave 5.1250 Ag.   77.300 per cent.
12.7170   "   9.8335 "   77.326   "

Mean, 77.313, ± .0088

In the third set of determinations, yellow phosphorus was oxidized by oxygen at reduced pressure, and the resulting P_2O_5 was weighed.

10.8230 grm. P gave 24.7925 P_2O_5.   Ratio, 2.29072
7.7624   "   17.7915 "   "   2.29201

As these figures fall within the range of Schrötter's, they may be averaged in with his series, the entire set of twelve determinations giving a mean of 2.28955, ± .00032.

From the following ratios an equal number of values for P may now be computed:

(1.) $2P : P_2O_5 :: 1.0 : 2.28955$, ± .00032
(2.) $3Ag : PCl_3 :: 100 : 42.4553$, ± .0036
(3.) $5Ag : P :: 100 : 5.7322$, ± .0045
(4.) $Ag_3PO_4 : 3Ag :: 100 : 77.313$, ± .0088

Starting with $O = 15.879$, ± .0003, $Ag = 107.108$, ± .0031, and $Cl = 35.179$, ± .0048, we have—

From (1)............................. $P = 30.784$, ± .0077
From (2)............................. " $= 30.882$, ± .0189
From (3)............................. " $= 30.698$, ± .0241
From (4)............................. " $= 30.774$, ± .0382

General mean................. $P = 30.789$, ± .0067

If $O = 16$, $P = 31.024$.

The highest of these figures is that from ratio number two, representing the work of Dumas. This is possibly due to the presence of oxychloride, in traces, in the trichloride taken. Such an impurity, if present, would tend to raise the apparent atomic weight of phosphorus.

---

*Compt. Rend., 100, 52. 1885.

VANADIUM.

Roscoe's determination of the atomic weight of vanadium was the first to have any scientific value. The results obtained by Berzelius* and by Czudnowicz† were unquestionably too high, the error being probably due to the presence of phosphoric acid in the vanadic acid employed. This particular impurity, as Roscoe has shown, prevents the complete reduction of V_2O_5 to V_2O_3 by means of hydrogen. All vanadium ores contain small quantities of phosphorus, which can only be detected with ammonium molybdate—a reaction unknown in Berzelius' time. Furthermore, the complete purification of vanadic acid from all traces of phosphoric acid is a matter of great difficulty, and probably never was accomplished until Roscoe undertook his researches.

In his determination of the atomic weight, Roscoe ‡ studied two compounds of vanadium, namely, the pentoxide, V_2O_5, and the oxychloride, $VOCl_3$. The pentoxide, absolutely pure, was reduced to V_2O_3 by heating in hydrogen, with the following results:

7.7397 grm. V_2O_5 gave 6.3827 grm. V_2O_3.		17.533 per cent. of loss.	
6.5819	" 5.4296 "	17.507	"
5.1895	" 4.2819 "	17.489	"
5.0450	" 4.1614 "	17.515	"
5.4296 grm. V_2O_3, reoxidized, gave 6.5814 grm. V_2O_5.		17.501 per cent. difference.	

Mean, 17.509, ± .005

Hence $V = 50.993, \pm .0219$.

Upon the oxychloride, $VOCl_3$, two series of experiments were made—one volumetric, the other gravimetric. In the volumetric series the compound was titrated with solutions containing known weights of silver, which had been purified according to the methods recommended by Stas. Roscoe publishes his weighings, and gives percentages deduced from them; his figures, reduced to a common standard, make the quantities of $VOCl_3$ given in the third column proportional to 100 parts of silver. He was assisted by two analysts:

Analyst A.

2.4322 grm. $VOCl_3$ = 4.5525 grm. Ag.		53.425
4.6840 "	8.7505 "	53.528
4.2188 "	7.8807 "	53.533
3.9490 "	7.3799 "	53.510
.9243 "	1.7267 "	53.530
1.4330 "	2.6769 "	53.532

---

* Poggend. Annal., 22, 14. 1831.
† Poggend. Annal., 120, 17. 1863.
‡ Journ. Chem. Soc., 6, pp. 330 and 344. 1869.

Analyst B.

2.8530 grm. VOCl₃	= 5.2853 grm. Ag.		53.980
2.1252 "	3.9535 "		53.755
1.4248 "	2.6642 "		53.479

Mean, 53.586, ± .039

The gravimetric series, of course, fixes the ratio between $VOCl_3$ and AgCl. If we put the latter at 100 parts, the proportion of $VOCl_3$ is as given in the third column:

Analyst A.

1.8521 grm. VOCl₃ gave	4.5932 grm. AgCl.		40.323
.7013 "	1.7303 "		40.531
.7486 "	1.8467 "		40.537
1.4408 "	3.5719 "		40.337
.9453 "	2.3399 "		40.399
1.6183 "	4.0282 "		40.174

Analyst B.

2.1936 grm. VOCl₃ gave	5.4039 grm. AgCl.		40.391
2.5054 "	6.2118 "		40.333

Mean, 40.378, ± .028

These two series give us two values for the molecular weight of $VOCl_3$:

From volumetric series.......... $VOCl_3 = 172.185, ± .1254$
From gravimetric series......... " $= 172.358, ± .1196$

General mean............ $VOCl_3 = 172.277, ± .0866$

Hence V = 50.881, ± .0877.
Combining the two values for V, we have:

From $VOCl_3$.................. V = 50.881, ± .0877
From V_2O_5.................. " = 50.993, ± .0219

General mean .... V = 50.986, ± .0212

If O = 16, V = 51.376. These values are calculated with O = 15.879, ± .0003; Cl = 35.179, ± .0048; Ag = 107.108, ± .0031, and AgCl = 142.287, ± .0037.

ARSENIC.

For the determination of the atomic weight of arsenic three compounds have been studied—the chloride, the trioxide, and sodium pyroarsenate. The bromide may also be considered, since it was analyzed by Wallace in order to establish the atomic weight of bromine. His series, in the light of more recent knowledge, may properly be inverted, and applied to the determination of arsenic.

In 1826 Berzelius[*] heated arsenic trioxide with sulphur in such a way that only SO_2 could escape. 2.203 grammes of As_2O_3, thus treated, gave a loss of 1.069 of SO_2. Hence As = 74.460.

In 1845 Pelouze[†] applied his method of titration with known quantities of pure silver to the analysis of the trichloride of arsenic, $AsCl_3$. Using the old Berzelian atomic weights, and putting Ag = 1349.01 and Cl = 443.2, he found in three experiments for As the values 937.9, 937.1, and 937.4. Hence 100 parts of silver balance the following quantities of $AsCl_3$:

56.029
56.009
56.016

Mean, 56.018, ± .004

Later, the same method was employed by Dumas,[‡] whose weighings, reduced to the foregoing standard, give the following results:

4.298 grm. $AsCl_3$	=	7.673 grm. Ag.	Ratio, 56.015
5.535	"	9.880	" 56.022
7.660	"	13.686	" 55.970
4.680	"	8.358	" 55.993

Mean, 56.000, ± .008

The two series of Pelouze and Dumas, combined, give a general mean of 56.014, ± .0035, as the amount of $AsCl_3$ equivalent to 100 parts of silver. Hence As = 74.450, ± .019, a value closely agreeing with that deduced from the single experiment of Berzelius.

The same process of titration with silver was applied by Wallace[§] to the analysis of arsenic tribromide, $AsBr_3$. This compound was repeatedly distilled to ensure purity, and was well crystallized. His weighings show that the quantities of bromide given in the third column are proportional to 100 parts of silver:

8.3246 grm. $AsBr_3$	=	8.58 grm. Ag.	97.023
4.4368	"	4.573	97.022
5.098	"	5.257	96.970

Mean, 97.005, ± .012

---
[*] Poggend. Annalen, 8, 1.
[†] Compt. Rend., 20, 1047.
[‡] Ann. Chim. Phys. (3), 55, 174; 1859.
[§] Phil. Mag. (4), 18, 279.

Hence As = 73.668, ± .0436. Why this value should be so much lower than that from the chloride is unexplained.

The volumetric work done by Kessler,* for the purpose of establishing the atomic weights of chromium and of arsenic, is described in the chromium chapter. In that investigation the amount of potassium dichromate required to oxidize 100 parts of As_2O_3 to As_2O_5 was determined and compared with the quantity of potassium chlorate necessary to produce the same effect. From the molecular weight of $KClO_3$, that of $K_2Cr_2O_7$ was then calculable.

From the same figures, the molecular weights of $KClO_3$ and of $K_2Cr_2O_7$ being both known, that of As_2O_3 may be easily determined. The quantities of the other compounds proportional to 100 parts of As_2O_3 are as follows:

$K_2Cr_2O_7$.	$KClO_3$.
98.95	41.156
98.94	41.116
99.17	41.200
98.98	41.255
99.08	41.201
99.15	41.086
	41.199
Mean, 99.045, ± .028	41.224
	41.161
	41.193
	41.149
	41.126
	Mean, 41.172, ± .009

Another series with the dichromate gave the following figures:

99.08
99.06
99.10
98.97
98.97

Mean, 99.036, ± .019
Previous series, 99.045, ± .028

General mean, 99.039, ± .016

Other defective series are given to illustrate the partial oxidation of the As_2O_3 by the action of the air. From Kessler's data we get two values for the molecular weight of As_2O_3, thus:

From $KClO_3$ series.............. As_2O_3 = 196.951, ± .0445
From $K_2Cr_2O_7$ series............. "   = 196.726, ± .0562

General mean............ As_2O_3 = 196.851, ± .0349

And As = 74.607, ± .0175.

---

* Poggend Annal., 95, 204. 1855. Also 113, 134. 1861.

ARSENIC. 215

The determinations made by Hibbs* are based upon an altogether different process from any of the preceding measurements. Sodium pyroarsenate was heated in gaseous hydrochloric acid, yielding sodium chloride. The latter was perfectly white, completely soluble in water, unfused, and absolutely free from arsenic. The vacuum weights are subjoined, with a column giving the percentage of chloride obtained from the pyroarsenate.

$Na_4As_2O_7$.	$NaCl$.	Percentage.
.02177	.01439	66.100
.04713	.03115	66.094
.05795	.03830	66.091
.40801	.26981	66.128
.50466	.33345	66.092
.77538	.51249	66.095
.82897	.54791	66.095
1.19124	.78731	66.092
1.67545	1.10732	66.091
3.22637	2.13267	66.101

Mean, 66.098, ± .0030

Hence As = 74.340, ± .0235.

In the calculation of the foregoing values for arsenic, the subjoined atomic weights have been assumed:

O = 15.879, ± .0003   K = 38.817, ± .0051
Ag = 107.108, ± .0031   Na = 22.881, ± .0046
Cl = 35.179, ± .0048   S = 31.828, ± .0015
Br = 79.344, ± .0062   Cr = 51.742, ± .0034

To the single determination by Berzelius we may arbitrarily assign a weight equal to that of the result from Wallace's bromide series. The general combination is then as follows:

From Berzelius' experiment .... As = 74.460, ± .0436
From $AsCl_3$. .... " = 74.450, ± .0190
From $AsBr_3$. .... " = 73.668, ± .0436
From As_2O_3 (Kessler).............. " = 74.607, ± .0175
From $Na_4As_2O_7$ .... " = 74.340, ± .0235

General mean................ As = 74.440, ± .0106

If O = 16, As = 75.007.

---

* Doctoral thesis, University of Pennsylvania, 1896. Work done under the direction of Professor E. F. Smith. In the fifth experiment the weight of NaCl is printed .33045. This is evidently a misprint, which I have corrected by comparison with the other data. The rejection of this experiment would not affect the final result appreciably.

ANTIMONY.

After some earlier, unsatisfactory determinations, Berzelius,* in 1826, published his final estimation of the atomic weight of antimony. He oxidized the metal by means of nitric acid, and found that 100 parts of antimony gave 124.8 of Sb_2O_4. Hence, if $O = 16$, $Sb = 129.03$. The value 129 remained in general acceptance until 1855, when Kessler, † by special volumetric methods, showed that it was certainly much too high. Kessler's results will be considered more fully further along, in connection with a later paper; for present purposes a brief statement of his earlier conclusions will suffice. Antimony and various compounds of antimony were oxidized partly by potassium dichromate and partly by potassium chlorate, and from the amounts of oxidizing agent required the atomic weight in question was deduced:

By oxidation of Sb_2O_3 from 100 parts of Sb..... $Sb = 123.84$
By oxidation of Sb with $K_2Cr_2O_7$.............. " $= 123.61$
By oxidation of Sb with $KClO_3 + K_2Cr_2O_7$..... " $= 123.72$
By oxidation of Sb_2O_3 with $KClO_3 + K_2Cr_2O_7$... " $= 123.80$
By oxidation of Sb_2S_3 with $K_2Cr_2O_7$ .... " $= 123.58$
By oxidation of tartar emetic .... " $= 119.80$

The figures given are those calculated by Kessler himself. A recalculation with our newer atomic weights for O, K, Cl, Cr, S, and C would yield lower values. It will be seen that five of the estimates agree closely, while one diverges widely from the others. It will be shown hereafter that the concordant values are all vitiated by constant errors, and that the exceptional figure is after all the best.

Shortly after the appearance of Kessler's first paper, Schneider ‡ published some results obtained by the reduction of antimony sulphide in hydrogen. The material chosen was a very pure stibnite from Arnsberg, of which the gangue was only quartz. This was corrected for, and corrections were also applied for traces of undecomposed sulphide carried off mechanically by the gas stream, and for traces of sulphur retained by the reduced antimony. The latter sulphur was estimated as barium sulphate. From 3.2 to 10.6 grammes of material were taken in each experiment. The final corrected percentages of S in Sb_2S_3 were as follows:

28.559
28.557
28.501
28.554
28.532

---
* Poggend. Annalen, 8, 1.
† Poggend. Annalen, 95, 215.
‡ Poggend. Annalen, 98, 293. 1856. Preliminary note in Bd. 97.

28.485
28.492
28.481

Mean, 28.520, ± .008

Hence, if $S = 32$, $Sb = 120.3$.

Immediately after the appearance of Schneider's memoir, Rose* published the result of a single analysis of antimony trichloride, previously made under his supervision by Weber. This analysis, if $Cl = 35.5$, makes $Sb = 120.7$, a value of no great weight, but in a measure confirmatory of that obtained by Schneider.

The next research upon the atomic weight of antimony was that of Dexter,† published in 1857. This chemist, having tried to determine the amount of gold precipitable by a known weight of antimony, and having obtained discordant results, finally resorted to the original method of Berzelius. Antimony, purified with extreme care, was oxidized by nitric acid, and the gain in weight was determined. From 1.5 to 3.3 grammes of metal were used in each experiment. The reduction of the weights to a vacuum standard was neglected as being superfluous. From the data obtained, we get the following percentages of Sb in Sb_2O_4:

79.268
79.272
79.255
79.266
79.253
79.271
79.264
79.260
79.286
79.274
79.232
79.395
79.379

Mean, 79.283, ± .009

Hence, if $O = 16$, $Sb = 122.46$.

The determinations of Dumas ‡ were published in 1859. This chemist sought to fix the ratio between silver and antimonious chloride, and obtained results for the atomic weight of antimony quite near to those of Dexter. The $SbCl_3$ was prepared by the action of dry chlorine upon pure antimony; it was distilled several times over antimony powder, and it seemed to be perfectly pure. Known weights of this preparation were added to solutions of tartaric acid in water, and the silver chloride was precipitated without previous removal of the antimony. Here, as

---

* Poggend. Annalen, 98, 455. 1856.
† Poggend. Annalen, 100, 363. 1857.
‡ Ann. Chim. Phys. (3), 55, 175.

Cooke has since shown, is a possible source of error, for under such circumstances the crystalline argento-antimonious tartrate may also be thrown down and contaminate the chloride of silver. But be that as it may, Dumas' weighings, reduced to a common standard, give as proportional to 100 parts of silver, the quantities of $SbCl_3$ which are stated in the third of the subjoined columns:

1.876 grm. $SbCl_3$ =	2.660 grm. Ag.		70.526
4.336 "	6.148 "		70.527
5.065 "	7.175 "		70.592
3.475 "	4.930 "		70.487
3.767 "	5.350 "		70.411
5.910 "	8.393 "		70.416
4.828 "	6.836 "		70.626

Mean, 70.512, ± .021

Hence, if Ag = 108, and Cl = 35.5, Sb = 122.

In 1861 Kessler's second paper* relative to the atomic weight of antimony appeared. Kessler's methods were somewhat complicated, and for full details the original memoirs must be consulted. A standard solution of potassium dichromate was prepared, containing 6.1466 grammes to the litre. With this, solutions containing known quantities of antimony or of antimony compounds were titrated, the end reaction being adjusted with a standard solution of ferrous chloride. In some cases the titration was preceded by the addition of a definite weight of potassium chlorate, insufficient for complete oxidation; the dichromate then served to finish the reaction. The object in view was to determine the amount of oxidizing agent, and therefore of oxygen, necessary for the conversion of known quantities of antimonious into antimonic compounds.

In the later paper Kessler refers to his earlier work, and shows that the values then found for antimony were all too high, except in the case of the series made with tartar emetic. That series he merely states, and subsequently ignores, evidently believing it to be unworthy of further consideration. For the remaining series he points out the sources of error. These need not be rediscussed here, as the discussion would have no value for present purposes; suffice it to say that in the series representing the oxidation of Sb_2O_3 with dichromate and chlorate, the material used was found to be impure. Upon estimating the impurity and correcting for it, the earlier value of Sb = 123.80 becomes Sb = 122.36, according to Kessler's calculations.

In the paper now under consideration four series of results are given. The first represents experiments made upon a pure antimony trioxide which had been sublimed, and which consisted of shining colorless needles. This was dissolved, together with some potassium chlorate, in

---

* Poggend. Annalen, 113, 145. 1861.

hydrochloric acid, and titrated with dichromate solution. Six experiments were made, but Kessler rejects the first and second as untrustworthy. The data for the others are as follows:

Sb_2O_3.	$KClO_3$.	$K_2Cr_2O_7$ sol. in cc.
1.7888 grm.	.4527 grm.	19.2 cc.
1.6523 "	.4506 "	3.9 "
3.2998 "	.8806 "	16.5 "
1.3438 "	.3492 "	10.2 "

From these figures Kessler deduces Sb = 122.16.

These data, reduced to a common standard, give the following quantities of oxygen needed to oxidize 100 parts of Sb_2O_3 to Sb_2O_5. Each cubic centimetre of the $K_2Cr_2O_7$ solution corresponds to one milligramme of O:

10.985
10.939
10.951
10.936

Mean, 10.953, ± .0075

In the second series of experiments pure antimony was dissolved in hydrochloric acid with the aid of an unweighed quantity of potassium chlorate. The solution, containing both antimonious and antimonic compounds, was then reduced entirely to the antimonious condition by means of stannous chloride. The excess of the latter was corrected with a strong hydrochloric acid solution of mercuric chloride, then, after diluting and filtering, a weighed quantity of potassium chlorate was added, and the titration with dichromate was performed as usual. Calculated as above, the percentages of oxygen given in the last column correspond to 100 parts of antimony:

Sb.	$KClO_3$.	$K_2Cr_2O_7$ sol. cc.	Per cent. O.
1.636 grm.	0.5000 grm.	18.3	13.088
3.0825 "	0.9500 "	30.2	13.050
4.5652 "	1.4106 "	45.5	13.098

Mean, 13.079, ± .0096

This series gave Kessler Sb = 122.34.

The third and fourth series of experiments were made with pure antimony trichloride, $SbCl_3$, prepared by the action of mercuric chloride upon metallic antimony. This preparation, in the third series, was dissolved in hydrochloric acid, and titrated. In one experiment solid $K_2Cr_2O_7$ in weighed amount was added before titration; in the other two estimations $KClO_3$ was taken as usual. The third column gives the percentages of oxygen corresponding to 100 parts of $SbCl_3$.

Per cent. O.

1.8576 grm. SbCl₃ needed	.5967 grm. K₂Cr₂O₇ and	33.4 cc. sol.	7.0338
1.9118 "	.3019 " KClO₃	" 16.2 "	7.0321
4.1235 "	.6801 " KClO₃	" 23.2 "	7.0222

Mean, 7.0294, ± .0024

The fourth set of experiments was gravimetric. The solution of SbCl₃, mixed with tartaric acid, was first precipitated by hydrogen sulphide, in order to remove the antimony. The excess of H₂S was corrected by copper sulphate, and then the chlorine was estimated as silver chloride in the ordinary manner. 100 parts of AgCl correspond to the amounts of SbCl₃ given in the third column.

1.8662 grm. SbCl₃ gave	3.483 grm. AgCl.	53.580
1.6832 "	3.141 "	53.588
2.7437 "	5.1115 "	53.677
2.6798 "	5.0025 "	53.569
5.047 "	9.411 "	53.629
3.8975 "	7.2585 "	53.696

Mean, 53.623, ± .015

The volumetric series with SbCl₃ gave Kessler values for Sb ranging from 121.16 to 121.47. The gravimetric series, on the other hand, yielded results from Sb = 124.12 to 124.67. This discrepancy Kessler rightly attributes to the presence of oxygen in the chloride; and, ingeniously correcting for this error, he deduces from both sets combined the value of Sb = 122.37.

The several mean results for antimony agree so fairly with each other, and with the estimates obtained by Dexter and Dumas, that we cannot wonder that Kessler felt satisfied of their general correctness, and of the inaccuracy of the figures published by Schneider. Still, the old series of data obtained by the titration of tartar emetic with dichromate contained no evident errors, and was not accounted for. This series,[*] if we reduce all of Kessler's figures to a single common standard, gives a ratio between $K_2Cr_2O_7$ and $C_4H_4KSbO_7.\tfrac{1}{2}H_2O$. 100 parts of the former will oxidize of the latter:

336.64
338.01
336.83
337.93
338.59
335.79

Mean, 337.30, ± .29

From this, if $K_2Cr_2O_7 = 292.271$, Sb = 118.024.

The newer atomic weights found in other chapters of this work will

---

[*] Poggend. Annalen, 95, 217.

be applied to the discussion of all these series further along. It may, however, be properly noted at this point that the probable errors assigned to the percentages of oxygen in three of Kessler's series are too low. These percentages are calculated from the quantities of $KClO_3$ involved in the several reactions, and their probable errors should be increased with reference to the probable error of the molecular weight of that salt. The necessary calculations would be more laborious than the importance of the figures would warrant, and accordingly, in computing the final general mean for antimony, Kessler's figures will receive somewhat higher weight than they are legitimately entited to.

Naturally, the concordant results of Dexter, Kessler, and Dumas led to the general acceptance of the value of 122 for antimony as against the lower figure, 120, of Schneider. Still, in 1871, Unger * published the results of a single analysis of Schlippe's salt, $Na_3SbS_4.9H_2O$. This analysis gave $Sb = 119.76$, if $S = 32$ and $Na = 23$, but no great weight could be attached to the determination. It served, nevertheless, to show that the controversy over the atomic weight of antimony was not finally settled.

More than ten years after the appearance of Kessler's second paper the subject of the atomic weight of antimony was again taken up, this time by Professor Cooke. His results appeared in the autumn of 1877 † and were conclusive in favor of the lower value, approximately 120. For full details the original memoir must be consulted; only a few of the leading points can be cited here.

Schneider analyzed a sulphide of antimony which was already formed. Cooke, reversing the method, effected the synthesis of this compound. Known weights of pure antimony were dissolved in hydrochloric acid containing a little nitric acid. In this solution weighed balls of antimony were boiled until the liquid became colorless; subsequently the weight of metal lost by the balls was ascertained. To the solution, which now contained only antimonious compounds, tartaric acid was added, and then, with a supersaturated aqueous sulphhydric acid, antimony trisulphide was precipitated. The precipitate was collected by an ingenious process of reverse filtration, converted into the black modification by drying at 210°, and weighed. After weighing, the Sb_2S_3 was dissolved in hydrochloric acid, leaving a carbonaceous residue unacted upon. This was carefully estimated and corrected for. About two grammes of antimony were taken in each experiment and thirteen syntheses were performed. In two of these, however, the antimony trisulphide was weighed only in the red modification, and the results were uncorrected by conversion into the black variety and estimation of the carbonaceous residue. In fact, every such conversion and correction was preceded by a weighing of the red modification of the Sb_2S_3. The mean result of these weighings, if $S = 32$, gave $Sb = 119.994$. The mean result of the cor-

---

* Archiv. der Pharmacie, 197, 194. Quoted by Cooke.
† Proc. Amer. Acad., 5, 13.

rected syntheses gave Sb = 120.295. In these eleven experiments the following percentages of S in Sb_2S_3 were established:

28.57
28.60
28.57
28.43
28.42
28.53
28.50
28.49
28.58
28.50
28.51

Mean, 28.5182, ± .0120

These results, confirmatory of the work of Schneider, were presented to the American Academy in 1876. Still, before publication, Cooke thought it best to repeat the work of Dumas, in order to detect the cause of the old discrepancy between the values Sb = 120 and Sb = 122. Accordingly, various samples of antimony trichloride were taken, and purified by repeated distillations. The final distillate was further subjected to several recrystallizations from the fused state; or, in one case, from a saturated solution in a bisulphide of carbon. The portions analyzed were dissolved in concentrated aqueous tartaric acid, and precipitated by silver nitrate, many precautions being observed. The silver chloride was collected by reverse filtration, and dried at temperatures from 110° to 120°. In one experiment the antimony was first removed by H_2S. Seventeen experiments were made, giving, if Ag = 108 and Cl = 35.5, a mean value of Sb = 121.94. If we reduce to a common standard, Cooke's analyses give, as proportional to 100 parts of AgCl, the quantities of $SbCl_3$ stated in the third column:

1.5974 grm. $SbCl_3$	gave	3.0124 grm. AgCl.		53.028
1.2533	"	2.3620	"	53.061
.8876	"	1.6754	"	52.978
.8336	"	1.5674	"	53.184
.5326	"	1.0021	"	53.148
.7270	"	1.3691	"	53.101
1.2679	"	2.3883	"	53.088
1.9422	"	3.6646	"	52.999
1.7702	"	3.3384	"	53.025
2.5030	"	4.7184	"	53.048
2.1450	"	4.0410	"	53.081
1.7697	"	3.3281	"	53.175
2.3435	"	4.4157	"	53.072
1.3686	"	2.5813	"	53.020
1.8638	"	3.5146	"	53.030
2.0300	"	3.8282	"	53.028
2.4450	"	4.6086	"	53.053

Mean, 53.066, ± .0096

ANTIMONY. 223

This mean may be combined with that of Kessler's series, as follows:

Kessler.................... .... 53.623, ± .015
Cooke..................... .... 53.066, ± .0096

General mean .... 53.2311, ± .008

The results thus obtained with $SbCl_3$ confirmed Dumas' determination of the atomic weight of antimony as remarkably as the syntheses of Sb_2S_3 had sustained the work of Schneider. Evidently, in one or the other series a constant error must be hidden, and much time was spent by Cooke in searching for it. It was eventually found that the chloride of antimony invariably contained traces of oxychloride, an impurity which tended to increase the apparent atomic weight of the metal under consideration. It was also found, in the course of the investigation, that hydrochloric acid solutions of antimonious compounds oxidize in the air during boiling as rapidly as ferrous compounds, a fact which explains the high values for antimony found by Kessler.

In order to render "assurance doubly sure," Professor Cooke also undertook the analysis of the bromide and the iodide of antimony. The bromide, $SbBr_3$, was prepared by adding the finely powdered metal to a solution of bromine in carbon disulphide. It was purified by repeated distillation over pulverized antimony, and by several recrystallizations from bisulphide of carbon. The bromine determinations resemble those of chlorine, and gave, if $Ag = 108$ and $Br = 80$, a mean value for antimony of $Sb = 120$. Reduced to a common standard, the fifteen analyses give the subjoined quantities of $SbBr_3$ proportional to 100 parts of silver bromide:

1.8621 grm.	$SbBr_3$ gave	2.9216 grm.	AgBr.	63.736
.9856	"	1.5422	"	63.909
1.8650	"	2.9268	"	63.721
1.5330	"	2.4030	"	63.795
1.3689	"	2.1445	"	63.833
1.2124	"	1.8991	"	63.841
.9417	"	1.4749	"	63.848
2.5404	"	3.9755	"	63.901
1.5269	"	2.3905	"	63.874
1.8604	"	2.9180	"	63.756
1.7298	"	2.7083	"	63.870
3.2838	"	5.1398	"	63.890
2.3589	"	3.6959	"	63.825
1.3323	"	2.0863	"	63.859
2.6974	"	4.2285	"	63.791

Mean, 63.830, ± .008

The iodide of antimony was prepared like the bromide, and analyzed in the same way. At first, discordant results were obtained, due to the presence of oxyiodide in the iodide studied. The impurity, however, was removed by subliming the iodide in an atmosphere of dry carbon dioxide. With this purer material, seven estimations of iodine were made, giving, if Ag = 108 and I = 127, a value for antimony of Sb = 120. Reduced to a uniform standard, Cooke's weighings give the following quantities of SbI_3 proportional to 100 parts of silver iodide:

1.1877 grm. SbI_3 gave	1.6727 grm. AgI.		71.005
.4610 "	.6497	"	70.956
3.2527 "	4.5716	"	71.150
1.8068 "	2.5389	"	71.165
1.5970 "	2.2456	"	71.117
2.3201 "	3.2645	"	71.071
.3496 "	.4927	"	70.956

Mean, 71.060, ± .023

Although Cooke's work was practically conclusive, as between the rival values for antimony, his results were severely criticised by Kessler,[*] who evidently had read Cooke's paper in a very careless way. On the other hand, Schneider published in Poggendorff's Annalen a friendly review of the new determinations, which so well vindicated his own accuracy. In reply to Kessler, Cooke undertook still another series of experiments with antimony bromide,[†] and obtained absolute confirmation of his previous results. To a solution of antimony bromide was added a solution containing a known weight of silver not quite sufficient to precipitate all the bromine. The excess of the latter was estimated by titration with a normal silver solution. Five analyses gave values for antimony ranging from 119.98 to 120.02, when Ag = 108 and Br = 80. Reduced to a common standard, the weights obtained gave the amounts of SbBr stated in the third column as proportional to 100 parts of silver:

2.5032 grm. $SbBr_3$ = 2.2528 grm. Ag.		111.115
2.0567 "	1.8509 "	111.119
2.6512 "	2.3860 "	111.115
3.3053 "	2.9749 "	111.106
2.7495 "	2.4745 "	111.113

Mean, 111.114, ± .0014

Schneider,[‡] also, in order to more fully answer Kessler's objections, repeated his work upon the Arnsberg stibnite. This he reduced in hydrogen as before, correcting scrupulously for impurities. The following percentages of sulphur were found:

28.546
28.534
28.542

Mean, 28.541, ± .0024

---

[*] Berichte d. Deutsch. Chem. Gesell., 12, 1044. 1879.
[†] Amer. Journ. Sci. and Arts, May, 1880. Berichte, 13, 951.
[‡] Journ. für Prakt. Chem. (2), 22, 131.

These figures confirm his old results, and may be fairly combined with them and with the percentages found by Cooke, as follows:

Schneider, early series	28.520,	± .008
Schneider, late series	28.541,	± .0024
Cooke	28.5182,	± .0120
General mean	28.5385,	± .0023

In 1881 Pfeifer* determined electrolytically the direct ratios between silver and antimony, and copper and antimony. With copper the following data were obtained:

$$Cu_3 : Sb_2 :: 100 : x.$$

1.412 grm. Sb =	1.1008 Cu.	128.270
1.902 "	1.4832 "	128.236
3.367 "	2.6249 "	128.272

Mean, 128.259, ± .0077

If Cu = 63.6, Sb = 122.36.
With silver he found—

$$Ag_3 : Sb :: 100 : x.$$

5.925 grm. Sb =	15.774 Ag.	37.562
6.429 "	17.109 "	37.577
10.116 "	26.972 "	37.506
4.865 "	13.014 "	37.383
4.390 "	11.697 "	37.531
9.587 "	25.611 "	37.433
4.525 "	12.097 "	37.406

Mean, 37.485, ± .0198

If Ag = 108, Sb = 121.45.

The latter ratio was also determined by Popper,† several years afterwards. The two metals were precipitated simultaneously by the same current; and in some experiments two portions of antimony were thrown down against one of silver. These are indicated in the subjoined table by suitable bracketing, and the ratio is given in the third column:

Sb.	Ag.	Ratio.
1.4856 }	3.9655	37.463
1.4788 }		37.292
2.0120 }	5.3649	37.503
2.0074 }		37.417
3.8882 }	10.3740	37.480
3.8903 }		37.500
4.1893 }	11.1847	37.455
4.1885 }		37.447

---

* Ann. Chem. Pharm., 209, 161.
† Ann. Chem., 233, 153.

226   THE ATOMIC WEIGHTS.

4.2710 } 4.2752 }	11.3868	37.507 37.545
5.6860 } 5.6901 }	15.1786	37.460 37.487
4.4117	11.8014	37.383
4.9999	13.3965	37.322
5.2409	14.0679	37.250

Mean, 37.434, ± .0149
Pfeifer found, 37.485, ± .0198

General mean, 37.452, ± .0119

If $Ag = 108$, Popper's figures give in mean $Sb = 121.3$.

I am inclined to attach slight importance to these electrolytic data, for the reasons that it would be very difficult to ensure the absolute purity and freedom from occlusions of the antimony as weighed, or to guarantee that no secondary reactions had modified the ratios.

The work done by Bongartz* in 1883 was quite different from any of the determinations which had preceded it. Carefully purified antimony was weighed as such, and then dissolved in a concentrated solution of potassium sulphide. From this, after strong dilution, antimony trisulphide was thrown down by means of dilute sulphuric acid. After thorough washing, this sulphide was oxidized by hydrogen peroxide, by Classen's method, and the sulphur in it was weighed as barium sulphate. The ratio measured, therefore, was $2Sb : 3BaSO_4$, and the data were as follows. The $BaSO_4$ equivalent to 100 parts of Sb is the ratio stated :

Sb Taken.	$BaSO_4$ Found.	Ratio.
1.4921	4.3325	290.362
.6132	1.7807	290.394
.5388	1.5655	290.553
1.2118	3.5205	290.518
.9570	2.7800	290.491
.6487	1.8855	290.349
.7280	2.1100	289.835
.9535	2.7655	290.036
1.0275	2.9800	290.024
.9635	2.7980	290.399
.9255	2.6865	290.275
.7635	2.2175	290.438

Mean, 290.306, ± .0436

We have now before us the following ratios, good and bad, from which to calculate the atomic weight of antimony. The single results obtained by Weber and by Unger, being unimportant, are not included :

---

* Ber. Deutsch. Chem. Gesell., 16, 1942. 1883.

ANTIMONY.

(1.) Percentage of S in Sb_2S_3, 28.5385, ± .0023
(2.) Percentage of Sb in Sb_2O_4, 79.283, ± .009
(3.) O needed to oxidize 100 parts $SbCl_3$, 7.0294, ± .0024
(4.) O needed to oxidize 100 parts Sb_2O_3, 10.953, ± .0075
(5.) O needed to oxidize 100 parts Sb, 13.079, ± .0096
(6.) $K_2Cr_2O_7$: tartar emetic : : 100 : 337.30, ± .29
(7.) Ag_3 : $SbCl_3$: : 100 : 70.512, ± .021
(8.) $3AgCl$: $SbCl_3$: : 100 : 53.2311, ± .008
(9.) Ag_3 : $SbBr_3$: : 100 : 111.114, ± .0014
(10.) $3AgBr$: $SbBr_3$: : 100 : 63.830, ± .008
(11.) $3AgI$: SbI_3 : : 100 : 71.060, ± .023
(12.) Cu_3 : Sb_2 : : 100 : 128.259, ± .0077
(13.) Ag_3 : Sb : : 100 : 37.452, ± .0119
(14.) Sb_2 : $3BaSO_4$: : 100 : 290.306, ± .0436

In the reduction of these ratios a considerable number of antecedent atomic weights are required, thus:

O = 15.879, ± .0003
Ag = 107.108, ± .0031
Cl = 35.179, ± .0048
Br = 79.344, ± .0062
I = 125.888, ± .0069
K = 38.817, ± .0051
S = 31.828, ± .0015
C = 11.920, ± .0004
Cu = 63.119, ± .0015
Ba = 136.392, ± .0086
Cr = 51.742, ± .0034
AgCl = 142.287, ± .0037
AgBr = 186.452, ± .0054
AgI = 232.996, ± .0062

Three of the ratios give the molecular weight of antimony trichloride, and two give corresponding values for the bromide. These values may be combined, as follows: First, for the chloride—

From (3) .... $SbCl_3$ = 225.894, ± .0771
From (7) .... " = 226.572, ± .0678
From (8) .... " = 227.223, ± .0347

General mean .... $SbCl_3$ = 226.924, ± .0286

Hence Sb = 121.387, ± .0321.

For the bromide we have—

From (9).................... $SbBr_3$ = 357.036, ± .0113
From (10).................... " = 357.037, ± .0250

General mean........... $SbBr_3$ = 357.036, ± .0103

Hence Sb = 119.005, ± .0212.

All the data yield eleven values for antimony, which are arranged below in the order of their magnitude:

THE ATOMIC WEIGHTS.

1. From tartar emetic, ratio (6).....	Sb = 118.024,	± .2827
2. From $SbBr_3$.	" = 119.005,	± .0212
3. From SbI_3, ratio (11)...........	" = 119.037,	± .1626
4. From Sb_2S_3, ratio (1)...........	" = 119.548,	± .0069
5. From ratio (14)................	" = 119.737,	± .0188
6. From ratio (13)................	" = 120.342,	± .0384
7. From ratio (4).................	" = 121.155,	± .1000
8. From $SbCl_3$....................	" = 121.387,	± .0321
9. From ratio (5).................	" = 121.408,	± .0891
10. From ratio (12)................	" = 121.434,	± .0078
11. From Sb_2O_4, ratio (2)..........	" = 121.542,	± .0546
General mean...............	Sb = 120.299,	± .0047

If $O = 16$, this becomes $Sb = 121.218$.

Among these figures the discordance is so great that the mathematical combination has no real value. We must base our judgment in this case mainly upon chemical evidence, and this, as shown in the investigations of Cooke and of Schneider, favors a lower rather than a higher value for the atomic weight of antimony. Dumas' work was affected by constant errors which are now known, and Dexter's data are also presumably in the wrong. A general mean of values 2, 3, 4, and 5 gives $Sb = 119.521$, ± .0062, or, if $O = 16$, $Sb = 120.432$. Even now the range of uncertainty is greater than it should be, but none of the four values combined can be accepted exclusively or rejected without more evidence. This result, therefore, should be adopted until new determinations, of a more conclusive nature, have been made.

BISMUTH.

Early in the century the combining weight of bismuth was approximately fixed through the experiments of Lagerhjelm.* Effecting the direct union of bismuth and sulphur, he found that ten parts of the metal yield the following quantities of trisulphide:

12.2520
12.2065
12.2230
12.2465
———
Mean, 12.2320

Hence Bi = 215 in round numbers, a value now known to be much too high. Lagerhjelm also oxidized bismuth with nitric acid, and, after ignition, weighed the trioxide thus formed. Ten parts of metal gave the following quantities of Bi_2O_3:

11.1382
11.1275
———
Mean, 11.13285

Hence, if $O = 16$, $Bi = 211.85$, a figure still too high.

In 1851 the subject of the atomic weight of bismuth was taken up by Schneider,† who, like Lagerhjelm, studied the oxidation of the metal with nitric acid. The work was executed with a variety of experimental refinements, by means of which every error due to possible loss of material was carefully avoided. For full details the original paper must be consulted; there is only room in these pages for the actual results, as follows. The figures represent the percentages of Bi in Bi_2O_3:

89.652
89.682
89.644
89.634
89.656
89.666
89.655
89.653
———
Mean, 89.6552, ± .0034

Hence, if $O = 16$, $Bi = 208.05$.

Next in order are the results obtained by Dumas.‡ Bismuth tri-

---
* Annals of Philosophy, 4, 358. 1814. Adopted by Berzelius.
† Poggend. Annalen, 82, 303. 1851.
‡ Ann. Chim. Phys. (3), 55, 176. 1859.

chloride was prepared by the action of dry chlorine upon bismuth, and repeatedly rectified by distillation over bismuth powder. The product was weighed in a closed tube, dissolved in water, and precipitated with sodium carbonate. In the filtrate, after strongly acidulating with nitric acid, the chlorine was precipitated by a known amount of silver. The figures in the third column show the quantities of $BiCl_3$ proportional to 100 parts of silver:

3.506 grm. $BiCl_3$ = 3.545 grm. Ag.		98.900
1.149 "	1.168 "	98.373
1.5965 "	1.629 "	98.005
2.1767 "	2.225 "	97.829
3.081 "	3.144 "	97.996
2.4158 "	2.470 "	97.806
1.7107 "	1.752 "	97.643
3.523 "	3.6055 "	97.712
5.241 . "	5.361 "	97.762

Mean, 98.003, ± .090

Hence, with Ag = 108 and Cl = 35.5, Bi = 211.03.

The first three of the foregoing experiments were made with slightly discolored material. The remaining six percentages give a mean of 97.791, whence, on the same basis as before, Bi = 110.79. Evidently these results are now of slight value, for it is probable that the chloride of bismuth, like the corresponding antimony compound, contained traces of oxychloride. This assumption fully accounts for the discordance between Dumas' determination and the determinations of Schneider and of still more recent investigators.

In 1883 Marignac* took up the subject, attacking the problem by two methods. His point of departure was commercial subnitrate of bismuth, which was purified by re-solution and reprecipitation, and from which he prepared the oxide. First, bismuth trioxide was reduced by heating in hydrogen, beginning with a moderate temperature and closing the operation at redness. The results were as follows, with the percentage of Bi in Bi_2O_3 added:

2.6460 grm. Bi_2O_3 lost 0.2730 grm. O.		89.683 per cent.
6.7057 "	.6910 "	89.696 "
3.6649 "	.3782 "	89.681 "
5.8024 "	.5981 "	89.692 "
5.1205 "	.5295 "	89.658 "
5.5640 "	.5742 "	89.680 "

Mean, 89.682, ± .0036

Hence, if O = 16, Bi = 208.60.

---

*Arch. Sci. Phys. et Nat. (3), 10, 10.

Marignac's second method of determination was by conversion of the oxide into the sulphate. The oxide was dissolved in nitric acid, and then sulphuric acid was added in slight excess from a graduated tube. The mass was evaporated to dryness with great care, and finally heated over a direct flame until fumes of SO_3 no longer appeared. The third column gives the sulphate formed from 100 parts of oxide:

2.6503 Bi_2O_3 gave	4.0218 $Bi_2(SO_4)_3$.	Ratio,	151.749
2.8025 "	4.2535 "	"	151.775
2.710 "	4.112 "	"	151.734
2.813 "	4.267 "	"	151.688
2.8750 "	4.3625 "	"	151.739
2.7942 "	4.2383 "	"	151.682

Mean, 151.728, ± .0099

Hence, with $O = 16$ and $S = 32.06$, $Bi = 208.16$.

This result needs to be studied in the light of Bailey's observation,[*] that bismuth sulphate has a very narrow range of stability. It loses the last traces of free sulphuric acid at 405°, and begins to decompose at 418°, so that the foregoing ratio is evidently uncertain. The concordance of the data, however, is favorable to it.

The next determination of this atomic weight was by Löwe,[†] who oxidized the metal with nitric acid, and reduced the nitrate to oxide by ignition. Special care was taken to prepare bismuth free from arsenic, and the oxide was fused before weighing. In the paper just quoted Bailey calls attention to the volatility of bismuth oxide, which doubtless accounts for the low results found in this investigation. The data are as follows:

Bi Taken.	Bi_2O_3 Found.	Per cent. Bi.
11.309	12.616	89.640
12.2776	13.694	89.656

Mean, 89.648, ± .0040

Hence, if $O = 16$, $Bi = 207.84$.

In Classen's[‡] work upon the atomic weight of bismuth, the metal itself was first carefully investigated. Commercial samples, even those which purported to be pure, were found to be contaminated with lead and other impurities, and these were not entirely removable by many successive precipitations as subnitrate. Finally, pure bismuth was obtained by an electrolytic process, and this was converted into oxide by means of nitric acid and subsequent ignition to incipient fusion. Results as follows, with the percentage of Bi in Bi_2O_3 added:

---

[*] Journ. Chem. Soc., 51, 676.
[†] Zeit. Anal. Chem., 24, 498.
[‡] Ber. Deutsch. Chem. Gesell., 23, 938. 1890.

THE ATOMIC WEIGHTS.

Bi Taken.	Bi_2O_3 Found.	Per cent. Bi.
25.0667	27.9442	89.703
21.0691	23.4875	89.7035
27.2596	30.3922	89.693
36.5195	40.7131	89.700
27.9214	31.1295	89.6944
32.1188	35.8103	89.692
30.1000	33.5587	89.694
26.4825	59.5257	89.693
19.8008	22.0758	89.695

Mean, 89.696, ± .0009

Hence, if O = 16, Bi = 208.92, or, reduced to vacuum standards, 208.90.

Classen's paper was followed by a long controversy between Schneider and Classen,* in which the former upheld the essential accuracy of the work done by Marignac and himself. Schneider had started out with commercial bismuth, and Classen found that the commercial bismuth which he met with was impure. Schneider, by various analyses, showed that other samples of bismuth were so nearly pure that the common modes of purification were adequate; but Classen replied that the original sample used by Schneider in his atomic weight investigation had not been reëxamined. Accordingly, Schneider published a new series of determinations † made by the old method, but with metal which had been scrupulously purified. Results as follows:

Bi.	Bi_2O_3.	Per cent. Bi.
5.0092	5.5868	89.661
3.6770	4.1016	89.648
7.2493	8.0854	89.659
9.2479	10.3142	89.662
6.0945	6.7979	89.653
12.1588	13.5610	89.660

Mean, 89.657, ± .0015

Hence with O = 16, Bi = 208.05, a confirmation of the earlier determinations.

Although the results so far are not final, a combination of the data relative to bismuth oxide is not without interest.

1. Lagerhjelm .... 89.865, ± .0650
2. Schneider, 1851 .... 89.655, ± .0034
3. Marignac .... 89.682, ± .0036
4. Löwe .... 89.648, ± .0040
5. Classen .... 89.696, ± .0009
6. Schneider, 1894 .... 89.657, ± .0015

General mean...................... 89.681, ± .0007

---

* Journ. für Prakt. Chem. (2), 42, 553; 43, 133; and 44, 23 and 411.
† Journ. für Prakt. Chem. (2), 50, 461. 1894.

Omitting the first and fifth means, the other data give a general mean percentage of 89.659, ± .0012.

The ratios now before us are as follows:

(1.) Percentage of Bi in Bi_2O_3, 89.681, ± .0007
(2.) $Bi_2O_3 : Bi_2(SO_4)_3 :: 100 : 151.728$, ± .0099
(3.) $3Ag : BiCl_3 :: 100 : 98.003$, ± .090

For computation we have—

$O = 15.879$, ± .0003 $Ag = 107.108$, ± .0031
$S = 31.828$, ± .0015 $Cl = 35.179$, ± .0048

Hence, reducing the ratios—

From (1) .... $Bi = 207.003$, ± .0150
From (2) .... " $= 206.613$, ± .0444
From (3) .... " $= 209.370$, ± .2847

General mean... .... $Bi = 206.971$, ± .0142

If $O = 16$, $Bi = 208.548$.

Classen's data alone give $Bi = 207.389$, or, with $O = 16$, 208.969. Omitting this set of determinations and rejecting Dumas', the remaining data give—

From Bi_2O_3 .... $Bi = 206.512$, ± .0244
From $Bi_2(SO_4)_3$ .... " $= 206.613$, ± .0444

General mean .... $Bi = 206.536$, ± .0214

If $O = 16$, this becomes $Bi = 208.11$. Between this figure and Classen's, future investigation must decide. The confirmation afforded by the sulphate series is in favor of the lower value.

COLUMBIUM.*

The atomic weight of this metal has been determined by Rose, Hermann, Blomstrand, and Marignac. Rose † analyzed a compound which he supposed to be chloride, but which, according to Rammelsberg, ‡ must have been nearly pure oxychloride. If it was chloride, then the widely varying results give approximately $Cb = 122$; if it was oxychloride, the value becomes nearly 94. If it was chloride, it was doubtless contaminated with tantalum compounds.

Hermann's § results seem to have no present value, and Blomstrand's ‖ are far from concordant. The latter chemist studied columbium pentachloride and sodium columbate. In the first case he weighed the columbium as columbium pentoxide, and the chlorine as silver chloride, the oxide being determined by several distinct processes. In some cases it was thrown down by water, in others by sulphuric acid, and in still others by sodium carbonate or ammonia jointly with sulphuric acid. The weights given are as follows:

$CbCl_5$.	Cb_2O_5.	$AgCl$.
.591	.294	.....
.8085	.401	2.085
.633	.317	.....
.195	.0974	.500
.507	.2505	1.302
.9415	.472	2.454
.563	.2796	.....
.9385	.4675	2.465
.4788	.2378	.....
.408	.204	1.067
.9065	.4515	.....

Hence the subjoined percentages, and the ratios $5AgCl : CbCl_5 :: 100 : x$, and $5AgCl : Cb_2O_5 :: 100 : x$.

Per cent. Cb_2O_5.	$AgCl : CbCl_5$.	$AgCl : Cb_2O_5$.
49.788	......	......
49.598	38.777	19.233
50.079	......	......
49.949	39.000	19.435
49.408	38.940	19.240
50.135	38.366	19.234

---

* This name has priority over the more generally accepted "niobium," and therefore deserves preference.
† Poggend. Annal., 104, 439. 1858.
‡ Poggend. Annal., 136, 353. 1869.
§ Journ. für Prakt. Chem., 68, 73. 1856.
‖ Acta Univ. Lund, 1864.

49.662	......	......
49.813	38.073	18.966
49.666	......	......
50.000	38.238	19.119
49.807	......	......
Mean, 49.806, ± .045	Mean, 38.566, ± .108	Mean, 19.205, ± .043

From these means the atomic weight of columbium may be computed, thus:

From $2CbCl_5 : Cb_2O_5$ .... $Cb = 95.397$
From $CbCl_5 : 5AgCl$ .... " $= 98.477$
From $5AgCl : Cb_2O_5$ .... " $= 96.933$,

when $O = 15.879$, $Ag = 107.108$, and $Cl = 35.179$.

The series upon sodium columbate, which salt was decomposed with sulphuric acid, both Cb_2O_5 and Na_2SO_4 being weighed, is too discordant for discussion. The exact nature of the salt studied is not clear, and the data given, when transformed into the ratio $Na_2SO_4 : Cb_2O_5 :: 100 : x$, give values for x ranging from 151.65 to 161.20. Further consideration of this series would therefore be useless. It seems highly probable that Blomstrand's materials were not entirely free from tantalum, however, since the atomic weight of columbium derived from his analyses of the chloride are evidently too high.

Marignac* made about twenty analyses of the potassium fluoxycolumbate, $CbOF_3.2KF.H_2O$. 100 parts of this salt give the following percentages:

Cb_2O_5............ Extremes 44.15 to 44.60   Mean, 44.36
K_2SO_4...:........   " 57.60 " 58.05
H_2O............   "   5.75 " 5.98
F................   " 30.62 " 32.22

From the mean percentage of Cb_2O_5, $Cb = 92.852$. If $O = 16$, this becomes 93.56.

From the mean between the extremes given for K_2SO_4, $Cb = 93.192$. If $O = 16$, this becomes 93.90.

As Deville and Troost's † results for the vapor density of the chloride and oxychloride agree fairly well with $Cb = 94$, we may adopt this value as approximately correct. The mean of the two values computed from Marignac's data is 93.022 when $H = 1$, and 93.73 when $O = 16$.

---
* Arch. Sci. Phys. Nat, (2), 23. 1865.
† Compt. Rend., 56, 891. 1863.

TANTALUM.

The results obtained for the atomic weight of this metal by Berzelius,[*] Rose,[†] and Hermann [‡] may be fairly left out of account as valueless. These chemists could not have worked with pure preparations, and their data are sufficiently summed up in Becker's "Digest."

Blomstrand's determinations, [§] as in the case of columbium, were made upon the pentachloride. His weights are as follows:

$TaCl_5$.	Ta_2O_5.	$AgCl$.
.9808	.598	......
1.4262	.867	2.906
2.5282	1.5375	5.0105
1.0604	.6455	2.156
2.581	1.577	......
.8767	.534	......

Hence the subjoined percentages of Ta_2O_5 from $TaCl_5$, and the ratios $5AgCl : TaCl_5 :: 100 : x$, and $5AgCl : Ta_2O_5 :: 100 : x$.

Per cent. Ta_2O_5.	$AgCl : TaCl_5$.	$AgCl : Ta_2O_5$.
60.971	......	......
60.791	49.078	29.835
60.814	50.458	30 685
60.873	49.297	29.940
60.960	......	......
60.924	......	......
Mean, 60.889, ± .0208	49.611, ± .289	30.153, ± .180

From these ratios we get for the atomic weight of tantalum:

From per cent. Ta_2O_5 .... Ta = 172.342
From $5AgCl : TaCl_5$ .... " = 177.055
From $5AgCl : Ta_2O_5$ .... " = 174.821

These results are too low. Probably Blomstrand's material still contained some columbium.

In 1866 Marignac's determinations appeared.[||] He made four analyses of a pure potassium fluotantalate, and four more experiments upon the ammonium salt. The potassium compound, K_2TaF_7, was treated with sulphuric acid, and the mixture was then evaporated to dryness. The potassium sulphate was next dissolved out by water, while the residue

---

[*] Poggend. Annalen, 4, 14. 1825.
[†] Poggend. Annalen, 99, 80. 1856.
[‡] Journ. für Prakt. Chem., 70, 193. 1857.
[§] Acta Univ. Lund, 1864.
[||] Arch. Sci. Phys. Nat. (2), 26, 89. 1866.

was ignited and weighed as Ta_2O_5. 100 parts of the salt gave the following quantities of Ta_2O_5 and K_2SO_4:

Ta_2O_5.	K_2SO_4.
56.50	44.37
56.75	44.35
56.55	44.22
56.56	44.24
Mean, 56.59, ± .037	Mean, 44.295, ± .026

From these figures, 100 parts of K_2SO_4 correspond to the subjoined quantities of Ta_2O_5:

127.338
127.960
128.178
127.848

Mean, 127.831, ± .120

The ammonium salt, $(NH_4)_2TaF_7$, ignited with sulphuric acid, gave these percentages of Ta_2O_5. The figures are corrected for a trace of K_2SO_4 which was always present:

63.08
63.24
63.27
63.42

Mean, 63.25, ± .047

Hence we have four values for Ta:

From potassium salt, per cent. Ta_2O_5 .... Ta = 182.336
From potassium salt, per cent. K_2SO_4 .... " = 180.496
From potassium salt, K_2SO_4 : Ta_2O_5 .... " = 181.422
From ammonium salt, per cent. Ta_2O_5 .... " = 181.559

Average .... Ta = 181.453

Or, if O = 16, Ta = 182.836.

These values are computed with O = 15.879, K = 38.817, S = 31.828, N = 13.935, and F = 18.912.

CHROMIUM.

Concerning the atomic weight of chromium there has been much discussion, and many experimenters have sought to establish the true value. The earliest work upon it having any importance was that of Berzelius,* in 1818 and 1826, which led to results much in excess of the correct figure. His method consisted in precipitating a known weight of lead nitrate with an alkaline chromate and weighing the lead chromate thus produced. The error in his determination arose from the fact that lead chromate, except when thrown down from very dilute solutions, carries with it minute quantities of alkaline salts, and so has its apparent weight notably increased. When dilute solutions are used, a trace of the precipitate remains dissolved, and the weight obtained is too low. In neither case is the method trustworthy.

In 1844 Berzelius' results were first seriously called in question. The figure for chromium deduced from his experiments was somewhat over 56; but Peligot† now showed, by his analyses of chromous acetate and of the chlorides of chromium, that the true number was near 52.5. Unfortunately, Peligot's work, although good, was published with insufficient details to be useful here. For chromous acetate he gives the percentages of carbon and hydrogen, but not the actual weights of salt, carbon dioxide, and water from which they were calculated. His figures vary considerably, moreover—enough to show that their mean would carry but little weight when combined with the more explicit data furnished by other chemists.

Jacquelain's ‡ work we may omit entirely. He gives an atomic weight for chromium which is notoriously too low (50.1), and prints none of the numerical details upon which his result rests. The researches which particularly command our attention are those of Berlin, Moberg, Lefort, Wildenstein, Kessler, Siewert, Baubigny, Rawson, and Meineke.

Among the papers upon the atomic weight under consideration that by Berlin is one of the most important.§ His starting point was normal silver chromate; but in one experiment the dichromate $Ag_2Cr_2O_7$ was used. These salts, which are easily obtained in a perfectly pure condition, were reduced in a large flask by means of hydrochloric acid and alcohol. The chloride of silver thus formed was washed by decantation, dried, fused, and weighed without transfer. The united washings were supersaturated with ammonia, evaporated to dryness, and the residue treated with hot water. The resulting chromic oxide was then collected upon a filter, dried, ignited, and weighed. The results were as follows:

---

* Schweigg. Journ., 22, 53, and Poggend. Annal., 8, 22.
† Compt. Rend., 19, 609, and 734; 20, 1187; 21, 74.
‡ Compt. Rend., 24, 679. 1847.
† Journ. für Prakt. Chem., 37, 509, and 38, 149. 1846.

4.6680 grm. Ag_2CrO_4 gave 4.027 grm. AgCl and 1.0754 grm. Cr_2O_3.
3.4568   "   2.983   "   .7960   "
2.5060   "   2.1605   "   .5770   "
2.1530   "   1.8555   "   .4945   "
4.3335 grm. $Ag_2Cr_2O_7$ gave 2.8692   "   1.5300   "

From these weighings three values are calculable for the atomic weight of chromium. The three ratios upon which these values depend we will consider separately, taking first that between the chromic oxide and the original silver salt. In the four analyses of the normal chromate the percentages of Cr_2O_3 deducible from Berlin's weighings are as follows:

23.037
23.027
23.025
22.968

Mean, 23.014, ± .011

And from the single experiment with $Ag_2Cr_2O_7$ the percentage of Cr_2O_3 was 35.306.

For the ratio between Ag_2CrO_4 and AgCl, putting the latter at 100, we have for the former:

115.917
115.883
115.992
116.033

Mean, 115.956, ± .023

In the single experiment with dichromate 100 AgCl is formed from 151.035 $Ag_2Cr_2O_7$.

Finally, for the ratio between AgCl and Cr_2O_3, the five experiments of Berlin give, for 100 parts of the former, the following quantities of the latter:

26.705
26.685
26.707
26.650
26.662

Mean, 26.682, ± .0076

These results will be discussed, in connection with the work of other investigators, at the end of this chapter.

In 1848 the researches of Moberg[*] appeared. His method simply consisted in the ignition of anhydrous chromic sulphate and of ammonium chrome alum, and the determination of the amount of chromic

---

[*] Journ. für Prakt. Chem., 43, 114.

oxide thus left as residue. In the sulphate, $Cr_2(SO_4)_3$, the subjoined percentages of Cr_2O_3 were found. The braces indicate two different samples of material, to which, however, we are justified in ascribing equal value:

.542 grm. sulphate gave	.212 grm. Cr_2O_3.	39.114 per cent.	⎫
1.337 "	.523 "	39.117 "	⎬
.5287 "	.207 "	39.153 "	⎭
1.033 "	.406 "	39.303 "	⎫
.868 "	.341 "	39.286 "	⎭

Mean, 39.1946, ± .0280

From the alum, $NH_4.Cr(SO_4)_2.12H_2O$, we have these percentages of Cr_2O_3. The first series represents a salt long dried under a bell jar at a temperature of 18°. The crystals taken were clear and transparent, but may possibly have lost traces of water,* which would tend to increase the atomic weight found for chromium. In the second series the salt was carefully dried between folds of filter paper, and results were obtained quite near those of Berlin. Both of these series are discussed together, neither having remarkable value:

1.3185 grm. alum gave	.213 grm. Cr_2O_3.	16.155 per cent.
.7987 "	.129 "	16.151 "
1.0185 "	.1645 "	16.151 "
1.0206 "	.1650 "	16.167 "
.8765 "	.1420 "	16.201 "
.7680 "	.1242 "	16.172 "
1.6720 "	.2707 "	16.190 "
.5410 "	.0875 "	16.174 "
1.2010 "	.1940 "	16.153 "
1.0010 "	.1620 "	16.184 "
.7715 "	.1235 "	16.007 "
1.374 "	.2200 "	16.012 "

Mean, 16.143, ± .0125

The determinations made by Lefort † are even less valuable than those by Moberg. This chemist started out from pure barium chromate, which, to thoroughly free it from moisture, had been dried for several hours at 250°. The chromate was dissolved in pure nitric acid, the barium thrown down by sulphuric acid, and the precipitate collected upon a filter, dried, ignited, and weighed in the usual manner. The natural objection to the process is that traces of chromium may be carried down with the sulphate, thus increasing its weight. In fact, Lefort's results are somewhat too high. Calculated from his weighings, 100 parts of $BaSO_4$ correspond to the amounts of $BaCrO_4$ given in the third column:

---

* This objection is suggested by Berlin in a note upon Lefort's paper. Journ. für Prakt. Chem., 71, 191.
† Journ. für Prakt. Chem., 51, 261. 1850.

CHROMIUM.

1.2615 grm. BaCrO₄ gave	1.1555 grm. BaSO₄.	109.174
1.5895 "	1.4580 "	109.019
2.3255 "	2.1340 "	108.974
3.0390 "	2.7855 "	109.101
2.3480 "	2.1590 "	108.754
1.4230 "	1.3060 "	108.708
1.1975 "	1.1005 "	108.814
3.4580 "	3.1690 "	109.119
2.0130 "	1.8430 "	109.224
3.5570 "	3.2710 "	108.744
1.6470 "	1.5060 "	109.363
1.8240 "	1.6725 "	109.058
1.6950 "	1.5560 "	108.933
2.5960 "	2.3870 "	108.756

Mean, $108.9815, \pm .0369$

Wildenstein,* in 1853, also made barium chromate the basis of his researches. A known weight of pure barium chloride was precipitated by a neutral alkaline chromate, and the precipitate allowed to settle until the supernatant liquid was perfectly clear. The barium chromate was then collected on a filter, washed with hot water, dried, gently ignited, and weighed. Here again arises the objection that the precipitate may have retained traces of alkaline salts, and again we find deduced an atomic weight which is too high. One hundred parts BaCrO₄ correspond to BaCl₂ as follows:

81.87	81.57
81.80	81.75
81.61	81.66
81.78	81.83
81.52	81.66
81.84	81.80
81.85	81.66
81.70	81.85
81.68	81.57
81.54	81.83
81.66	81.71
81.55	81.63
81.81	81.56
81.86	81.58
81.54	81.67
81.68	81.84

Mean, $81.702, \pm .014$

Next in order we have to consider two papers by Kessler, who employed a peculiar volumetric method entirely his own. In brief, he compared the oxidizing power of potassium dichromate with that of the chlorate, and from his observations deduced the ratio between the molecular weights of the two salts.

---

† Journ. für Prakt. Chem., 59, 27.

In his earlier paper* the mode of procedure was about as follows: The two salts, weighed out in quantities having approximate chemical equivalency, were placed in two small flasks, and to each was added 100 cc. of a ferrous chloride solution and 30 cc. hydrochloric acid. The ferrous chloride was added in trifling excess, and, when action ceased, the amount unoxidized was determined by titration with a standard solution of dichromate. As in each case the quantity of ferrous chloride was the same, it became easy to deduce from the data thus obtained the ratio in question. I have reduced all of his somewhat complicated figures to a simple common standard, and give below the amount of chromate equivalent to 100 of chlorate:

120.118
120.371
120.138
120.096
120.241
120.181

Mean, 120.191; ± .028

In his later paper † Kessler substituted arsenic trioxide for the iron solution. In one series of experiments the quantity of dichromate needed to oxidize 100 parts of the arsenic trioxide was determined, and in another the latter substance was similarly compared with the chlorate. The subjoined columns give the quantity of each salt proportional to 100 of As_2O_3:

$K_2Cr_2O_7$.	$KClO_3$.
98.95	41.156
98.94	41.116
99.17	41.200
98.98	41.255
99.08	41.201
99.15	41.086
	41.199
Mean, 99.045, ± .028	41.224
	41.161
	41.193
	41.149
	41.126

Mean, 41.172, ± .009

Reducing the later series to the standard of the earlier, the two combine as follows:

(1) $2KClO_3 : K_2Cr_2O_7 :: 100 : 120.191, ± .028$
(2) $2KClO_3 : K_2Cr_2O_7 :: 100 : 120.282, ± .043$

General mean...... 120.216, ± .0235

---

* Poggend. Annalen, 95, 208 1855.
† Poggend. Annalen, 113, 137. 1861.

Siewert's determinations, which do not seem to have attracted general attention, were published in 1861.* He, reviewing Berlin's work, found that upon reducing silver chromate with hydrochloric acid and alcohol, the chromic chloride solution always retained traces of silver chloride dissolved in it. These could be precipitated by dilution with water; but, in Berlin's process, they naturally came down with the chromium hydroxide, making the weight of the latter too high; hence too large a value for the atomic weight of chromium. In order to find a more correct value Siewert resorted to the analysis of sublimed, violet, chromic chloride. This salt he fused with sodium carbonate and a little nitre, treated the fused mass with water, and precipitated from the resulting solution the chlorine by silver nitrate in presence of nitric acid. The weight of the silver chloride thus obtained, estimated after the usual manner, gave means for calculating the atomic weight of chromium. His figures, reduced to a common standard, give, as proportional to 100 parts of chloride of silver, the quantities of chromic chloride stated in the third of the subjoined columns:

.2367 grm. $CrCl_3$	gave	.6396 grm.	AgCl.		37.007
.2946	"	.7994	"		36.853
.2593	"	.7039	"		36.838
.4935	"	1.3395	"		36.842
.5850	"	1.5884	"		36.830
.6511	"	1.76681	"		36.852
.5503	"	1.49391	"		36.836

$$\text{Mean, } 36.865, \pm .0158$$

The first of these figures varies so widely from the others that we are justified in rejecting it, in which case the mean becomes $36.842, \pm .0031$.

Siewert also made two analyses of silver dichromate by the following process. The salt, dried at 120°, was dissolved in nitric acid. The silver was then thrown down by hydrochloric acid, and, in the filtrate, chromium hydroxide was precipitated by ammonia. Reduced to a uniform standard, we find from his results, corresponding to 100 parts of AgCl, $Ag_2Cr_2O_7$ as in the last column:

.7866 grm. $Ag_2Cr_2O_7$	gave	.52202 AgCl	and	.2764 Cr_2O_3.	150.684
1.089	"	.72249	"	.3840 "	150.729

Berlin's single determination of this ratio gave 151.035. Taking all three values together as one series, they give a mean of $150.816, \pm .074$.

Siewert's percentages of Cr_2O_3 obtained from $Ag_2Cr_2O_7$ are as follows, calculated from the above weighings:

$$35.139$$
$$35.262$$

$$\text{Mean, } 35.2005, \pm .0415$$

---

* Zeit. Gesammt. Wissenschaften, 17. 530.

Combining, as before, with Berlin's single result, giving the latter equal weight with one of these, we have a general mean of 35.236, ± .0335.

For the ratio between silver chloride and chromic oxide, Siewert's two analyses of the dichromate come out as follows. For 100 parts of AgCl we have of Cr_2O_3:

$$52.948$$
$$53.150$$

Mean, 53.049, ± .068

This figure, reduced to the standard of Berlin's work on the monochromate, becomes 26.525, ± .034. Berlin's mean was 26.682, ± .0076. The two means, combined, give a general mean of 26.676, ± .074.

By Baubigny* we have only three experiments upon the calcination of anhydrous chromic sulphate, as follows:

1.989 grm. $Cr_2(SO_4)_3$ gave	.7715 grm. Cr_2O_3.	38.788 per cent.
3.958 "	1.535 "	38.782 "
2.6052 "	1.0115 "	38.826 "

Mean, 38.799, ± .0092

Moberg found for the same ratio the percentage 39.195, ± .028. The general mean of both series, Moberg's and Baubigny's, is 38.838, ± .0087.

In Rawson's work † ammonium dichromate was the substance studied. Weighed quantities of this salt were dissolved in water, and then reduced by hydrochloric acid and alcohol. After evaporation to dryness the mass was treated with water and ammonia, reëvaporated, dried five hours at 140°, and finally ignited in a muffle. The residual chromic oxide was bright green, and was tested to verify its purity. The corrected weights are as follows:

$Am_2Cr_2O_7$.	Cr_2O_3.	Per cent. Cr_2O_3.
1.01275	.61134	60.365
1.08181	.65266	60.330
1.29430	.78090	60.334
1.13966	.68799	60.368
.98778	.59595	60.332
1.14319	.68987	60.346

Mean, 60.346, ± .0046

Latest in time and most elaborate of all, we come to the determinations of the atomic weight of chromium made by Meineke,‡ who studied the chromate and ammonio-chromate of silver, and also the dichromates of potassium and ammonium. For the latter salt he measured the same ratio that Rawson determined, but by a different method. He precipi-

---

*Compt. Rend., 98, 146.
†Journ. Chem. Soc., 55, 213.
‡Ann. d. Chem., 261, 339. 1891.

tated its solution with mercurous nitrate, and ignited the precipitate, with the subjoined results. Vacuum weights are given:

$Am_2Cr_2O_7$.	Cr_2O_3.	Per cent. Cr_2O_3.
2.0416	1.2316	60.325
2.1618	1.3040	60.320
2.0823	1.2562	60.328
2.1913	1.3221 *	60.335
2.0970	1.2656	60.353

Mean, 60.332, ± .0037
Rawson found, 60.346, ± .0046

General mean, 60.337, ± .0029

The chromate of silver, Ag_2CrO_4, and the ammonio-chromate, $Ag_2CrO_4 \cdot 4NH_3$, both prepared with all necessary precautions to insure purity, were first treated essentially as in Berlin's experiments, except that the traces of silver chloride held in solution by the chromic chloride were thrown out by sulphuretted hydrogen, estimated, and their amount added to the main portion. Thus the chief error in Berlin's work was avoided. I subjoin the data obtained, with vacuum standards, as usual. All of Meineke's results are so corrected:

Ag_2CrO_4.	$AgCl$.	Cr_2O_3.
2.7826	2.4047	.6384
3.2627	2.8199	.7480
3.6362	3.1416	.8338
4.6781	4.0414	1.0726
3.2325	2.7930	.7411
3.9137	3.3805	.8976

Hence we have the following ratios, as in the case of Berlin's data:

Per cent. Cr_2O_3.	$100 AgCl : Ag_2CrO_4$.	$100 AgCl : Cr_2O_3$.
22.943	115.715	26.548
22.926	115.703	26.526
22.931	115.744	26.602
22.928	115.754	26.601
22.924	115.736	26.531
22.935	115.773	26.552

Mean, 22.931, ± .0019   Mean, 115.737, ± .0072   Mean, 26.560, ± .0093
Berlin, 23.014, ± .0110   Berlin, 115.956, ± .0230

General mean, 22.934, ± .0018   General mean, 115.760, ± .0069

With the ammonio-chromate Meineke found as follows:

$Ag_2CrO_4 \cdot 4NH_3$.	$AgCl$.	Cr_2O_3.
4.1518	2.9724	.7904
4.2601	3.0592	.8125
5.9348	4.2654	1.1317

---

* Calculated back from Meineke's value for Cr, to replace an evident misprint in the original.

And the ratios become—

Per cent. Cr_2O_3.	$100 AgCl : Salt$.	$100 AgCl : Cr_2O_3$.
19.037	139.679	26.591
19.072	139.255	26.559
19.059	139.138	26.532
Mean, 19.059, ± .0074	Mean, 139.357, ± .1109	Mean, 26.561, ± .0115

The first of these three analyses is rejected by Meineke as suspicious, but for the present I shall allow it to remain. The data in the third column may now be combined with the corresponding figures from the normal chromate, as found by Meineke and his predecessors.

Berlin	26.682, ± .0076
Siewert, from $Ag_2Cr_2O_7$	26.525, ± .0340
Meineke, from Ag_2CrO_4	26.560, ± .0093
Meineke, from $Ag_2CrO_4.4NH_3$	26.561, ± .0115
General mean	26.620, ± .0052

$$4AgCl : Cr_2O_3 :: 100 : 26.620, \pm .0052$$

Obviously, this mean is vitiated by the known error in Berlin's work, the ultimate effect of which is yet to be considered.

In all four of the salts studied by Meineke he determined volumetrically the oxygen in excess of the normal oxides by measuring the amount of iodine liberated in acid solutions. With the silver salts the process was essentially as follows: A weighed quantity of the chromate was dissolved in weak ammonia, and the solution was precipitated with potassium iodide. After the silver iodide had been filtered off, five or six grammes of potassium iodide were added to the filtrate, which was then acidulated with phosphoric acid and a little sulphuric. The liberated iodine was then titrated with sodium thiosulphate solution, which had been standardized by means of pure iodine, prepared by Stas' method. From the iodine thus measured the excessive oxygen was computed, and from that datum the atomic weight of chromium was found. For present purposes, however, the data may be used more directly, as giving the ratios $I_2 : Ag_2CrO_4$ and $I_2 : Ag_2CrO_4.4NH_3$. Thus treated, the weights are as follows, reduced to a vacuum. Reckoning the salt as 100, the third column gives the percentage of iodine liberated:

Ag_2CrO_4.	I Set Free.	Percentage.
.43838	.50251	114.628
.90258	1.03432	114.595
.89858	1.02980	114.603
.89868	1.03072	114.693

Mean, 114.630, ± .015

CHROMIUM. 247

The next series, obviously, gives the ratio $I_3 : Ag_2CrO_4.4NH_3$.

$Ag_2CrO_4.4NH_3$.	I Set Free.	Percentage.*
.54356	.51784	95.267
.54856	.52046	94.877
.54926	.52322	95.258
.54906	.52376	95.392
.54466	.51910	95.307
.54536	.51891	95.150

Mean, 95.208, ± .0497

In dealing with the two dichromates Meineke used the acid potassium iodate in place of potassium iodide, the chromate and the iodate reacting in the molecular ratio of 2 : 1. The thiosulphate was standardized by means of the acid iodate, so that we have direct ratios between the latter and the two chromates. The data are as follows, with the amount of iodate proportional to one hundred parts of the dichromate in the third column:

$K_2Cr_2O_7$.	KHI_2O_6.	Percentage.
.25090	.16609	66.198
.25095	.16613	66.200
.25078	.16601	66.197
.24979	.16541	66.220
.24987	.16540	66.192
.24966	.16543	66.262
.25015	.16559	66.196
.25012	.16559	66.204
.24977	.16546	66.245
.25034	.16572	66.198
.25025	.16567	66.202
.25015	.16568	66.234

Mean, 66.212, ± .0044

$Am_2Cr_2O_7$.	KHI_2O_6.	Percentage.
.21457	.16584	77.290
.21465	.16588	77.279
.21464	.16584	77.264
.21416	.16543	77.246
.21447	.16564	77.232
.21427	.16559	77.281
.22196	.17152	77.272
.22194	.17151	77.278
.22180	.17139	77.272

Mean, 77.268, ± .0041

* These figures are not wholly in accord with the percentages of oxygen computed by Meineke. I suspect that there is a misprint among his data as published, probably in the second experiment, but I cannot trace it with certainty.

THE ATOMIC WEIGHTS.

The following ratios are now available for computing the atomic weight of chromium:

(1.) Percentage Cr_2O_3 from Ag_2CrO_4, 22.934, ± .0018
(2.) Percentage Cr_2O_3 from $Ag_2Cr_2O_7$, 35.236, ± .0335
(3.) $2AgCl : Ag_2CrO_4 :: 100 : 115.760$, ± .0069
(4.) $2AgCl : Ag_2Cr_2O_7 :: 100 : 150.816$, ± .074
(5.) $4AgCl : Cr_2O_3 :: 100 : 26.620$, ± .0052
(6.) Percentage Cr_2O_3 in $Cr_2(SO_4)_3$, 38.838, ± .0087
(7.) Percentage Cr_2O_3 in $AmCr(SO_4)_2.12H_2O$, 16.143, ± .0125
(8.) $BaSO_4 : BaCrO_4 :: 100 : 108.9815$, ± .0369
(9.) $BaCrO_4 : BaCl_2 :: 100 : 81.702$, ± .014
(10.) $3AgCl : CrCl_3 :: 100 : 36.842$, ± .0031
(11.) $2KClO_3 : K_2Cr_2O_7 :: 100 : 120.216$, ± .0235
(12.) Percentage Cr_2O_3 in $Ag_2CrO_4.4NH_3$, 19.059, ± .0074
(13.) $2AgCl : Ag_2CrO_4.4NH_3 :: 100 : 139.357$, ± .1109
(14.) Percentage Cr_2O_3 in $Am_2Cr_2O_7$, 60.337, ± .0029
(15.) $Ag_2CrO_4 : 3I :: 100 : 114.630$, ± .015
(16.) $Ag_2CrO_4.4NH_3 : 3I :: 100 : 95.208$, ± .0497
(17.) $2K_2Cr_2O_7 : KHI_2O_6 :: 100 : 66.212$, ± .0044
(18.) $2Am_2Cr_2O_7 : KHI_2O_6 :: 100 : 77.268$, ± .0041

The antecedent values to use in the reduction are—

$O = 15.879$, ± .0003 $S = 31.828$, ± .0015
$Ag = 107.108$, ± .0031 $N = 13.935$, ± .0021
$Cl = 35.179$, ± .0048 $Ba = 136.392$, ± .0086
$I = 125.888$, ± .0069 $AgCl = 142.287$, ± .0037
$K = 38.817$, ± .0051

For the molecular weight of Cr_2O_3, seven values are now calculable, as follows:

From (1) .... $Cr_2O_3 = 151.120$, ± .0130
From (2) .... " $= 151.105$, ± .1636
From (5) .... " $= 151.507$, ± .0299
From (6) .... " $= 151.384$, ± .0341
From (7) .... " $= 153.756$, ± .1205
From (12) .... " $= 151.478$, ± .0606
From (14) .... " $= 151.190$, ± .0110

General mean............ $Cr_2O_3 = 151.229$, ± .0039

For silver chromate there are two values—

From (3).................... $Ag_2CrO_4 = 329.423$, ± .0195
From (15)................... " $= 329.464$, ± .0467

General mean........... $Ag_2CrO_4 = 329.430$, ± .0180

And for the ammonio-chromate we have—

From (13) .... $Ag_2CrO_4.4NH_3 = 396.574$, ± .3158
From (16)............. " $= 396.673$, ± .2082

General mean.... $Ag_2CrO_4.4NH_3 = 396.647$, ± .1738

CHROMIUM.   249

From (4) .... $Ag_2Cr_2O_7 = 429.177, \pm .2109$
From (10) .... $CrCl_3 = 157.266, \pm .0113$
From (18) .... $Am_2Cr_2O_7 = 250.341, \pm .0164$

For the molecular weights of $K_2Cr_2O_7$ and $BaCrO_4$ there are two estimates each, as given below:

From (11) .... $K_2Cr_2O_7 = 292.433, \pm .0189$
From (17) .... " $= 292.143, \pm .0224$

General mean.......... $K_2Cr_2O_7 = 292.311, \pm .0144$

From (8) .... $BaCrO_4 = 252.549, \pm .0966$
From (9) .... " $= 253.054, \pm .0377$

General mean.......... $BaCrO_4 = 252.985, \pm .0351$

Finally, from these molecular weights, eight independent values are obtained for the atomic weight of chromium:

From Cr_2O_3............... $Cr = 51.796, \pm .0039$
From Ag_2CrO_4............. " $= 51.698, \pm .0191$
From $Ag_2CrO_4, 4NH_3$....... " $= 51.175, \pm .1741$
From $Ag_2Cr_2O_7$............ " $= 51.904, \pm .1055$
From $Am_2Cr_2O_7$............ " $= 51.659, \pm .0085$
From $K_2Cr_2O_7$............. " $= 51.762, \pm .0102$
From $CrCl_3$................. " $= 51.729, \pm .0183$
From $BaCrO_4$................ " $= 53.077, \pm .0362$

General mean............ $Cr = 51.778, \pm .0032$

If $O = 16$, $Cr = 52.172$.

Rejecting the last of the eight values, that from barium chromate, the mean becomes—

$Cr = 51.767, \pm .0032.$

Even this result is probably too high, for it includes ratios which are certainly erroneous, and which yet exert appreciable weight. From the ratios which are reasonably concordant a better mean is derivable, as follows:

From (1)..................... $Cr = 51.741, \pm .0065$
From (2)..................... " $= 51.734, \pm .0818$
From (14).................... " $= 51.776, \pm .0055$
From (3) and (15)............ " $= 51.698, \pm .0191$
From (4)..................... " $= 51.904, \pm .1055$
From (10).................... " $= 51.729, \pm .0183$
From (18).................... " $= 51.659, \pm .0085$
From (11) and (17)........... " $= 51.762, \pm .0102$

General mean............ $Cr = 51.742, \pm .0034$

If $O = 16$, this becomes 52.136, a value which is probably not very far from the truth.

MOLYBDENUM.

If we leave out of account the inaccurate determination made by Berzelius,* we shall find that the data for the atomic weight of molybdenum lead to two independent estimates of its value—one near 92, the other near 96. The earlier results found by Berlin and by Svanberg and Struve lead to the lower number; the more recent investigations, together with considerations based upon the periodic law, point conclusively to the higher.

The earliest investigation which we need especially to consider is that of Svanberg and Struve.† These chemists tried a variety of different methods, but finally based their conclusions upon the two following: First, molybdenum trioxide was fused with potassium carbonate, and the carbon dioxide which was expelled was estimated; secondly, molybdenum disulphide was converted into the trioxide by roasting, and the ratio between the weights of the two substances was determined.

By the first method it was found that 100 parts of MoO_3 will expel the following quantities of CO_2:

$$31.4954$$
$$31.3749$$
$$31.4705$$

Mean, $31.4469, \pm .0248$

The carbon dioxide was determined simply from the loss of weight when the weighed quantities of trioxide and carbonate were fused together. It is plain that if, under these circumstances, a little of the trioxide should be volatilized, the total loss of weight would be slightly increased. A constant error of this kind would tend to bring out the atomic weight of molybdenum too low.

By the second method, the conversion by roasting of MoS_2 into MoO_3, Svanberg and Struve obtained these results. Two samples of artificial disulphide were taken, A and B, and yielded for each hundred parts the following of trioxide:

$$\left.\begin{array}{l}89.7919\\89.7291\end{array}\right\} A.$$

$$\left.\begin{array}{l}89.6436\\89.7082\\89.7660\\89.7640\\89.8635\end{array}\right\} B.$$

Mean, $89.7523, \pm .0176$

Three other experiments in series B gave divergent results, and, although published, are rejected by the authors themselves. Hence it is

---

* Poggend. Annalen, 8, 1. 1826.
† Journ. für Prakt. Chem., 44, 301. 1848.

not necessary to cite them in this discussion. We again encounter in these figures the same source of constant error which apparently vitiates the preceding series, namely, the possible volatilization of the trioxide. Here, also, such an error would tend to reduce the atomic weight of molybdenum.

From the CO_2 series .... Mo = 91.25
From the MoS_2 series........................ Mo = 92.49

Berlin,* a little later than Svanberg and Struve, determined the atomic weight of molybdenum by igniting a molybdate of ammonium and weighing the residual MoO_3. Here, again, a loss of the latter by volatilization may (and probably does) lead to too low a result. The salt used was $(NH_4)_4Mo_5O_{17}.3H_2O$, and in it these percentages of MoO_3 were found:

. 81.598
81.612
81.558
81.555
———
Mean, 81.581, ± .0095

Hence Mo = 91.559.

Until 1859 the value 92 was generally accepted on the basis of the foregoing researches, but in this year Dumas † published some figures tending to sustain a higher number. He prepared molybdenum trioxide by roasting the disulphide, and then reduced it to metal by ignition in hydrogen. At the beginning the hydrogen was allowed to act at a comparatively low temperature, in order to avoid volatilization of trioxide; but at the end of the operation the heat was raised sufficiently to insure a complete reduction. From the weighings I calculate the percentages of metal in MoO_3:

.448 grm. MoO_3 gave .299 grm. Mo.   66.741 per cent.
.484   "   .323   "   66.736   "
.484   "   .322   "   66.529   "
.498   "   .332   "   66.667   "
.559   "   .373   "   66.726   "
.388   "   .258   "   66.495   "
                                        ————
                                    Mean, 66.649, ± .030

In 1868 the same method was employed by Debray.‡ His trioxide was purified by sublimation in a platinum tube. His percentages are as follows:

5.514 grm. MoO_3 gave 3.667 grm. Mo.   66.503 per cent.
7.910   "   5.265   "   61.561   "
9.031   "   6.015   "   66.604   "
                                         ————
                                    Mean, 66.556, ± .020

---

* Journ. für Prakt. Chem., 49, 444. 1850.
† Ann. Chem. Pharm., 105, 84, and 113, 23.
‡ Compt. Rend., 66, 734.

For the same ratio we have also a single experiment by Rammelsberg,* who, closely following Dumas' method, found in molybdenum trioxide 66.708 per cent. of metal. As this figure falls within the limits of Dumas' series, we may assign it equal weight with one experiment in the latter.

Debray also made two experiments upon the precipitation of molybdenum trioxide in ammoniacal solution by nitrate of silver. In his results, as published, there is curious discrepancy, which, I have no doubt, is due to a typographical error. These results I am therefore compelled to leave out of consideration. They could not, however, exert a very profound influence upon the final discussion.

In 1873, Lothar Meyer † discussed the analyses made by Liechti and Kemp ‡ of four chlorides of molybdenum, and in the former edition of this work the same data were considered in detail. The analyses, however, were not intended as determinations of atomic weight, and since good determinations have been more recently published, the work on the chlorides will be omitted from further consideration. It is enough to state here that they gave values for Mo ranging near 96, both above and below that number, with an extreme range of over eight-tenths of a unit.

In 1893 the determinations by Smith and Maas appeared, § representing an entirely new method. Sodium molybdate, purified by many recrystallizations and afterwards dehydrated, was heated in a current of pure, dry, gaseous hydrochloric acid. The compound $MoO_3.2HCl$ was thus distilled off, and the sodium molybdate was quantitatively transformed into sodium chloride. The latter salt was afterwards carefully examined, and proved to be free from molybdenum. The data, with all weights reduced to a vacuum standard, are subjoined:

Na_2MoO_4.	$NaCl$.	Per cent. $NaCl$.
1.14726	.65087	56.733
.89920	.51023	56.743
.70534	.40020	56.739
.70793	.40182	56.760
1.26347	.71695	56.745
1.15217	.65367	56.734
.90199	.51188	56.750
.81692	.46358	56.747
.65098	.36942	56.748
.80563	.45717	56.747

Mean, 56.745, ± .0017

In 1895, Seubert and Pollard ∥ determined the atomic weight of mo-

---

* Berlin Monatsbericht, 1877, p. 574.
† Ann. Chem. Pharm., 169, 365. 1873.
‡ Ann. Chem. Pharm., 169, 344.
§ Journ. Amer. Chem. Soc., 15, 397. 1893.
∥ Zeitsch. Anorg. Chem., 8, 434. 1895.

MOLYBDENUM.   253

lybdenum by two methods. First, the carefully purified trioxide, in weighed amounts, was dissolved in an excess of a standard solution of caustic soda. This solution was standardized by means of hydrochloric acid, which in turn had been standardized gravimetrically as silver chloride. Hence, indirectly, the ratio $2AgCl : MoO_3$ was measured. Sulphuric acid and lime water were also used in the titrations, so that the entire process was rather complicated. Ignoring the intermediate data, the end results, in weights of MoO_3 and AgCl, were as follows. The third column gives the MoO_3 proportional to 100 parts of AgCl:

MoO_3.	$AgCl$.	Ratio.
3.6002	7.1709	50.206
3.5925	7.1569	50.196
3.7311	7.4304	50.214
3.8668	7.7011	50.211
3.9361	7.8407	50.201
3.8986	7.7649	50.208
3.9630	7.8941	50.202
3.9554	7.8806	50.192
3.9147	7.7999	50.189
3.8543	7.6767	50.208
3.9367	7.8437	50.190

Mean, 50.202, ± .0018

The second method adopted by Seubert and Pollard was the old one of reducing the trioxide to metal by heating in a current of hydrogen. The weights and percentages of metal are subjoined:

MoO_3.	Mo.	Per cent.
1.8033	1.2021	66.661
1.9345	1.1564	66.670
3.9413	2.6275	66.666
1.5241	1.0160	66.662
4.0533	2.7027	66.679

Mean, 66.668, ± .0022

This mean may be combined with the results of previous investigators, thus:

Dumas .... 66.649, ± .0300
Debray .... 66.556, ± .0200
Rammelsberg .... 66.708, ± .0680
Seubert and Pollard.................... 66.668, ± .0022

General mean .... 66.665, ± .0022

Here the data of Seubert and Pollard alone exert any appreciable influence.

Neglecting all determinations made previous to 1859, there are now three ratios from which to compute the atomic weight of molybdenum, viz:

(1.) Percentage Mo in MoO_3, 66.665, ± .0022.
(2.) $2AgCl : MoO_3 :: 100 : 50.202$, ± .0018
(3.) $2NaCl : Na_2MoO_4 : 56.745$, ± .0017 : 100.

These involve the following values:

$O = 15.879$, ± .0003 $AgCl = 142.287$, ± .0037
$Na = 22.881$, ± .0046 $NaCl = 58.060$, ± .0017

Hence for the atomic weight in question—

From (1) .... $Mo = 95.267$, ± .0072
From (2) .... " $= 95.225$, ± .0064
From (3) .... " $= 95.357$, ± .0126

General mean .... $Mo = 95.259$, ± .0045

With $O = 16$, $Mo = 95.985$.

This value is essentially that derived from Seubert and Pollard's data alone. Reducing the latter to a vacuum would affect the result very slightly—so slightly that the correction may be ignored.

TUNGSTEN.

The atomic weight of tungsten has been determined from analyses of the trioxide, the hexchloride, and the tungstates of iron, silver, and barium.

The composition of the trioxide has been the subject of many investigations. Malaguti* reduced this substance to the blue oxide, and from the difference between the weights of the two compounds obtained a result now known to be considerably too high. In general, however, the method of investigation has been to reduce WO_3 to W in a stream of hydrogen at a white heat, and afterwards to reoxidize the metal, thus getting from one sample of material two results for the percentage of tungsten. This method is probably accurate, provided that the trioxide used be pure.

The first experiments which we need consider are, as usual, those of Berzelius.† 899 parts WO_3 gave, on reduction, 716 of metal. 676 of metal, reoxidized, gave 846 WO_3. Hence these percentages of W in WO_3:

79.644, by reduction.
79.905, by oxidation.

Mean, 79.7745, ± .0880

These figures are far too high, the error being undoubtedly due to the presence of alkaline impurity in the trioxide employed.

Next in order of time comes the work of Schneider, ‡ who with characteristic carefulness, took every precaution to get pure material. His percentages of tungsten are as follows:

Reduction Series.

79.336
79.254
79.312
79.326
79.350

Mean, 79.3156

Oxidation Series.

79.329
79.324
79.328

Mean, 79.327
Mean of all, 79.320, ± .0068

---
* Journ. für Prakt. Chem., 8, 179. 1836.
† Poggend. Annalen, 8, 1. 1825.
‡ Journ. für Prakt. Chem., 50, 152. 1850.

Closely agreeing with these figures are those of Marchand,* published in the following year:

Reduction Series.

79.307
79.302

Mean, 79.3045

Oxidation Series.

79.321
79.352

Mean, 79.3365
Mean of all, 79.3205, ± .0073

The figures obtained by v. Borch † agree in mean tolerably well with the foregoing. They are as follows:

Reduction Series.

79.310
79.212
79.289
79.313
79.225
79.290
79.302

Mean, 79.277

Oxidation Series.

79.359
79.339

Mean, 79.349
Mean of all, 79.293, ± .0108

Dumas ‡ gives only a reduction series, based upon trioxide obtained by the ignition of a pure ammonium tungstate. The reduction was effected in a porcelain boat, platinum being objectionable on account of the tendency of tungsten to alloy with it. Dumas publishes only weighings, from which I have calculated the percentages:

2.784 grm. WO_3 gave	2.208 grm. W.	79.310 per cent.
2.994 "	2.373 "	79.259 "
4.600 "	3.649 "	79.326 "
.985 "	.781 "	79.289 "
.917 "	.727 "	79.280 "
.917 "	.728 "	79.389 "
1.717 "	1.362 "	79.324 "
2.988 "	2.370 "	79.317 "

Mean, 79.312, ± .009

---

* Ann. Chem. Pharm., 77, 261. 1851.
† Journ. für Prakt. Chem., 54, 254. 1851.
‡ Ann. Chem. Pharm., 113, 23. 1860.

TUNGSTEN. 257

The data furnished by Bernoulli * differ widely from those just given. This chemist undoubtedly worked with impure material, the trioxide having a greenish tinge. Hence the results are too high. These are the percentages of W:

Reduction Series.

79.556
79.526
79.553
79.558
79.549
78.736

Mean, 79.413

Oxidation Series.

79.558
79.656
79.555
79.554

Mean, 79.581
Mean of all, 79.480, ± .056

Two reduction experiments by Persoz † give the following results:

1.7999 grm. WO_3 gave 1.4274 grm. W.   79.304 per cent.
2.249   "   1.784   "   79.324   "

Mean, 79.314, ± .007

Next in order is the work done by Roscoe.‡ This chemist used a porcelain boat and tube, and made six weighings, after successive reductions and oxidations, with the same sample of 7.884 grammes of trioxide. These weighings give me the following five percentages, which, for the sake of uniformity with foregoing series, I have classified under the usual, separate headings:

Reduction Series.

79.196
79.285
79.308

Mean, 79.263

Oxidation Series.

79.230
79 299

Mean, 79.2645
Mean of all, 79.264, ± .0146

---

* Poggend. Annalen, 111, 573. 1860.
† Zeit. Anal. Chem., 3, 260. 1864.
‡ Ann. Chem. Pharm., 162, 368. 1872.

17

258   THE ATOMIC WEIGHTS.

In Waddell's experiments* especial precautions were taken to procure tungstic oxide free from silica and molybdenum. Such oxide, elaborately purified, was reduced in hydrogen, with the following results:

1.4006 grm. WO_3 gave	1.1115 W.		79.359 per cent.	
.9900	"	.7855 "	79.343	"
1.1479	"	.9110 "	79.362	"
.9894	"	.7847 "	79.311	"
4.5639	"	3.6201 "	79.320	"

79.339, ± .0069

The investigation by Pennington and Smith † started from the supposition that the tungsten compounds studied by their predecessors had not been completely freed from molybdenum. Accordingly, tungstic oxide, carefully freed from all other impurities, was heated in a stream of gaseous hydrochloric acid, so as to volatilize all molybdenum as the compound $MoO_3.2HCl$. The residual WO_3, was then reduced in pure hydrogen, and the tungsten so obtained was oxidized in porcelain crucibles. Care was taken to exclude reducing gases, and the trioxide was finally cooled in vacuum desiccators over sulphuric acid. The oxidation data are as follows, with the usual percentage column added. The weights are reduced to a vacuum:

Tungsten.	Oxygen Gained.	Percentage.
.862871	.223952	79.394
.650700	.168900	79.392
.597654	.155143	79.390
.666820	.173103	79.391
.428228	.111168	79.390
.671920	.174406	79.392
.590220	.153193	79.394
.568654	.147588	79.394
1.080973	.280600	79.392

Mean, 79.392, ± .0004

With O = 16, this series gives W = 184.92.

The very high value for tungsten found by Pennington and Smith, nearly a unit higher than that which was commonly accepted, seems to have at once attracted the attention of Schneider,‡ who criticised the paper somewhat fully, and gave some new determinations of his own. The tungsten trioxide employed in this new investigation was heated in gaseous hydrochloric acid, and the absence of molybdenum was proved. The data obtained, both by reduction and by oxidation, are as follows:

---

*Am. Chem. Journ., 8, 280. 1886.
† Read before the Amer. Philos. Soc., Nov. 2, 1894.
‡ Journ. für Prakt. Chem. (2), 53, 288. 1896.

Reduction Series.

2.0738 grm. WO_3 gave	1.6450 W.	79.323 per cent.
4.0853 "	3.2400 "	79.309 "
6.1547 "	4.8811 "	79.307 "

Oxidation Series.

1.5253 grm. W gave	1.9232 WO_3.	79.311 per cent.
3.1938 "	4.0273 "	79.304 "
4.7468 "	5.9848 "	79.314 "

Mean of all, 79.311, ± .0018

Hence with $O = 16$, $W = 184.007$.

In order to account for the difference between this result and that of Pennington and Smith, an impurity of molybdenum trioxide amounting to about one per cent. would be necessary. Schneider suggests that the quantities of material used by Pennington and Smith were too small, and that there may have been mechanical loss of small particles during the long heatings. Such losses would tend to raise the atomic weight computed from the experiments. On the other hand, the losses could hardly have been uniform in extent, and the extremely low probable error of Pennington and Smith's series renders Schneider's supposition improbable. The error, if error exists, must be accounted for otherwise.

Since Schneider's paper appeared, another set of determinations by Shinn* has been published from Smith's laboratory. Attempts to verify the results obtained by Smith and Desi having proved abortive, and other experiments having failed, Shinn resorted to the oxidation method and gives the subjoined data. The percentage column is added by myself:

.22297 grm. W gave .28090 WO_3.	79.377
.17200 " .21664 "	79.394
.10989 " .13844 "	79.377
.10005 " .12598 "	79.417

Mean, 79.391, ± .0066

This figure is very close to that found in Pennington and Smith's series, and therefore serves as a confirmation. The discordance between these results and Schneider's is still to be explained.

There are still other experiments by Riche,† which I have not been able to get in detail. They cannot be of any value, however, for they give to tungsten an atomic weight of about ten units too low. We may therefore neglect this series, and go on to combine the others:

Berzelius	79.7745,	± .0880
Schneider, 1850	79.320,	± .0068
Marchand	79.3205,	± .0073
v. Borch	79.293,	± .0108
Dumas	79.312,	± .0090

---

* Doctoral thesis., University of Pennsylvania, 1896. "The atomic mass of tungsten."
† Journ. für Prakt. Chem., 69, 10. 1857.

260   THE ATOMIC WEIGHTS.

Bernoulli	79.480,	± .0560
Persoz	79.314,	± .0070
Roscoe	79.264,	± .0146
Waddell	79.339,	± .0069
Pennington and Smith	79.392,	± .0004
Schneider, 1896	79.311,	± .0018
Shinn	79.391,	± .0066
General mean	79.388,	± .00039

Here the work of Pennington and Smith vastly outweighs everything else; and if their supposition as to the presence of molybdenum in all the previous investigations is correct, this result is to be accepted.

The rejection of the figures given by Berzelius and by Bernoulli would exert an unimportant influence upon the final result. There is, therefore, no practical objection to retaining them in the discussion.

In 1861 Scheibler* deduced the atomic weight of tungsten from analyses of barium metatungstate, $BaO.4WO_3.9H_2O$. In four experiments he estimated the barium as sulphate, getting closely concordant results, which were, however, very far too low. These, therefore, are rejected. But from the percentage of water in the salt a better result was attained. The percentages of water are as follows:

13.053
13.054
13.045
13.010
13.022

Mean, 13.0368, ± .0060

The work of Zettnow,† published in 1867, was somewhat more complicated than any of the foregoing researches. He prepared the pure tungstates of silver and of iron, and from their composition determined the atomic weight of tungsten.

In the case of the iron salt the method of working was this: The pure, artificial $FeWO_4$ was fused with sodium carbonate, the resulting sodium tungstate was extracted by water, and the thoroughly washed residual ferric oxide was dissolved in hydrochloric acid. This solution was then reduced by zinc, and titrated for iron with potassium permanganate. Corrections were applied for the drop in excess of permanganate needed to produce distinct reddening, and for the iron contained in the zinc. 11.956 grammes of the latter metal contained iron corresponding to 0.6 cc. of the standard solution. The permanganate was standardized by comparison with pure ammonium-ferrous sulphate, $Am_2Fe(SO_4)_2.6H_2O$, so that, in point of fact, Zettnow establishes directly only the ratio between that salt and the ferrous tungstate. From Zettnow's four experiments in standardizing I find that 1 cc. of his solution

---
* Journ. für Prakt. Chem., 83, 324.
† Poggend. Annalen, 130, 30.

corresponds to 0.0365457 gramme of the double sulphate, with a probable error of ± .0000012.

Three sets of titrations were made. In the first a quantity of ferrous tungstate was treated according to the process given above; the iron solution was diluted to 500 cc., and four titrations made upon 100 cc. at a time. The second set was like the first, except that three titrations were made with 100 cc. each, and a fourth upon 150 cc. In the third set the iron solution was diluted to 300 cc., and only two titrations upon 100 cc. each were made. In sets one and two thirty grammes of zinc were used for the reduction of each, while in number three but twenty grammes were taken. Zettnow's figures, as given by him, are quite complicated; therefore I have reduced them to a common standard. After applying all corrections the following quantities of tungstate, in grammes, correspond to 1 cc. of permanganate solution :

.028301 ⎫
.028291 ⎬ First set.
.028311 ⎪
.028301 ⎭

.028367 ⎫
.028368 ⎬ Second set.
.028367 ⎪
.028367 ⎭

.028438 ⎱ Third set.
.028438 ⎰

Mean, .0283549, ± .0000115

With the silver tungstate, Ag_2WO_4, Zettnow employed two methods. In two experiments the substance was decomposed by nitric acid, and the silver thus taken into solution was titrated with standard sodium chloride. In three others the tungstate was treated directly with common salt, and the residual silver chloride collected and weighed. Here again, on account of some complexity in Zettnow's figures, I am compelled to reduce his data to a common standard. To 100 parts of AgCl the following quantities of Ag_2WO_4 correspond:

By First Method.
161.665
161.603

Mean, 161.634, ± .021

By Second Method.
161.687
161.651
161.613

Mean, 161.650, ± .014
General mean from both series, 161.645, ± .012

For tungsten hexchloride we have two analyses by Roscoe, published in the same paper with his results upon the trioxide. In one experiment the chlorine was determined as AgCl; in the other the chloride was reduced by hydrogen, and the residual tungsten estimated. By bringing both results into one form of expression we have for the percentage of chlorine in WCl_6:*

53.588
53.632

Mean, 53.610, ± .015

The work done by Smith and Desi † probably ought to be considered in connection with that of Pennington and Smith on the trioxide. Smith and Desi started with tungsten trioxide, freed from molybdenum by means of gaseous hydrochloric acid. This material was reduced in a stream of carefully purified hydrogen, and the water formed was collected in a calcium chloride tube and weighed. To the results found I add the percentage of water obtained from 100 parts of WO_3. Vacuum weights are given.

WO_3.	H_2O.	Per cent. H_2O.
.983024	.22834	23.228
.998424	.23189	23.226
1.008074	.23409	23.221
.911974	.21184	23.229
.997974	.23179	23.226
1.007024	.23389	23.226

Mean, 23.226, ± .0008

There are now six ratios from which to calculate the atomic weight of tungsten:

(1.) Percentage of W in WO_3, 79.388, ± .00039
(2.) Percentage of H_2O in $BaO.4WO_3.9H_2O$, 13.0368, ± .0060
(3.) $WO_3 : 3H_2O :: 100 : 23.226$, ± .0008
(4.) $Am_2Fe(SO_4)_2.6H_2O : FeWO_4 :: .0365457$, ± .0000012 : .0283549, ± .0000115
(5.) $2AgCl : Ag_2WO_4 :: 100 : 161.645$, ± .012
(6.) Percentage of Cl in WCl_6, 53.610, ± .015

These are reduced with—

O = 15.879, ± .0003   S = 31.828, ± .0015
Ag = 107.108, ± .0031   Ba = 136.392, ± .0086
Cl = 35.179, ± .0048   Fe = 55.597, ± .0023
N = 13.935, ± .0021   AgCl = 142.287, ± .0037

---

* The actual figures are as follows:

19.5700 grm. WCl_6 gave 42.4127 grm. AgCl.
10.4326 " 4.8374 grm. tungsten.

† Read before Amer. Philos. Soc., Nov. 2, 1894.

Hence there are six values for the atomic weight of tungsten, as follows:

From (1) .... W = 183.485, ± .0051
From (2) .... " = 182.638, ± .1248
From (3) .... " = 183 298, ± .0088
From (4) .... " = 183.035, ± .1229
From (5) .... " = 182.268, ± .0663
From (6) .... " = 182.647, ± .0820

General mean .... W = 183.429, ± .0044

If O = 16, W = 184.827. The rejection of all values except the first and third raises the mean by 0.009; that is, four of the ratios count for almost nothing, and the work done in Smith's laboratory dominates all the rest. The questions raised by Schneider in his latest determination, however, are not yet answered, and farther investigation is required in order to fully establish the true atomic weight of tungsten.

URANIUM.

The earlier attempts to determine the atomic weight of uranium were all vitiated by the erroneous supposition that the uranous oxide was really the metal. The supposition, of course, does not affect the weighings and analytical data which were obtained, although these, from their discordance with each other and with later and better results, have now only a historical value.

For present purposes the determinations made by Berzelius,* by Arfvedson,† and by Marchand ‡ may be left quite out of account. Berzelius employed various methods, while the others relied upon estimating the percentage of oxygen lost upon the reduction of U_3O_8 to UO_2. Rammelsberg's § results also, although very suggestive, need no full discussion. He analyzed the green chloride, UCl_4; effected the synthesis of uranyl sulphate from uranous oxide; determined the amount of residue left upon the ignition of the sodio and bario-uranic acetates; estimated the quantity of magnesium uranate formed from a known weight of UO_3, and attempted also to fix the ratio between the green and the black oxides. His figures vary so widely that they could count for little in the establishing of any general mean; and, moreover, they lead to estimates of the atomic weight which are mostly below the true value. For instance, twelve lots of U_3O_8 from several different sources were reduced to UO_2 by heating in hydrogen. The percentages of loss varied from 3.83 to 4.67, the mean being 4.121. These figures give values for the atomic

---
* Schweigg. Journ., 22, 336. 1818. Poggend. Annalen, 1, 359. 1825.
† Poggend. Annalen, 1, 245. Berz. Jahr., 3, 120. 1822.
‡ Journ. für Prakt. Chem., 23, 497. 1841.
§ Poggend. Annalen, 55, 318, 1842; 56, 125, 1842; 59, 9, 1843; 66, 91, 1845. Journ. für Prakt. Chem., 29, 324.

weight of uranium ranging from 184.33 to 234.05, or, in mean, 214.53. Such discordance is due partly to impurity in some of the material studied, and illustrates the difficulties inherent in the problem to be solved. Some of the uranoso-uranic oxide was prepared by calcining the oxalate, and retained an admixture of carbon. Many such points were worked up by Rammelsberg with much care, so that his papers should be scrupulously studied by any chemist who contemplates a redetermination of the atomic weight of uranium.

In 1841 and 1842 Peligot published certain papers * showing that the atomic weight of uranium must be somewhere near 240. A few years qater the same chemist published fuller data concerning the constant in luestion, but in the time intervening between his earlier and his final researches other determinations were made by Ebelmen and by Wertheim. These investigations we may properly discuss in chronological order. For present purposes the early work of Peligot may be dismissed as only preliminary in character. It showed that what had been previously regarded as metallic uranium was in reality an oxide, but gave figures for the atomic weight of the metal which were merely approximations.

Ebelmen's † determinations of the atomic weight of uranium were based upon analyses of uranic oxalate. This salt was dried at 100°, and then, in weighed amount, ignited in hydrogen. The residual uranous oxide was weighed, and in some cases converted into U_3O_8 by heating in oxygen. The following weights are reduced to a vacuum standard:

10.1644 grm. oxalate gave 7.2939 grm. UO_2.
12.9985   "   9.3312   "   Gain on oxidation, .3685
11.8007   "   8.4690   "   "   .3275
9.9923   "   7.1731   "   "   .2812
11.0887   "   7.9610   "   "   .3105
10.0830   "   7.2389   "
6.7940   "   4.8766   "
16.0594   "   11.5290   "   "   .4531

Reducing these figures to percentages, we may present the results in two columns. Column A gives the percentages of UO_2 in the oxalate, while B represents the amount of U_3O_8 formed from 100 parts of UO_2:

A.	B.
71.924	......
71.787	103.949
71.767	103.867
71.621	103.920
71.794	103.900
71.793	
71.778	......
71.790	103 930
Mean, 71.782, ± .019	Mean, 103.913, ± .009

---
* Compt. Rend., 12, 735. 1841. Ann. Chim. Phys. (3), 55. 1842.
† Journ. für Prakt. Chem., 27, 385. 1842.

Wertheim's* experiments were even simpler in character than those of Ebelmen. Sodio-uranic acetate, carefully dried at 200°, was ignited, leaving the following percentages of sodium uranate:

$$67.51508$$
$$67.54558$$
$$67.50927$$

Mean, $67.52331, \pm .0076$

The final results of Peligot's † investigations appeared in 1846. Both the oxalate and the acetate of uranium were studied and subjected to combustion analysis. The oxalate was scrupulously purified by repeated crystallizations, and thirteen analyses, representing different fractions, were made. Seven of these gave imperfect results, due to incomplete purification of the material; six only, from the later crystallizations, need to be considered. In these the uranium was weighed as U_3O_8, and the carbon as CO_2. From the ratio between the CO_2 and U_3O_8 the atomic weight of uranium may be calculated without involving any error due to traces of moisture possibly present in the oxalate. I subjoin Peligot's weighings, and give, in the third column, the U_3O_8 proportional to 100 parts of CO_2:

CO_2.	U_3O_8.	Ratio.
1.456 grm.	4.649 grm.	319.299
1.369 "	4.412 "	322.279
2.209 "	7.084 "	320.688
1.019 "	3.279 "	321.786
1.069 "	3.447 "	322.461
1.052 "	3.389 "	322.148

Mean, $321.443, \pm .338$

From the acetate, $UO_2(C_2H_3O_2)_2 \cdot 2H_2O$, the following percentages of U_3O_8 were obtained:

5.061 grm. acetate gave 3.354 grm. U_3O_8.		66.2715 per cent.
4.601 "	3.057 "	66.4421 "
1.869 "	1.238 "	66.2386 "
3.817 "	2.541 "	66.5706 "
10.182 "	6.757 "	66.3622 "
4.393 "	2.920 "	66.4694 "
2.868 "	1.897 "	66.1437 "

Mean, $66.3569, \pm .038$

The acetate also yielded the subjoined percentages of carbon and of water. Assuming that the figures for carbon were calculated from known

---

* Journ. für Prakt. Chem., 29, 209. 1843.
† Compt. Rend., 22, 487. 1846.

weights of dioxide, with $C = 12$ and $O = 16$, I have added a third column, in which the carbon percentages are converted into percentages of CO_2:

$H_2O.$	$C.$	$CO_2.$
21.60	11.27	41.323
21.16	11.30	41.433
21.10	11.30	41.433
21.20	11.10	40.700
Mean, 21.265, ± .187	Mean, 11.24	Mean, 41.222, ± .092

From these data we get the following values for the molecular weight of uranyl acetate:

From percentage of U_3O_8 .... 423.183, ± .4781
From percentage of CO_2 .... 423.842, ± .9462
From percentage of H_2O .... 420.386, ± 2.9033

General mean................ 423.257, ± .4222

In the posthumous paper of Zimmermann, edited by Krüss and Alibegoff,* the atomic weight of uranium is determined by two methods. First, UO_2, prepared by several methods, is converted into U_3O_8 by heating in oxygen. To begin with, U_3O_8 was prepared, and reduced to UO_2 by ignition in hydrogen. When the reduction takes place at moderate temperatures, the UO_2 is somewhat pyrophoric, but if the operation is performed over the blast lamp this difficulty is avoided. After weighing the UO_2, the oxidation is effected, and the gain in weight observed. The preliminary U_3O_8 was derived from the following sources: A, from uranium tetroxide; B, from the oxalate; C, from uranyl nitrate; D, by precipitation with mercuric oxide. The full data, lettered as indicated above, are subjoined:

	$UO_2.$	$U_3O_8.$	Per cent. of Gain.
A.	8.9363	9.2872	3.927
	7.9659	8.2789	3.929
	12.4385	12.9270	3.927
B.	12.8855	13.3913	3.925
	5.7089	5.9331	3.927
	9.6270	10.0051	3.928
C.	13.1855	13.7036	3.929
	9.9973	10.3901	3.929
D.	15.8996	16.5242	3.928
	7.4326	7.7245	3.927

Mean, 3.9276, ± .0003
Ebelmen found, 3.913, ± .009

General mean, 3.9276, ± .0003

In short, Ebelmen's mean vanishes when combined with Zimmermann's.

* Ann. d. Chem., 232, 299. 1886.

Zimmermann's second method was essentially that of Wertheim, namely, the ignition of the double acetate $UO_2(C_2H_3O_2)_2.NaC_2H_3O_2$, the residue being sodium uranate, $Na_2U_2O_7$.

Double Acetate.	Uranate.	Per cent. Uranate.
4.272984	2.886696	67.557
5.272094	3.560770	67.540
2.912283	1.967428	67.556
3.181571	2.149309	67.555

Mean, 67.552, ± .0027
Wertheim found, 67.523, ± .0076

General mean, 67.549, ± .0025

All the data for uranium now sum up thus:

(1.) Per cent. UO_2 from uranyl oxalate, 71.782, ± .019
(2.) $6CO_2 : U_3O_8 : : 100 : 321.443$, ± .338
(3.) Molecular weight of uranyl acetate, 423.842, ± .4222
(4.) $3UO_3 : U_3O_8 : : 100 : 103.9276$, ± .0003
(5.) Per cent. $Na_2U_2O_7$ from $UO_2.Na(C_2H_3O_2)_3$, 67.549, ± .0025

Computing with $O = 15.879$, ± .0003; $C = 11.920$, ± .0004, and $Na = 22.881$, ± .0046, we have—

From (1).................... $U = 235.948$, ± .1938
From (2).................... " $= 238.462$, ± .2953
From (3).................... " $= 238.541$, ± .4223
From (4).................... " $= 237.770$, ± .0055
From (5).................... " $= 237.902$, ± .0283

General mean............... $U = 237.774$, ± .0054

If $O = 16$, $U = 239.586$.

In this case Zimmermann's data control the final result. All the other determinations might be rejected without appreciable effect.

SELENIUM.

The atomic weight of this element was first determined by Berzelius,[*] who, saturating 100 parts of selenium with chlorine, found that 179 of chloride were produced. Further on these figures will be combined with similar results by Dumas.

We may omit, as unimportant for present purposes, the analyses of alkaline selenates made by Mitscherlich and Nitzsch,[†] and pass on to the experiments published by Sacc[‡] in 1847. This chemist resorted to a variety of methods, some of which gave good results, while others were unsatisfactory. First, he sought to establish the exact composition of SeO_2, both by synthesis and by analysis. The former plan, according to which he oxidized pure selenium by nitric acid, gave poor results; better figures were obtained upon reducing SeO_2 with ammonium bisulphite and hydrochloric acid, and determining the percentage of selenium set free:

.6800 grm. SeO_2 gave	.4828 grm. Se.		71.000 per cent.	
3.5227 "	2.5047 "		71.102 "	
4.4870 "	3.1930 "		71.161 "	
			Mean, 71.088, ± .032	

In a similar manner Sacc also reduced barium selenite, and weighed the resulting mixture of barium sulphate and free selenium. This process gave discordant results, and a better method was found in calcining $BaSeO_3$ with sulphuric acid, and estimating the resulting quantity of $BaSO_4$. In the third column I give the amounts of $BaSO_4$ equivalent to 100 of $BaSeO_3$:

.5573 grm. $BaSeO_3$ gave	.4929 grm. $BaSO_4$.	88.444	
.9942 "	.8797 "	88.383	
.2351 "	.2080. "	88.473	
.9747 "	.8621 "	88.448	
		Mean, 88.437, ± .013	

Still other experiments were made with the selenites of silver and lead; but the figures were subject to such errors that they need no further discussion here.

A few years after Sacc's work was published, Erdmann and Marchand made with their usual care a series of experiments upon the atomic weight under consideration.[§] They analyzed pure mercuric selenide, which had been repeatedly sublimed and was well crystallized. Their

---

[*] Poggend. Annalen, 8, 1. 1826.
[†] Poggend. Annalen, 9, 623. 1827.
[‡] Ann. d. Chim. et d. Phys. (3), 21, 119.
[§] Jour. für Prakt. Chem., 55, 202. 1852.

method of manipulation has already been described in the chapter upon mercury. These percentages of Hg in HgSe were found:

71.726
71.731
71.741

Mean, 71.7327, ± .003

The next determinations were made by Dumas,* who returned to the original method of Berzelius. Pure selenium was converted by dry chlorine into $SeCl_4$, and from the gain in weight the ratio between Se and Cl was easily deducible. I include Berzelius' single experiment, which I have already cited, and give in a third column the quantity of chlorine absorbed by 100 parts of selenium:

1.709 grm. Se absorb 3.049 grm. Cl.		178.409
1.810 " 3.219 "		177.845
1.679 " 3.003 "		178.856
1.498 , " 2.688 "		179.439
1.944 " 3.468 "		178.395
1.887 " 3.382 "		179.226
1.935 " 3.452 "		178.398
		179.000—Berzelius.

Mean, 178.696, ± .125

The question may here be properly asked, whether it would be possible thus to form $SeCl_4$, and be certain of its absolute purity? A trace of oxychloride, if simultaneously formed, would increase the apparent atomic weight of selenium. In point of fact, this method gives a higher value for Se than any of the other processes which have been adopted, and that value has the largest probable error of any one in the entire series. A glance at the table which summarizes the discussion at the end of this chapter will render this point sufficiently clear.

Still later, Ekman and Pettersson † investigated several methods for the determination of this atomic weight, and finally decided upon the two following:

First, pure silver selenite, Ag_2SeO_3 was ignited, leaving behind metallic silver, which, however, sometimes retained minute traces of selenium. The data obtained were as follows:

Ag_2SeO_3.	Ag.	Per cent. Ag.
5.2102	3.2787	62.93
5.9721	3.7597	62.95
7.2741	4.5803	62.97
7.5390	4.7450	62.94
6.9250	4.3612	62.98
7.3455	4.6260	62.98
6.9878	4.3992	62.95

Mean, 62.957, ± .005

* Ann. Chem. Pharm., 113, 32. 1860.
† Ber. d. Deutsch. Chem. Gesell., 9, 1210. 1876. Published in detail by the society at Upsala.

Secondly, a warm aqueous solution of selenious acid was mixed with HCl, and reduced by a current of SO_2. The reduced Se was collected upon a glass filter, dried, and weighed.

SeO_2.	Se.	Per cent. Se.
11.1760	7.9573	71.199
11.2453	8.0053	71.185
24.4729	17.4232	71.193
20.8444	14.8383	71.187
31.6913	22.5600	71.191

Mean, 71.191, ± .0016
Sacc found, 71.088, ± .0320

General mean, 71.1907, ± .0016

There are now five series of figures from which to deduce the atomic weight of selenium:

(1.) Per cent. of Se in SeO_2, 71.1907, ± .0016
(2.) $BaSeO_3 : BaSO_4 :: 100 : 88.437$, ± .013
(3.) Per cent. of Hg in HgSe, 71.7327, ± .003
(4.) $Se : Cl_4 :: 100 : 178.696$, ± .125
(5.) Per cent. of Ag in Ag_2SeO_3, 62.957, ± .005

From these, computing with—

O = 15.879, ± .0003   S = 31.828, ± .0015
Ag = 107.108, ± .0031   Ba = 136.392, ± .0086
Cl = 35.179, ± .0048   Hg = 198.491, ± .0083,

five values for Se are calculable, as follows:

From (1)...................... Se = 78.477, ± .0049
From (2)...................... " = 78.006, ± .0410
From (3)...................... " = 78.217, ± .0095
From (4)...................... " = 78.740, ± .0561
From (5)...................... " = 78.405, ± .0201

General mean............... Se = 78.419, ± .0042

If O = 16, this becomes Se = 79.016.

TELLURIUM.

Particular interest attaches to the atomic weight of tellurium on account of its relations to the periodic law. According to that law, tellurium should lie between antimony and iodine, having an atomic weight greater than 120 and less than 126. Theoretically, Mendelejeff assigns it a value of Te = 125, but all of the best determinations lead to a mean number higher than is admissible under the currently accepted hypotheses. Whether theory or experiment is at fault remains to be discovered.

The first, and for many years the only, determinations of the constant in question were made by Berzelius.* By means of nitric acid he oxidized tellurium to the dioxide, and from the increase in weight deduced a value for the metal. He published only his final results, from which, if O = 100, Te = 802.121. The three separate experiments give Te = 801.74, 801.786, and 802.838, whence we can calculate the following percentages of metal in the dioxide:

$$80.057$$
$$80.036$$
$$80.034$$

Mean, $80.042, \pm .005$

The next determinations were made by von Hauer,† who resorted to the analysis of the well crystallized double salt $TeBr_4.2KBr$. In this compound the bromine was estimated as silver bromide, the values assumed for Ag and Br being respectively 108.1 and 80. Recalculating, with our newer atomic weights for the above-named elements, we get from von Hauer's analyses, for 100 parts of the salt, the quantities of AgBr which are put in the third column:

2.000 grm.	K_2TeBr_6 gave	69.946	per cent. Br.		164.460
6.668	"	69.8443	"		164.221
2.934	"	69.9113	"		164.379
3.697	"	70.0163	"		164.626
1.000	"	69.901	"		164.355

Mean, $164.408, \pm .045$

From Berzelius' series we may calculate Te = 127.366, and from von Hauer's Te = 126.454. Dumas,‡ by a method for which he gives absolutely no particulars, found Te = 129.

In 1879, with direct reference to Mendelejeff's theory, the subject of the atomic weight of tellurium was taken up by Wills. § The methods

---

* Poggend. Annalen, 28, 395. 1833.
† Sitzungsb. Wien Akad., 25, 142.
‡ Ann. Chim. Phys. (3), 55, 129. 1859.
§ Journ. Chem. Soc., Oct., 1879, p. 704.

of Berzelius and von Hauer were employed, with various rigid precautions in the way of testing balance and weights, and to ensure purity of material. In the first series of experiments tellurium was oxidized by nitric acid to form TeO_2. The results gave figures ranging from Te = 125.64 to 128.66:

2.21613 grm. Te gave	2.77612 grm. TeO_2.	79.828 per cent. Te.
1.45313 "	1.81542 "	80.044 "
2.67093 "	3.33838 "	80.007 "
4 77828 "	5.95748 "	80.207 "
2.65029 "	3.31331 "	79.989 "

Mean, 80.015, ± .041

In the second series tellurium was oxidized by aqua regia to TeO_2, with results varying from Te = 127.10 to 127.32:

2.85011 grm. Te gave	3.56158 grm. TeO_2.	80.024 per cent. Te.
3.09673 "	3.86897 "	80.040 "
5.09365 "	6.36612 "	80.012 "
3.26604 "	4.08064 "	80.037 "

Mean, 80.028, ± .004

By von Hauer's process, the analysis of $TeBr_4.2KBr$, Will's figures give results ranging from Te = 125.40 to 126.94. Reduced to a common standard, 100 parts of the salt yield the quantities of AgBr given in the third column:

1.70673 grm. K_2TeBr_6 gave	2.80499 grm. AgBr.	164.349
1.75225 "	2.88072 "	164.398
2.06938 "	3.40739 "	164.657
3.29794 "	5.43228 "	164.717
2.46545 "	4 05742 "	164.571

Mean, 164.538, ± .048

Combined with von Hauer's mean, 164.408, ± .045, this gives a general mean of 164.468, ± .033. Hence Te = 126.502.

The next determinations in order of time were those of Brauner.[*] This chemist tried various unsuccessful methods for determining the atomic weight of tellurium, among them being the synthetic preparation of silver, copper, and gold tellurides, and the basic sulphate, Te_2SO_7. None of these methods gave sufficiently concordant results, and they were therefore abandoned. The oxidation of tellurium to dioxide by means of nitric acid was also unsatisfactory, but a series of oxidations with aqua regia gave data as follows. The third column contains the percentage of tellurium in the dioxide:

---

[*] Journ. Chem. Soc., 55, 382. 1889.

TELLURIUM.

Te.	TeO_2.	Per cent. Te.
2.3092	2.9001	79.625
2.8153	3.5332	79.681
4.0176	5.0347	79.798
3.1613	3.9685	79.660
.8399	1.0526	79.793

Mean, 79.711, ± .0239

Hence Te = 124.709.

In a single analysis of the dioxide, by reduction with SO_2, 2.5489 grammes TeO_2 gave 2.0374 of metal. If we give this experiment the weight of one observation in the synthetic series, the percentage of tellurium found by it becomes—

79.932, ± .0534.

Hence Te = 126.494.

Brauner's best results were obtained from analyses of tellurium tetrabromide, prepared from pure tellurium and pure bromine, and afterwards sublimed in a vacuum. This compound was titrated with standard solutions of silver, and three series of experiments, made with samples of bromide of different origin, gave results as follows. The $TeBr_4$ equivalent to 100 parts of silver appears in the third column:

First Series.

$TeBr_4$.	Ag.	Ratio.
2.14365	2.06844	103.636
1.76744	1.70531	103.643
1.47655	1.42477	103.634
1.23354	1.19019	103.642

Second Series.

$TeBr_4$.	Ag.	Ratio.
3.07912	2.97064	103.651
5.47446	5.28157	103.652
3.30927	3.19313	103.637
7.26981	7.01414	103.645
3.52077	3.39667	103.654

Third Series.

$TeBr_4$.	Ag.	Ratio.
2.35650	2.27363	103.645
1.51931	1.46564	103.662
1.43985	1.38942	103.630

Mean of all as one series, 103.644, ± .0018

Hence Te = 126.668, ± .0290. A reduction of the weighings to a vacuum raises this by 0.07 to 126.738.

Still another series of analyses, made with fractionated material, gave values for tellurium running up to as high as 137. These experiments led Brauner to believe that he had found in tellurium a higher homologue of that element, a view which he has since abandoned.* Brauner also made a series of analyses of tellurium dibromide, but the results were unsatisfactory.

In the series of determinations by Gooch and Howland† an alkaline solution of tellurium dioxide was oxidized by means of standard solutions of potassium permanganate. This was added in excess, the excess being measured, after acidification with sulphuric acid, by back titration with oxalic acid and permanganate. Two series are given, varying in detail, but for present purposes they may be treated as one. The ratio $TeO_2 : O :: 100 : x$ is given in the third column.

TeO_2 Taken.	O Required.	Ratio.
.1200	.01202	10.017
.0783	.00785	10.026
.0931	.00940	10.097
.1100	.01119	10.149
.0904	.00909	10.055
.1065	.01078	10.122
.0910	.00915	10.055
.0910	.00910	10.000
.0911	.00924	10.143
.0913	.00915	10.022
.0912	.00915	10.033
.0914	.00923	10.098

Mean, 10.068, ± .0100

Hence Te = 125.96.

In Staudenmaier's‡ determinations of the atomic weight of tellurium, crystallized telluric acid, H_6TeO_6 was the starting point. By careful heating in a glass bulb this compound can be reduced to TeO_2, and by heating in hydrogen, to metal. In the latter case finely divided silver was added to prevent volatilization of tellurium. The telluric acid was fractionally crystallized, but the different fractions gave fairly constant results. I therefore group Staudenmaier's data so as to bring them into series more suitable for the present discussion.

---

* Journ. Chem. Soc., 67, 549. 1895.
† Amer. Journ. Sci., 58, 375. 1894. Some misprints in the original publication have been kindly corrected by Professor Gooch; hence the differences between these data and the figures formerly given.
‡ Zeitsch. Anorg. Chem., 10, 189. 1895.

TELLURIUM.   275

First. H_6TeO_6 to TeO_2.

H_6TeO_6.	Loss in Weight.	Per cent. TeO_2.
1.7218	.5260	69.451
2.8402	.8676	69.453
4.0998	1.2528	69.442
3.0916	.9450	69.433
1.1138	.3405	69.429
4.9843	1.5236	69.432
4.6716	1.4278	69.437

Mean, 69.440, ± .0024

Hence Te = 126.209.

Second. H_6TeO_6 to Te.

H_6TeO_6.	Loss in Weight.	Per cent. Te.
1.2299	.5471	55.517
1.0175	.4526	55.518
2.5946	1.1549	55.488

Mean, 55.508, ± .0068

Hence Te = 126.303.

Staudenmaier also gives four reductions of TeO_2 to Te, in presence of finely divided silver. The data are as follows:

TeO_2.	Loss in Weight.	Per cent. Te.
.9171	.1839	79.948
1 9721	.3951	79.966
2.4115	.4835	79 950
1.0172	.2041	79.935

Mean, 79.950, ± .0043

Hence Te = 126.636.

The last series, giving the percentage of tellurium in the dioxide, combines with previous series thus:

Berzelius,	80.042, ± .0050
Wills, first series	80.015, ± .0410
Wills, second series	80.028, ± .0040
Brauner, synthesis	79.711, ± .0239
Brauner, analysis	79.932, ± .0534
Staudenmaier	79.950, ± .0043
General mean	80.001, ± .0025

The very recent determinations by Chikashigé[*] were made by Brauner's method, giving the ratio between silver and $TeBr_4$. In all essential particulars the work resembles that of Brauner, except that the tellurium,

[*] Journ. Chem. Soc., 69, 881. 1896.

instead of being extracted from metallic tellurides, was derived from Japanese native sulphur, in which it exists as an impurity. This difference of origin in the material studied gives the chief interest to the investigation. The data are as follows:

$TeBr_4$.	Ag.	Ratio.
4.1812	4.0348	103.628
4.3059	4.1547	103.639
4.5929	4.4319	103.633

Mean, 103.633, ± .0023
Brauner found, 103.644, ± .0018

General mean, 103.640, ± .0014

Now, to sum up, the subjoined ratios are available for computing the atomic weight of tellurium:

(1.) Percentage Te in TeO_2, 80.001, ± .0025
(2.) Percentage Te in H_6TeO_6, 55.508, ± .0068
(3.) Percentage TeO_2 in H_6TeO_6, 64.440, ± .0024
(4.) Ag_4 : $TeBr_4$ :: 100 : 103.640, ± .0014
(5.) K_2TeBr_6 : $6AgBr$ :: 100 : 164.468, ± .0330
(6.) TeO_2 : O :: 100 : 10.068, ± .0100

To reduce these ratios we have—

O = 15.879, ± .0003   K = 38.817, ± .0051
Ag = 107.108, ± .0031   AgBr = 186.452, ± .0054
Br = 79.344, ± .0062

For the atomic weight of tellurium six values appear, as follows:

From (1)............ Te = 127.040, ± .0165
From (4)............ " = 126.650, ± .0302
From (5)............ " = 126.502, ± .1430
From (2)............ " = 126.303, ± .0246
From (3)............ " = 126.209, ± .0138
From (6)............ " = 125.960, ± .1574

General mean........... Te = 126.523, ± .0092

If O = 16, Te = 127.487.

A careful consideration of the foregoing figures, and of the experimental methods by which they were obtained, will show that they are not absolutely conclusive with regard to the place of tellurium under the periodic law. The atomic weight of iodine, calculated in a previous chapter, is 125.888. Wills' values for Te, rejecting his first series as relatively unimportant, range from 125.40 to 127.32; that is, some of them fall below the atomic weight of iodine, although none descend quite to the 125 assumed by Mendelejeff.

Some of Brauner's data fall even lower; and the same thing is true in

Gooch and Howland's series, of which the mean gives Te = 125.96, a value very little above that of iodine.

In considering the experimental methods, reference may properly be made to the controversy regarding the atomic weight of antimony. It will be seen that Dexter, estimating the latter constant by the conversion of the metal into Sb_2O_4, obtained a value approximately of Sb = 122. Dumas, working with $SbCl_3$, obtained nearly the same value. Schneider and Cooke, on the other hand, have established an atomic weight for antimony near 120, and Cooke in particular has traced out the constant errors which lurked unsuspected in the work of Dumas. Now in their physical aspects tellurium and antimony are quite similar. The oxidation of tellurium to dioxide resembles in many particulars that of antimony, and may lead to error in the same way. In each of the six tellurium ratios there is still uncertainty, and a positive measurement, free from objections, of the constant in question is yet to be made.

FLUORINE.

The atomic weight of fluorine has been chiefly determined by one general method, namely, by the conversion of fluorides into sulphates. The work of Christensen, however, is on different lines. Excluding the early results of Davy,* we have to consider first the experiments of Berzelius, Louyet, Dumas, De Luca, and Moissan with reference to the fluorides of calcium, sodium, potassium, barium, and lead.

The ratio between calcium fluoride and sulphate has been determined by the five investigators above named, and by one general process. The fluoride is treated with strong sulphuric acid, the resulting sulphate is ignited, and the product weighed. In order to insure complete transformation special precautions are necessary, such, for instance, as repeated treatment with sulphuric acid, and so on. For details like these the original papers must be consulted.

The first experiments in chronological order are those of Berzelius,† who operated upon an artificial calcium fluoride. He found, in three experiments, for one part of fluoride the following of sulphate:

1.749
1.750
1.751
———
Mean, 1.750, ± .0004

Louyet's researches ‡ were much more elaborate than the foregoing. He began with a remarkably concordant series of results upon fluor spar,

---

* Phil. Trans., 1814, 64.
† Poggend. Annalen, 8, 1. 1826.
‡ Ann. Chim. Phys. (3), 25, 300. 1849.

in which one gramme of the fluoride yielded from 1.734 to 1.737 of sulphate. At first he regarded these as accurate, but he soon found that particles of spar had been coated with sulphate, and had therefore escaped action. In the following series this source of error was guarded against.

Starting with fluor spar, Louyet found of sulphate as follows:

$$1.742$$
$$1.744$$
$$1.745$$
$$1.744$$
$$1.7435$$
$$1.7435$$

Mean, $1.7437, \pm .0003$

A second series, upon artificial fluoride, gave:

$$1.743$$
$$1.741$$
$$1.741$$

Mean, $1.7417, \pm .0004$

Dumas[*] published but one result for calcium fluoride. .495 grm. gave .864 grm. sulphate, the ratio being $1 : 1.7455$.

De Luca[†] worked with a very pure fluor spar, and published the following results. The ratio between $CaSO_4$ and one gramme of CaF_2 is given in the third column:

.9305 grm. CaF_2 gave	1.630 grm. $CaSO_4$.		1.7518
.836 "	1.459 "		1.7452
.502 "	.8755 "		1.7440
.3985 "	.6945 "		1.7428

If we include Dumas' single result with these, we get a mean of $1.7459, \pm .0011$.

Moissan[‡] unfortunately gives no details nor weighings, but merely states that four experiments with calcium fluoride gave values for F ranging from 19.02 to 19.08. To S he assigned the value 32.074, and probably Ca was taken as $= 40$. With these data his extreme values as given may be calculated back into uniformity with the ratio as stated above, becoming—

$$1.7444$$
$$1.7410$$

Mean, 1.7427.

---

[*] Ann. Chem. Pharm., 113, 28. 1860.
[†] Compt. Rend., 51, 299. 1860.
[‡] Compt. Rend., 111, 570. 1890.

FLUORINE.

If we assign this equal weight with Berzelius' series, the data for this ratio combine thus:

Berzelius	1.7500,	± .0004
Louyet, first series	1.7437,	± .0003
Louyet, second series	1.7417,	± .0004
De Luca with Dumas	1.7459,	± .0011
Moissan	1.7427,	± .0004
General mean	1.7444,	± .00018

For the ratio between the two sodium salts we have experiments by Dumas, Louyet, and Moissan. According to Louyet, one gramme of NaF gives of Na_2SO_4,—

$$1.686$$
$$1.683$$
$$1.685$$

Mean, 1.6847, ± .0006

The weighings published by Dumas are as follows:

.777 grm. NaF give 1.312 grm. Na_2SO_4. Ratio, 1.689
1.737 " 2.930 " " 1.687

Mean, 1.688, ± .0007

Moissan says only that five experiments with sodium fluoride gave $F = 19.04$ to 19.08. This was calculated with $Na = 23.05$ and $S = 32.074$. Hence, reckoning backward, the two values give for the standard ratio—

$$1.6889$$
$$1.6873$$

Mean, 1.6881

Giving this equal weight with Dumas' mean, we have—

Louyet	1.6847,	± .0006
Dumas	1.688,	± .0007
Moissan	1.6881,	± .0007
General mean	1.6867,	± .00038

Dumas also gives experiments upon potassium fluoride. The quantity of sulphate formed from one gramme of fluoride is given in the last column:

1.483 grm. KF give 2.225 grm. K_2SO_4. 1.5002
1.309 " 1.961 " 1.4981

Mean, 1.4991, ± .0007

The ratio between barium fluoride and barium sulphate was measured by Louyet and Moissan. According to Louyet, one gramme of BaF_2 gives of $BaSO_4$,—

$$1.332$$
$$1.331$$
$$1.330$$

Mean, $1.331, \pm .0004$

Moissan, in five experiments, found $F = 19.05$ to 19.09. Assuming that he put $Ba = 137$, and $S = 32.074$ as before, these two extremes become—

$$1.3311$$
$$1.3305$$

Mean, 1.3308

Giving this equal weight with Louyet's mean, we get the subjoined combination:

Louyet.................................... $1.331, \pm .0004$
Moissan................................... $1.3308, \pm .0004$
General mean.......................... $1.3309, \pm .00028$

The experiments with lead fluoride are due to Louyet, and a new method of treatment was adopted. The salt was fused, powdered, dissolved in nitric acid, and precipitated by dilute sulphuric acid. The evaporation of the fluid and the ignition of the sulphate was then effected without transfer. Five grammes of fluoride were taken in each operation, yielding of sulphate:

$$6.179$$
$$6.178$$
$$6.178$$

Mean, $6.1783, \pm .0002$

In Christensen's determinations[*] we find a method adopted which is radically unlike anything in the work of his predecessors. He started out with the salt $(NH_4)_2MnF_6$. When this is added to a mixture, in solution, of potassium iodide and hydrochloric acid, iodine is set free, and may be titrated with sodium thiosulphate. One molecule of the salt (as written above), liberates one atom of iodine. In four experiments Christensen obtained the following data:

3.1199 grm. Am_2MnF_6 gave	2.12748 I.	68.191 per cent.	
3.9190	"	2.67020 "	68.135 "
3.5005	"	2.38429 "	68.113 "
1.2727	"	.86779 "	68.185 "

Mean, $68.156, \pm .0128$

---

[*] Journ. für Prakt. Chem. (2), 35, 541. Christensen assigns to the salt double the formula here given.

The ratios from which to compute the atomic weight of fluorine are now—

(1.) $CaF_2 : CaSO_4 :: 1.0 : 1.7444, \pm .00018$
(2.) $2NaF : Na_2SO_4 :: 1.0 : 1.6867, \pm .00038$
(3.) $2KF : K_2SO_4 :: 1.0 : 1.4991, \pm .0007$
(4.) $BaF_2 : BaSO_4 :: 1.0 : 1.3309, \pm .00028$
(5.) $PbF_2 : PbSO_4 :: 5.0 : 6.1783, \pm .0002$
(6.) $Am_2MnF_6 : I :: 100 : 68.156, \pm .0128$

To reduce them we have—

$O = 15.879, \pm .0003$ $K = 38.817, \pm .0051$
$S = 31.828, \pm .0015$ $Ca = 39.764, \pm .0045$
$N = 13.935, \pm .0021$ $Ba = 136.392, \pm .0086$
$I = 125.888, \pm .0069$ $Pb = 205.358, \pm .0040$
$Na = 22.881, \pm .0046$ $Mn = 54.571, \pm .0013$

And the values derived for fluorine are as follows:

From (1)............................ $F = 18.844, \pm .0048$
From (2)............................ " $= 18.948, \pm .0108$
From (3)............................ " $= 18.877, \pm .0276$
From (4)............................ " $= 18.869, \pm .0192$
From (5)............................ " $= 18.997, \pm .0047$
From (6)............................ " $= 18.853, \pm .0073$

General mean.................... $F = 18.912, \pm .0029$

If $O = 16$, $F = 19.056$.

In all probability these values for fluorine average a trifle too high. It is difficult to be certain that a fluoride has been completely converted into sulphate, and an incomplete conversion tends to raise the apparent atomic weight of fluorine. This possible source of error exists in all of the ratios except the last one, but the fair concordance of the results obtained seems to indicate that the uncertainty cannot be very large.

MANGANESE.

The earliest experiments of Berzelius* and of Arfvedson† gave values for Mn ranging between 56 and 57, and therefore need no farther consideration here. The first determinations to be noticed are those of Turner‡ and a later measurement by Berzelius.§ who both determined gravimetrically the ratio between the chlorides of manganese and silver. The manganese chloride was fused in a current of dry hydrochloric acid, and afterwards precipitated with a silver solution. I give the $MnCl_2$ equivalent to 100 parts of AgCl in the third column:

4.20775 grm. $MnCl_2 =$	9.575 grm. AgCl.	43.945	Berzelius.
3.063 "	= 6.96912 "	43.950	
12.47 grains $MnCl_2 =$	28.42 grains AgCl.	43.878	—Turner.

Mean, 43.924, ± .015

Many years later Dumas ‖ also made the chloride of manganese the starting point of some atomic weight determinations. The salt was fused in a current of hydrochloric acid, and afterwards titrated with a standard solution of silver in the usual way. One hundred parts of Ag are equivalent to the quantities of $MnCl_2$ given in the third column:

3.3672 grm. $MnCl_2 =$	5.774 grm. Ag.		58.317
3.0872 "	5.293 "		58.326
2.9671 "	5.0875 "		58.321
1.1244 "	1.928 "		58.320
1.3134 "	2.251 "		58.321

Mean, 58.321, ± .001

An entirely different method of investigation was followed by von Hauer,¶ who, as in the case of cadmium, ignited the sulphate in a stream of sulphuretted hydrogen, and determined the quantity of sulphide thus formed. I subjoin his weighings, and also the percentage of MnS in $MnSO_4$ as calculated from them:

4.0626 grm. $MnSO_4$ gave 2.3425 grm. MnS.		57.660 per cent.
4.9367 "	2.8442 "	57.613 "
5.2372 "	3.0192 "	57.649 "
7.0047 "	4.0347 "	57.600 "
4.9175 "	2.8297 "	57.543 "
4.8546 "	2.7955 "	57.585 "
4.9978 "	2.8799 "	57.625 "
4.6737 "	2.6934 "	57.629 "
4.7240 "	2.7197 "	57.572 "

Mean, 57.608, ± .008

---

* Poggend. Annalen, 8, 185. 1826.
† Berz. Jahresbericht, 9, 136. 1829.
‡ Trans. Roy. Soc. Edinb., 11, 143. 1831.
§ Lehrbuch, 5 Aufl., 3. 1224.
‖ Ann. Chem. Pharm., 113, 25. 1860.
¶ Journ. für Prakt. Chem., 72, 360. 1857.

This method of von Hauer, which seemed to give good results with cadmium, is, according to Schneider,* inapplicable to manganese, for the reason that the sulphide of the latter metal is liable to be contaminated with traces of oxysulphide. Such an impurity would bring the atomic weight out too high. The results of two different processes, one carried out by himself and the other in his laboratory by Rawack, are given by Schneider in this paper.

Rawack reduced manganoso-manganic oxide to manganous oxide by ignition in a stream of hydrogen, and weighed the water thus formed. From his weighings I get the values in the third column, which represent the Mn_3O_4 equivalent to one gramme of water:

4.149 grm. Mn_3O_4 gave	0.330 grm. H_2O.		12.5727
4.649 "	.370 "		12.5643
6.8865 "	.5485 "		12.5552
7.356 "	.5855 "		12.5636
8.9445 "	.7135 "		12.5361
11.584 "	.9225 "		12.5572

Mean, 12.5582, ± .0034

Here the most obvious source of error lies in the possible loss of water. Such a loss, however, would increase the apparent atomic weight of manganese; but we see that the value found is much lower than that obtained either by Dumas or von Hauer.

Schneider himself effected the combustion of manganous oxalate with oxide of copper. The salt was not absolutely dry, so that it was necessary to collect both water and carbon dioxide. Then, upon deducting the weight of water from that of the original material, the weight of anhydrous oxalate was easily ascertained. Subtracting from this the CO_2, we get the weight of Mn. If we put $CO_2 = 100$, the quantities of manganese equivalent to it will be found in the last column:

1.5075 grm. oxalate gave	.306 grm. H_2O and	.7445 grm. CO_2.	61.3835
2.253 "	.4555 "	1.1135 "	61.4291
3.1935 "	.652 "	1.5745 "	61.4163
5.073 "	1.028 "	2.507 "	61.3482

Mean, 61.3943, ± .0122

Up to this point the data give two distinct values for Mn—one near 54, the other approximately 55—and with no sure guide to preference between them. The higher value, however, has been confirmed by later testimony.

In 1883 Dewar and Scott † published the results of their work upon silver permanganate. This salt is easily obtained pure by recrystallization, and has the decided advantage of not being hygroscopic. Two sets

---

* Poggend. Annalen, 107, 605.
† Proc. Roy. Soc., 35, 44. 1883.

of experiments were made. First, the silver permanganate was heated to redness in a glass bulb, first in air, then in hydrogen. Before weighing, the latter gas was replaced by nitrogen. The data are as follows:

$AgMnO_4$.	$Ag + MnO$.	Per cent. $Ag + MnO$.
5.8696	4.63212	78.917
5.4988	4.33591	78.852
7.6735	6.05395	78.894
13.10147	10.31815	78.756
12.5799	{ 9.91065	78.782
	9.91435	78.811

Mean, 78.835, ± .0174

The duplication of the last weighing is not explained.

In the second series the permanganate was dissolved in dilute nitric acid, reduced by sulphur dioxide, potassium nitrite, or sodium formate, and titrated with potassium bromide. The $AgMnO_4$ equivalent to 100 KBr appears in the third column.

$AgMnO_4$.	KBr.	Ratio.
6.5289	3.42385	190.686
7.5378	3.9553	190.575
6.1008	3.20166	190.559
5.74647	3.00677	191.117
6.16593	3.23602	190.540
5.11329	2.6828	190.596
5.07438	2.66204	190.624
13.4484	7.05602	190.604
12.5799	6.60065	190.588
12.27025	6.43808	190.584

Mean, 190.647, ± .0361

Vacuum weights are given throughout. To the first series of experiments the authors attach little importance, and numbers 1 and 4 of the second series they also regard as questionable. These experiments represent the use of sulphur dioxide as the reducing agent, and were attended by the formation of an insoluble residue, apparently of a sulphide. Excluding them, the remaining eight experiments of the second series give in mean—

$KBr : AgMnO_4 :: 100 : 190.584, ± .0062$,

which will be used for the present calculation. Dewar and Scott also made determinations with manganese chloride and bromide. With the first salt they found $Mn = 54.91$, and with the second, $Mn = 54.97$; but they give no details.

Marignac's work upon the atomic weight of manganese also appeared in 1883.* He prepared the oxide, MnO, by ignition of the oxalate and

---
*Arch. Sci. Phys. et Nat. (3), 10, 21, 1883.

subsequent reduction of the resulting Mn_3O_4 in hydrogen. The oxide, with various precautions, was then converted into sulphate. The percentage of MnO in $MnSO_4$ is appended:

2.6587 grm. MnO gave 5.6530 $MnSO_4$.		47.032 per cent.	
2.5185	" 5.3600	" 46.987	"
2.5992	" 5.5295	" 47.006	"
2.8883	" 6.1450	" 47.002	"

Mean, 47.007, ± .0025

J. M. Weeren, in 1890,[*] published determinations made by two methods, the one Marignac's, the other von Hauer's. From manganese sulphate he threw down the hydrated peroxide electrolytically, and the latter compound was then reduced in hydrogen which had been proved to be free from oxygen. The resulting monoxide was cooled in a stream of purified nitrogen. After the oxide had been treated with sulphuric acid, converted into sulphate, and weighed, a few drops of sulphuric acid and a little sulphurous acid were added to it, after which it was reheated and weighed again. This process was repeated until four successive weighings absolutely agreed. The results of this set of experiments were as follows, with vacuum standards:

15.2349 grm. MnO gave 32.4142 $MnSO_4$.		47.005 per cent.	
13.9686	" 29.7186	" 47.004	"
13.7471	" 29.2493	" 47.000	"
15.5222	" 33.0246	" 47.001	"
14.9824	" 31.8755	" 47.002	"
14.6784	" 31.2304	" 47.000	"

Mean, 47.002, ± .0006

Marignac's mean, combined with this, hardly affects either the percentage itself or its probable error. Fortunately, both Marignac and Weeren are completely in agreement as to the ratio, and either set of measurements would be valid without the other. In order, therefore, to give Marignac's work some proper recognition, we can assume a géneral mean of 47.004, ± .0006, without danger of serious error.

The manganese sulphate produced in the foregoing series of experiments was used, with many precautions, for the next series carried out by von Hauer's method. It was transferred to a porcelain boat, dried at 260° to avoid errors due to retention of water taken up in the process of transfer, and then heated to constant weight in a stream of hydrogen sulphide. Before weighing, the sulphide was heated to redness in hydrogen and cooled in the same gas. The results, with vacuum weights, were as follows:

---

[*] Atom-Gewichtsbestimmung des Mangans. Inaugural Dissertation, Halle, 1890.

16.0029 grm.	MnSO₄ gave	9.2228	MnS =	57.632 per cent.
16.3191	"	9.4048	"	57.631 "
15.9307	"	9.1817	"	57.634 "
15.8441	"	9.1315	"	57.634 "
16.2783	"	9.3819	"	57.635 "
17.0874	"	9.8477	"	57.633 "

Mean, 57.633, ± .0004
von Hauer found, 57.608, ± .0080

Hence the general mean is identical with Weeren's to the third decimal place, which is unaffected by combination with von Hauer's data. We have now to consider the following ratios for manganese:

(1.) $2AgCl : MnCl_2 :: 100 : 41.924, \pm .0150$
(2.) $2Ag : MnCl_2 :: 100 : 58.321, \pm .0010$
(3.) $H_2O : Mn_3O_4 :: 100 : 1255.82, \pm .340$
(4.) $2CO_2 : Mn :: 100 : 61.3943, \pm .0122$
(5.) $AgMnO_4 : Ag + MnO :: 100 : 78.835, \pm .0174$
(6.) $KBr : AgMnO_4 :: 100 : 190.584, \pm .0062$
(7.) $MnSO_4 : MnO :: 100 : 47.004, \pm .0006$
(8.) $MnSO_4 : MnS :: 100 : 57.633, \pm .0004$

Computing with the subjoined preliminary data—

O = 15.879, ± .0003		K = 38.817, ± .0051
Ag = 107.108, ± .0031		C = 11.920, ± .0004
Cl = 35.179, ± .0048		S = 31.828, ± .0015
Br = 79.344, ± .0062		AgCl = 142.287, ± .0037

these ratios reduce as follows:

First, for the molecular weight of manganese chloride, two values are deducible.

From (1) .... $MnCl_2$ = 124.996, ± .0428
From (2) .... " = 124.933, ± .0042

General mean .... $MnCl_2$ = 124.934, ± .0042

Hence Mn = 54.576, ± .0075.

For manganese there are seven independent values, as follows:

From molecular weight $MnCl_2$........ Mn = 54.576, ± .0075
From (3)..................... " = 53.667, ± .0203
From (4)..................... " = 53.633, ± .0107
From (5)..................... " = 54.450, ± .1511
From (6)..................... " = 54.572, ± .0173
From (7)..................... " = 54.601, ± .0018
From (8)..................... " = 54.575, ± .0022

General mean............... Mn = 54.571, ± .0013

If O = 16, this becomes Mn = 54.987.

In this case five of the separate values are well in accord, and the rejection of the two aberrant values, which have high probable errors, is

IRON. 287

not necessary. Their influence is imperceptible. Weeren's marvelously-concordant data seem to receive undue weight, but they are abundantly confirmed by the evidence of other experimenters. In short, the atomic weight of manganese appears to be quite well determined.

IRON.

The atomic weight of iron has been mainly determined from the composition of ferric oxide, with some rather scanty data relative to other compounds.

Most of the earlier data relative to the percentage of metal and oxygen in ferric oxide we may reject at once, as set aside by later investigations. Among this no longer valuable material there is a series of experiments by Berzelius, another by Döbereiner, and a third by Capitaine. The work done by Stromeyer and by Wackenroder was probably good, but I am unable to find its details. The former found 30.15 per cent. of oxygen in the oxide under consideration, while Wackenroder obtained figures ranging from a minimum of 30.01 to a maximum of 30.38 per cent.*

In 1844 Berzelius † published two determinations of the ratio in question. He oxidized iron by means of nitric acid, and weighed the oxide thus formed. He thus found that when $O = 100$ $Fe = 350.27$ and 350.369.

Hence the following percentages of Fe in Fe_2O_3:

70.018
70.022

Mean, 70.020, ± .0013

About the same time Svanberg and Norlin ‡ published two elaborate series of experiments; one relating to the synthesis of ferric oxide, the other to its reduction. In the first set pure piano-forte wire was oxidized by nitric acid, and the amount of oxide thus formed was determined. The results were as follows:

1.5257 grm. Fe gave	2.1803 grm. Fe_2O_3.			69.977 per cent. Fe.	
2.4051	"	3.4390	"	69.936	"
2.3212	"	3.3194	"	69.928	"
2.32175	"	3.3183	"	69.968	"
2.2772	"	3.2550	"	69.960	"
2.4782	"	3.5418	"	69.970	"
2.3582	"	3.3720	"	69.935	"

Mean, 69.9534, ± .0050

---

* For additional details concerning these earlier papers I must refer to Oudemans' monograph, pp. 140, 141.
† Ann. Chem. Pharm., 30, 432. Berz. Jahresb., 25, 43.
‡ Berzelius' Jahresbericht, 25, 42.

In the second series ferric oxide was reduced by ignition in a current of hydrogen, yielding the subjoined percentages of metal:

2.98353 grm. Fe_2O_3 gave	2.08915 grm. Fe.		70.025 per cent.	
2.41515 "	1.6910	"	70.015	"
2.99175 "	2.09455	"	70.014	"
3.5783 "	2.505925	"	70.030	"
4.1922 "	2.9375	"	70.072	"
3.1015 "	2.17275	"	70.056	"
2.6886 "	1.88305	"	70.036	"

Mean, 70.0354, ± .0055

It is evident that one or both of these series must be vitiated by constant errors, and that these probably arise from impurities in the materials employed. Impurities in the wire taken for the oxidation series could hardly have been altogether avoided, and in the reduction series it is possible that weighable traces of hydrogen may have been retained by the iron. At all events, it is probable that the errors of both series are in contrary directions, and therefore in some measure compensatory.

In 1844 there was also published an important paper by Erdmann and Marchand.* These chemists prepared ferric oxide by the ignition of pure ferrous oxalate, and submitted it to reduction in a stream of hydrogen. Two sets of results were obtained with two different samples of ferrous oxalate, prepared by two different methods. For present purposes, however, it is not necessary to discuss these sets separately. The percentages of iron in Fe_2O_3 are as follows:

$$\left. \begin{array}{l} 70.013 \\ 69.962 \\ 69.979 \\ 70.030 \\ 69.977 \end{array} \right\} A.$$

$$\left. \begin{array}{l} 70.044 \\ 70.015 \\ 70.055 \end{array} \right\} B.$$

Mean, 70.0094, ± .0080

In 1850 Maumené's† results appeared. He dissolved pure iron wire in aqua regia, precipitated with ammonia, filtered off the precipitate, washed thoroughly, ignited, and weighed, after the usual methods of quantitative analysis. The percentages of Fe in Fe_2O_3 are given in the third column:

1.482 grm. Fe gave	2.117 grm. Fe_2O_3.		70.005 per cent.	
1.452 "	2.074	"	70.010	"
1.3585 "	1.941	"	69.990	"
1.420 "	2.0285	"	70.002	"
1.492 "	2.1315	"	69.998	"
1.554 "	2.220	"	70.000	"

Mean, 70.0008, ± .0019

---

* Journ. für Prakt. Chem., 33, 1. 1844.
† Compt. Rend., Oct. 17, 1850.

IRON. 289

Two more results, obtained by Rivot* through the reduction of ferric oxide in hydrogen, remain to be noticed. The percentages are:

$$69.31$$
$$69.35$$

Mean, $69.33, \pm .013$

We have thus before us six series of results, which we may now combine:

Berzelius,	70.020, ± .0013
Erdmann and Marchand,	70.0094, ± .0080
Svanberg and Norlin, oxidation,	69.9534, ± .0050
Svanberg and Norlin, reduction,	70.0354, ± .0055
Maumené,	70.0008, ± .0019
Rivot,	69.33, ± .013
General mean	70.0075, ± .0010

From this we get Fe = 55.596.

Dumas'† results, obtained from the chlorides of iron, are of so little weight that they might safely be omitted from our present discussion. For the sake of completeness, however, they must be included.

Pure ferrous chloride, ignited in a stream of hydrochloric acid gas, was dissolved in water and titrated with a silver solution in the usual way. One hundred parts of silver are equivalent to the amounts of $FeCl_2$ given in the third column:

3.677 grm. $FeCl_2$ = 6.238 grm. Ag.		58.945
3.924 " = 6.675 "		58.787

Mean, $58.866, \pm .053$

Ferric chloride, titrated in the same way, gave these results:

1.179 grm. $FeCl_3$ = 2.3475 grm. Ag.		50.224
1.742 " = 2.471 "		50.263

Mean, $50.2435, \pm .0132$

These give us two additional values for Fe, as follows:

From $FeCl_2$ .... Fe = 55.742
From $FeCl_3$ .... " = 55.907

A series of determinations of the equivalent of iron, made by students by measuring the hydrogen evolved when the metal is dissolved in an acid, was published by Torrey in 1888.‡ The data have, of course, slight

---

* Ann. Chem. Pharm., 78, 214. 1851.
† Ann. Chem. Pharm., 113, 26. 1860.
‡ Am. Chem. Journ., 10, 74.

value, but may be considered as being in some measure confirmatory. They are as follows:

56.40
55.60
55.38
55.56
55.48
55.50
55.86
56.06
56.22
55.80
55.78
55.60
55.70
55.94

Mean, 55.777, ± .0532

These values undoubtedly depend on Regnault's value for the weight of hydrogen. Correcting by the later value, as found in the chapter of this work relating to the density ratio H : O, the mean becomes Fe = 55.608, ± .0532. Here the probable error in the weight of the hydrogen is ignored, as being of no practical significance.

The four ratios for iron are now as follows:

(1.) Per cent. Fe in Fe_2O_3, 70.0075, ± .0010
(2.) Ag_2 : $FeCl_2$ :: 100 : 58.866, ± .0530
(3.) Ag_3 : $FeCl_3$ :: 100 : 50.2435, ± .0132
(4.) H : Fe :: 1 : 55.608, ± .0532

Reducing these with—

O = 15.879, ± .0003
Ag = 107.108, ± .0031
Cl = 35.179, ± .0048

we have—

From (1).......................... Fe = 55.596, ± .0023
From (2).......................... " = 55.742, ± .1140
From (3).......................... " = 55.907, ± .0450
From (4).......................... " = 55.608, ± .0532

General mean............... Fe = 55.597, ± .0023

If O = 16, then Fe = 56.021. Here all the values are absorbed practically by the first, the other three having no real significance.

NICKEL AND COBALT.

On account of the close similarity of these metals to each other, their atomic weights, approximately if not actually identical, have received of late years much attention.

The first determinations, and the only ones up to 1852, were made by Rothhoff,* each with but a single experiment. For nickel 188 parts of the monoxide were dissolved in hydrochloric acid; the solution was evaporated to dryness, the residue was dissolved in water, and precipitated by silver nitrate. 718.2 parts of silver chloride were thus formed; whence $Ni = 58.613$. The same process was applied also to cobalt, 269.2 parts of the oxide being found equivalent to 1029.9 of AgCl; hence $Co = 58.504$. These values are so nearly equal that their differences were naturally ascribable to experimental errors. They are, however, entitled to no special weight at present, since it cannot be certain from any evidence recorded that the oxide of either metal was absolutely free from traces of the other.

In 1852 Erdmann and Marchand † published some results, but without details, concerning the atomic weight of nickel. They reduced the oxide by heating in a current of hydrogen, and obtained values ranging from 58.2 to 58.6, when $O = 16$. Their results were not very concordant, and the lowest was probably the best.

In 1856, incidentally to other work, Deville ‡ found that 100 parts of pure metallic nickel yielded 262 of sulphate; whence $Ni = 58.854$.

To none of the foregoing estimations can any importance now be attached. The modern discussion of the atomic weights under consideration began with the researches of Schneider § in 1857. This chemist examined the oxalates of both metals, determining carbon by the combustion of the salts with copper oxide in a stream of dry air. The carbon dioxide thus formed was collected as usual in a potash bulb, which, in weighing, was counterpoised by a similar bulb, so as to eliminate errors due to the hygroscopic character of the glass. The metal in each oxalate was estimated, first by ignition in a stream of dry air, followed by intense heating in hydrogen. Pure nickel or cobalt was left behind in good condition for weighing. Four analyses of each oxalate were made, with the results given below. The nickel salt contained three molecules of water, and the cobalt salt two molecules:

---

* Cited by Berzelius. Poggend. Annalen, 8, 184. 1826.
† Journ. für Prakt. Chem., 55, 202. 1852.
‡ Ann. Chim. Phys. (3), 46, 182. 1856.
† Poggend. Annalen, 101, 387. 1857.

$NiC_2O_4.3H_2O.$

1.1945 grm. gave	.528 grm. CO_2.	44.203 per cent.
2.5555 "	1.12625 "	44.072 "
3.199 "	1.408 "	44.014 "
5.020 "	2.214 "	44.104 "

Mean, 44.098, ± .027

The following percentages of nickel were found in this salt:

29.107
29.082
29.066
29.082

Mean, 29.084, ± .006

$CoC_2O_4.2H_2O.$

1.6355 grm. gave	.781 grm. CO_2.	47.753 per cent.
1.107 . "	.5295 "	47.832 "
2.309 "	1.101 "	47.683 "
3.007 "	1.435 "	47.722 " .

Mean, 47.7475, ± .0213

The following were the percentages found for cobalt:

32.552
32.619
32.528
32.523

Mean, 32.5555, ± .0149

In a later paper[*] Schneider also gives some results obtained with a nickel oxalate containing but two molecules of water. This gave him 47.605 per cent. of CO_2, and the following percentages of nickel:

31.4115
31.4038

Mean, 31.4076, ± .0026

The conclusion at which Schneider arrived was that the atomic weights of cobalt and nickel are not identical, being about 60 and 58 respectively. The percentages given above will be discussed at the end of this chapter in connection with all the other data relative to the constants in question.

The next chemist to take up the discussion of these atomic weights was Marignac, in 1858.[†] He worked with the chlorides and sulphates

---

[*] Poggend. Annalen, 107, 616.
[†] Arch. des Sci. Phys et Nat. (nouv. serie). 1, 372. 1858.

of nickel and cobalt, using various methods, but publishing few details, as he did not consider the determinations final. The sulphates, taken as anhydrous, were calcined to oxides. From the ratio $NiSO_4 : NiO$, he found $Ni = 58.4$ to 59.0, and from five measurements of the ratio $CoSO_4 : Co$, $Co = 58.64$ to 58.76. If oxygen is taken as 16, these give for the percentages of oxide in sulphate:

CoO in $CoSO_4$.	NiO in $NiSO_4$.
48.267	48.187
48.307	48.387
Mean, 48.287, ± .0135	Mean, 48.287, ± .0675

The chlorides were dried at $100°$, but found to retain water; and in most cases were then either fused in a stream of chlorine or of dry, gaseous hydrochloric acid, or else calcined gently with ammonium chloride. The determinations were then made by titration with a standard solution of silver in nitric acid. Three experiments with an-hydrous $CoCl_2$ gave $Co = 58.72$ to 58.84. Three more with $CoCl_2$ dried at $100°$ gave $Co = 58.84$ to 59.02. Three with anhydrous $NiCl_2$ gave $Ni = 58.80$ to 59.00. If the calculations were made with $Ag = 108$ and $Cl = 35.5$, then these data give as proportional to 100 parts of silver:

$NiCl_2$.	$CoCl_2$.
60.093	60.056
60.185	60.111
	60.111
Mean, 60.139, ± .0310	60.194
	Mean, 60.118, ± .0192

In one more experiment $NiCl_2$ was precipitated with a known quantity of silver. The filtrate was calcined, yielding NiO; hence the ratio $Ag_2 : NiO$, giving $Ni = 59.29$. This experiment needs no farther attention.

In short, according to Marignac, and contrary to Schneider's views, the two atomic weights are approximately the same. Marignac criticises Schneider's earlier paper, holding that the nickel oxalate may have contained some free oxalic acid, and that the cobalt salt was possibly contaminated with carbonate or with basic compounds. In his later papers Schneider rejects these suggestions as unfounded, and in turn criticises Marignac. The purity of anhydrous $NiSO_4$ is not easy to guarantee, and, according to Schneider, the anhydrous chlorides of cobalt and nickel are liable to be contaminated with oxides. This is the case even when the chlorides are heated in chlorine, unless the gas is carefully freed from all traces of air and moisture.

Dumas'* determinations of the two atomic weights were made with the chlorides of nickel and cobalt. The pure metals were dissolved in aqua regia, the solutions were repeatedly evaporated to dryness, and the residual chlorides were ignited in dry hydrochloric acid gas. The last two estimations in the nickel series were made upon $NiCl_2$ formed by heating the spongy metal in pure chlorine. In the third column I give the $NiCl_2$ or $CoCl_2$ equivalent to 100 parts of silver:

.9123 grm.	$NiCl_2$ =	1.515 grm.	Ag.	60.218
2.295	"	3.8115	"	60.212
3.290	"	5.464	"	60.212
1.830	"	3.041	"	60.178
3.001	"	4.987	"	60.176

Mean, 60.1992, ± .0062

2.352 grm.	$CoCl_2$ =	3.9035 grm.	Ag.	60.254
4.210	"	6.990	"	60.229
3.592	"	5.960	"	60.268
2.492	"	4.1405	"	60.186
4.2295	"	7.0255	"	60.202

Mean, 60.2278, ± .011

These results give values for Co and Ni differing by less than a tenth of a unit; here, as elsewhere, the figure for Ni being a trifle the lower. Combining these data with Marignac's, we have—

$$Ag_2 : NiCl_2 :: 100 : x.$$

Marignac.................................. 60.139, ± .0310
Dumas......... .... 60.199, ± .0062

General mean .... 60.194, ± .0061

$$Ag_2 : CoCl_2 :: 100 : x.$$

Marignac.................................. 60.118, ± .0192
Dumas......,..:........................... 60.228, ± .0110

General mean.................... 60.200, ± .0095

In 1863 † the idea that nickel and cobalt have equal atomic weights was strengthened by the researches of Russell. He found that the black oxide of cobalt, by intense heating in an atmosphere of carbon dioxide, became converted into a brown monoxide of constant composition. The ordinary oxide of nickel, on the other hand, was shown to be convertible into a definite monoxide by simple heating over the blast lamp. The pure oxides of the two metals, thus obtained, were reduced by ignition in hydrogen, and their exact composition thus ascertained.

---
*Ann. Chem. Pharm., 113, 25. 1860.
† Journ. Chem. Soc. (2), 1, 51. 1863.

Several samples of each oxide were taken, yielding the following data. The separate samples are indicated by lettering:

Nickel.

	NiO.	Ni.	Per cent. Ni.
A.	2.0820	1.6364	78.597
	2.0956	1.6468	78.584
	2.0148	1.5838	78.608
B.	2.2069	1.7342	78.581
	2.2843	1.7952	78.589
	2.1329	1.6761	78.583
C.	2.2783	1.7911	78.616
	2.1434	1.6845	78.590
	2.4215	1.9030	78.588
D.	2.1859	1.7179	78.590
	2.0088	1.5788	78.594
	2.0839	1.6379	78.597
	2.6560	2.0873	78.588

Mean, 78.593, ± .0018

Cobalt.

	CoO.	Co.	Per cent. Co.
A.	2.1211	1.6670	78.591
	2.0241	1.5907	78.588
	2.1226	1.6673	78.550
	1.9947	1.5678	78.598
	3.0628	2.4078	78.614
B.	2.1167	1.6638	78.603
	1.7717	1.3924	78.591
	1.7852	1.4030	78.591
C.	1.6878	1.3264	78.588
	2.2076	1.7350	78.592
D.	2.6851	2.1104	78.597
	2.1461	1.6868	78.598
E.	3.4038	2.6752	78.595
	2.2778	1.7901	78.589
	2.1837	1.7163	78.596

Mean, 78.592, ± .0023

These percentages are practically identical, and lead to essentially the same mean value for each atomic weight.

In a later paper Russell[*] confirmed the foregoing results by a different process. He dissolved metallic nickel and cobalt in hydrochloric acid and measured the hydrogen evolved. Thus the ratio between the metal and the ultimate standard was fixed without the intervention of any other element. About two-tenths of a gramme of metal, or less, was

---

[*] Journ. Chem. Soc. (2), 7, 494. 1867.

taken in each experiment. The data obtained were as follows; the last column giving the weight of hydrogen, computed from its volume, yielded by 100 parts of cobalt or nickel:

Nickel.

	Wt. Ni.	Vol. H in cc.	Ratio.
A.	.0906	153.62	3.420
	.1017	172.32	3.418
	.1990	337.06	3.416
	.0997	168.93	3.417
	.1891	319.86	3.412
	.1859	314.75	3.415
	.1838	311.25	3.416
B.	.1892	318.75	3.398
	.1806	305.28	3.409
	.2026	333.81	3.404
C.	.1933	325.93	3.401
D.	.1890	319.77	3.412
	.1942	328.15	3.408
	.1781	301.09	3.410

Mean, 3.411, ± .001

Cobalt.

	Wt. Co.	Vol. H in cc.	Ratio.
A.	.1958	321.36	3.395
	.1905	312.95	3.398
	.1946	319.63	3.397
	.2002	328.96	3.398
B.	.1996	328.43	3.403
	.2000	329.55	3.401
	.1721	290.17	3.401
C.	.1877	308.97	3.404
	.1935	318.60	3.405
D.	.1909	314.73	3.410
	.1834	305.40	3.407

Mean, 3.4017, ± .0009

The weight of the hydrogen in these determinations was doubtless computed from Regnault's data concerning the density of that gas. Correcting by the new value for the weight of a litre of hydrogen, .089872 gramme, the ratios become:

For nickel .... 3.4211, ± .0010
For cobalt.. .... 3.4112, ± .0009

Some time after the publication of Russell's first paper, but before the appearance of his second, some other investigations were made known.

Of these the first was by Sommaruga,* whose results, obtained by novel methods, closely confirmed those of Schneider and antagonized those of Dumas, Marignac, and Russell. The atomic weight of nickel Sommaruga deduced from analyses of the nickel potassium sulphate, $K_2Ni(SO_4)_2.6H_2O$, which, dried at 100°, has a perfectly definite composition. In this salt the sulphuric acid was determined in the usual way as barium sulphate, a process to which there are obvious objections. In the third column are given the quantities of the nickel salt proportional to 100 parts of $BaSO_4$:

0.9798 grm. gave	1.0462 grm. $BaSO_4$.		93.653
1.0537	"	1.1251 "	93.654
1.0802	"	1.1535 "	93.645
1.1865	"	1.2669 "	93.654
3.2100	"	3.4277 "	93 649
3.2124	"	3.4303 "	93.648

Mean, 93.6505, ± .001

For cobalt Sommaruga used the purpureocobalt chloride of Gibbs and Genth. This salt, dried at 110°, is anhydrous and stable. Heated hotter, $CoCl_2$ remains. The latter, ignited in hydrogen, yields metallic cobalt. In every experiment the preliminary heating must be carried on cautiously until ammoniacal fumes no longer appear:

.6656 grm. gave	.1588 grm. Co.		23.858 per cent.
1.0918	"	.2600 "	23.814 "
.9058	"	.2160 "	23.846 "
1.5895	"	.3785 "	23.813 "
2.9167	"	.6957 "	23.847 "
1.8390	"	.4378 "	23.806 "
2.5010	"	.5968 "	23.808 "

Mean, 23.827, ± .006

Further along this series will be combined with a similar one by Lee. It may here be said that Sommaruga's paper was quickly followed by a critical essay from Schneider,† endorsing the former's work and objecting to the results of Russell.

In 1867 still another new process for the estimation of these atomic weights was put forward by Winkler,‡ who determined the amount of gold which pure metallic nickel and cobalt could precipitate from a neutral solution of sodio-auric chloride.

In order to obtain pure cobalt Winkler prepared purpureocobalt chloride, which, having been four or five times recrystallized, was ignited in hydrogen. His nickel was repeatedly purified by precipitation with sodium hypochlorite. From material thus obtained pure nickel chloride

---

* Sitzungsb. Wien. Akad., 54, 2 Abth., 50. 1866.
† Poggend. Annalen, 130, 310.
‡ Zeit. Anal. Chem., 6, 18. 1867.

was prepared, which, after sublimation in dry chlorine, was also reduced by hydrogen. One hundred parts of gold are precipitated by the quantities of nickel and cobalt given in the third columns respectively. In the cobalt series I include one experiment by Weselsky, which was published by him in a paper presently to be cited:

.4360 grm. nickel precipitated .9648 grm. gold.   45.191
.4367   "   .9666   "   45.179
.5189   "   1.1457   "   45.291
.6002   "   1.3286   "   45.175

Mean, 45.209, ± .019

.5890 grm. cobalt precipitated 1.3045 grm. gold.   45.151
.3147   "   .6981   "   45.080
.5829   "   1.2913   "   45.141
.5111   "   1.1312   "   45.182
.5821   "   1.2848   "   45.307
.559   "   1.241   "   45.044—Weselsky.

Mean, 45.151, ± .025

Weselsky's paper,* already quoted, relates only to cobalt. He ignited the cobalticyanides of ammonium and of phenylammonium in hydrogen, and from the determinations of cobalt thus made deduced its atomic weight. His results are as follows:

.7575 grm. $(NH_4)_6Co_2Cy_{12}$ gave .166 grm. Co.   21.914 per cent.
.5143   "   .113   "   21.972   "

Mean, 21.943, ± .029

.8529 grm. $(C_8H_9N)_6Co_2Cy_{12}$ gave .1010 grm. Co.   11.842 per cent.
.6112   "   .0723   "   11.829   "
.7140   "   .0850   "   11.905   "
.9420   "   .1120   "   11.890   "

Mean, 11.8665, ± .0124

Next in order is the work done by Lee† in the laboratory of Wolcott Gibbs. Like Weselsky, Lee ignited certain cobalticyanides and also nickelocyanides in hydrogen and determined the residual metal. The double cyanides chosen were those of strychnia and brucia, salts of very high molecular weight, in which the percentages of metal are relatively low. A series of experiments with purpureocobalt chloride was also carried out. In order to avoid admixture of carbon in the metallic residues, the salts were first ignited in air, and then in oxygen. Reduction by hydrogen followed. The salts were in each case covered by a porous septum of earthenware, through which the hydrogen diffused, and which served to prevent the mechanical carrying away of solid particles; fur-

---

* Ber. d. Deutsch. Chem. Gesell., 2, 592. 1868.
† Am. Journ. Sci. and Arts (3), 2, 44. 1871.

NICKEL AND COBALT. 299

thermore, heat was applied from above. The results attained were very satisfactory, and assign to nickel and cobalt atomic weights varying from each other by about a unit; Ni being nearly 58, and Co about 59, when $O = 16$. The exact figures will appear later. The cobalt results agree remarkably well with those of Weselsky. The following are the data obtained:

Brucia nickelocyanide, $Ni_3Cy_{12}(C_{23}H_{26}N_2O_4)_6H_6.10H_2O$.

Salt.	Ni.	Per cent. Ni.
.3966	.0227	5.724
.5638	.0323	5.729
.4000	.0230	5.750
.3131	.01795	5.733
.4412	.0252	5.712
.4346	.0249	5.729

Mean, 5.7295, ± .0034

Strychnia nickelocyanide, $Ni_3Cy_{12}(C_{21}H_{22}N_2O_2)_6H_6.8H_2O$.

Salt.	Ni.	Per cent. Ni.
.5358	.0354	6.607
.5489	.0363	6.613
.3551	.0234	6.589
.4495	.0297	6.607
.2530	.0166	6.561
.1956	.0129	6.595

Mean, 6.595, ± .005

Brucia cobalticyanide, $Co_3Cy_{12}(C_{23}H_{26}N_2O_4)_6H_6.20H_2O$.

Salt.	Co.	Per cent. Co.
.4097	.0154	3.759
.3951	.0147	3.720
.5456	.0204	3.739
.4402	.0165	3.748
.4644	.0174	3.747
.4027	.0151	3.749

Mean, 3.7437, ± .0036

Strychnia cobalticyanide, $Co_3Cy_{12}(C_{21}H_{22}N_2O_2)_6H_6.8H_2O$.

Salt.	Co.	Per cent. Co.
.4255	.0195	4.583
.4025	.0185	4.596
.3733	.0170	4.554
.4535	.0207	4.564
.2753	.0126	4.577
.1429	.0065	4.549

Mean, 4.5705, ± .005

Purpureo-cobalt chloride, $Co_2(NH_3)_{10}Cl_6$.

Salt.	Co.	Per cent. Co.
.9472	.2233	23.575
.8903	.2100	23.587
.6084	.1435	23.586
.6561	.1547	23.579
.6988	.1647	23.569
.7010	.1653	23.581

Mean, 23.5795, ± .0019

The last series may be combined with Sommaruga's, thus:

Sommaruga.......................... 23.817, ± .006
Lee .... 23.5795, ± .0019

General mean................... 23.6045, ± .0018

Baubigny's* determinations of the atomic weight of nickel are limited to two experiments upon the calcination of nickel sulphate, and his data are as follows:

6.2605 grm. NiSO₄ gave 3.9225 NiO.   48.279 per cent.
4.4935   "   2.1695   "   48.281   "

Mean, 48.280

Zimmermann's work, published after his death by Krüss and Alibegoff,† was based, like Russell's, upon the reduction of cobalt and nickel oxides in hydrogen. The materials used were purified with great care, and the results were as follows:

Nickel.

NiO.	Ni.	Per cent. Ni.
6.0041	4.7179	78.578
6.4562	5.0734	78.582
8.5960	6.7552	78.585
4.7206	3.7096	78.583
8.2120	6.4536	78.587
9.1349	7.1787	78.585
10.0156	7.8702	78.579
4.6482	3.6526	78.580
8.9315	7.0184	78.580
10.7144	8.4196	78.582
3.0036	2.3602	78.579

Mean, 78.582, ± .0006

* Compt. Rend., 97, 951. 1883.
† Ann. der Chem., 232, 324. 1886.

Cobalt.

CoO.	Co.	Per cent. Co.
6.3947	5.0284	78.634
6.6763	5.2501	78.638
5.6668	4.4560	78.633
2.9977	2.3573	78.637
8.7446	6.8763	78.635
3.2625	2.5655	78.636
6.3948	5.0282	78.630
8.2156	6.4606	78.638
9.4842	7.4580	78.636
9.9998	7.8630	78.632

Mean, 78.635, ± .0002

Shortly after the discovery of nickel carbonyl, NiC_4O_4, Mond, Langer, and Quincke* made use of it with reference to the atomic weight of nickel. The latter was purified by distillation as nickel carbonyl, then converted into oxide, and that was reduced by hydrogen in the usual way.

NiO.	Ni.	Per cent. Ni.
.2414	.1896	78.542
.3186	.2503	78.562
.3391	.2663	78.531

Mean, 78.545, ± .0061

Schutzenberger's experiments,† published in 1892, were also few in number. First, nickel sulphate, dehydrated at 440°, was calcined to oxide.

3.505 grm. $NiSO_4$ gave 1.690 NiO.   48.217 per cent.
2.6008   "   1.2561 "   48.297   "

Mean, 48.257, ± .027

Second, nickel oxide was reduced in hydrogen, as follows:

1.6865 grm. NiO gave 1.3245 Ni.   78.535 per cent.
1.2527   "   .9838 "   78.533   "

Mean, 78.534

In one experiment with cobalt oxide, 3.491 grm. gave 2.757 Co, or 78.975 per cent. In view of the many determinations of this ratio by other observers, this single estimation may be neglected. The experiments on nickel sulphate, however, should be combined with those of Marignac and Baubigny, giving the latter equal weight with Schutzenberger's, thus:

---

* Journ. Chem. Soc., 57, 753. 1890.
† Compt. Rend., 114, 1149. 1892.

THE ATOMIC WEIGHTS.

Marignac.............................. 48.287, ± .0675
Baubigny.............................. 48.280, ± .027
Schutzenberger....................... 48.257, ± .027

General mean.................... 48.269, ± .018

From this point on the determination of these atomic weights is complicated by the questions raised by Krüss as to the truly elementary character of nickel and cobalt. If that which has been called nickel really contains an admixture of some other hitherto unknown element, then all the determinations made so far are worthless, and the investigations now to be considered bear directly upon that question. First in order comes Remmler's research upon cobalt.* This chemist, asking whether cobalt is homogeneous, prepared cobaltic hydroxide in large quantity, and made a series of successive ammoniacal extracts from it, twenty-five in all. Each extract represented a fraction, from which, by a long series of operations, cobalt monoxide was prepared, and the latter was reduced in hydrogen after the manner of Russell. The actual determinations began with the second fraction, and the data are subjoined, the number of the fraction being given with each experiment:

	CoO.	Co.	Per cent. Co.
2	.09938	.07837	78.859
3	.15021	.11814	78.650
4	.22062	.17360	78.687
5	.39011	.30681	78.647
6	.28820	.22661	78.629
7	.34304	.26968	78.615
8	.43703	.34321	78.532
9	.91477	.71864	78.560
10	.63256	.49661	78.508
11	.32728	.25701	78.529
12	.38042	.29899	78.595
13	.16580	.13027	78.571
14	1.01607	.79873	78.610
15	1.31635	1.03545	78.661
16	.91945	.72315	78.650
17	.53100	.41773	78.668
18	.82381	.64728	78.572
19	.81139	.63754	78.574
20	.76698	.60292	78.610
21	1.13693	.89412	78.643
22	2.00259	1.57495	78.646
23	1.04629	.82185	78.549
24	.48954	.38466	78.576
25	.69152	.54326	78.560

Mean, 78.613, ± .0099

---

* Zeit. Anorg. Chem., 2, 221. Also more fully in an Inaugural Dissertation. Erlangen, 1891.

NICKEL AND COBALT. 303

Considered with reference to the purpose of the investigation, this mean and its probable error have no real significance. But it is very close to the means of other experimenters, and a study of the variations represented by the several fractions seems to indicate fortuity rather than system. Remmler regards his results as indicating lack of homogeneity in his material; but it seems more probable that such differences as exist are due to experimental errors and to impurities acquired in the long process of purification to which each fraction was submitted, rather than to any uncertainty regarding the nature of cobalt itself. For either interpretation the data are inconclusive, and I therefore feel justified in treating the mean like other means, and in combining it finally with them.

From the same point of view—that is, with reference to the supposed heterogeneity of nickel—Krüss and Schmidt * carried out a series of fractionations of the metal by distillation in a stream of carbon monoxide. Nickel oxide, free from obnoxious impurities, was first reduced to metal by heating in hydrogen, after which the current of carbon monoxide was allowed to flow. The latter, carrying its small charge of nickel tetracarbonyl was then passed through a Winkler's absorption apparatus containing pure aqua regia, from which, by evaporation, nickel chloride was obtained, and from that, by reduction in hydrogen, the nickel. Ten such fractions were successively prepared and studied; first, by preparation of NiO and its reduction in hydrogen; and, secondly, in some cases, by the reoxidation of the reduced metal, so as to give a synthetic value for the ratio Ni:O. The data obtained are as follows, the successive fractions being numbered:

Reduction of NiO.

	NiO.	Ni.	Per cent. Ni.
1.	.3722	.2926	78.614
	.7471	.5870	78.571
2.	.7659	.60085	78.450
	.7606	.5961	78.372
	1.0175	.7984	78.467
3.	1.2631	.99065	78.430
	1.2582	.9868	78.429
4.	.5193	.4076	78.490
	.9200	.7215	78.424
5.	.4052	.3179	78.455
	.6518	.5111	78.414
6.	.5623	.4399	78.232
	.5556	.4350	78.294
	.9831	.7724	78.568
7.	.9765	.7646	78.300
	.9639	.7557	78.400

* Zeit. Anorg. Chem., 2, 235. 1892.

8.	.5756	.4538	78.839
	.56765	.4451	78.411
	.5663	.4438	78.368
	.5449	.4272	78.400
9.	.3174	.2491	78.481
	.3148	.2467	78.367
10.	.4976	.3904	78.457
	.4961	.3891	78.432

Mean, 78.444, ± .0166

Oxidation of Ni.

	Ni.	NiO.	Per cent. Ni.
1.	.5870	.7471	78.571
2.	.6011	.7659	78.372
	.5961	.7606	78.359
3.	.7988	1.0175	78.506
	.9913	1.2631	78.482
	.9868	1.2582	78.429
4.	.4093	.5193	78.818
	.7216	.9200	78.435
5.	.3194	.4052	78.825
	.5111	.6518	78.414
6.	.4415	.5623	78.517
	.4350	.5556	78.294
7.	.7752	.9831	78.853
	.7667	.9765	78.515
	.7558	.9639	78.411
8.	.4555	.5756	79.135
	.4456	.56765	78.499
	.44415	.5663	78.430
	.4423	.5642	78.394
9.	.2508	.3174	79.015
	.2467	.3148	78.367
10.	.3918	.4976	78.738
	.3891	.4961	78.432

Mean, 78.557, ±.0319

To these data of Krüss and Schmidt the remarks already made concerning Remmler's work seem also to apply. The variations appear to be fortuitous, and not systematic, although the authors seem to think that they indicate a compositeness in that substance which has been hitherto regarded as elementary nickel. There is doubtless something to be said on both sides of the question; but if Krüss and Schmidt are right, all previous atomic weight determinations for cobalt and nickel are invalidated. In view of all the evidence, therefore, I prefer to regard their varying estimations as affected by accidental errors, and to treat their means like others. On this basis, their work combines with previous work as follows, Schutzenberger's measurements of the ratio NiO : Ni being assigned equal weight with those of Mond, Langer, and Quincke:

Russell..................................... 78.593, ± .0018
Zimmermann................................. 78.582, ± .0006
Mond, Langer, and Quincke .... 78.545, ± .0061
Schutzenberger............................. 78.534, ± .0061
Krüss and Schmidt, reduction series .... 78.444, ± .0166
Krüss and Schmidt, oxidation series .... 78.557, ± .0319

General mean.................... 78.570, ± .0006

In 1889 Winkler * published a short paper concerning the gold method for determining the atomic weights in question, but gave in it no actual measurements. In 1893 † he returned to the problem with a new line of attack, and at the same time he takes occasion to criticise Krüss and Schmidt somewhat severely. He utterly rejects the notion that either nickel or cobalt contain any hitherto unknown element, and ascribes the peculiar results obtained by Krüss and Schmidt to impurities derived from the glass apparatus used in their experiments. For his own part he now works with pure nickel and cobalt precipitated electrolytically upon platinum, and avoids the use of glass or porcelain vessels so far as possible. With material thus obtained he operates by two distinct but closely related methods, both starting with the metal, nickel or cobalt, converting it next into neutral chloride, and then measuring the chloride gravimetrically in one process, volumetrically in the other.

After precipitation in a platinum dish, the nickel or cobalt is washed with water, rinsed with alcohol and ether, and then weighed. It is next dissolved in pure hydrochloric acid, properly diluted, and by evaporation to dryness and long heating to 150° converted into anhydrous chloride. The nickel chloride thus obtained dissolves perfectly in water, but the cobalt salt always gave a slight residue in which the metal was electrolytically determined and allowed for. In the redissolved chloride, by precipitation with silver nitrate, silver chloride is obtained, giving a direct ratio between that compound and the nickel or cobalt originally taken. The gravimetric data are as follows, with the metal equivalent to 100 parts of silver chloride given in a final column :

Nickel.

$Ni.$	$AgCl.$	Ratio.
.3011	1.4621	20.594
.2242	1.0081	20.605
.5166	2.5108	20.570
.4879	2.3679	20.605
.3827	1.8577	20.601
.3603	1.7517	20.568

Mean, 20.590, ± .0049

---

* Ber. Deutsch. Chem. Gesell., 22, 891. 1889.
† Zeit. Anorg. Chem., 4, 10. 1893.

Cobalt.

Co.	AgCl.	Ratio.
.3458	1.6596	20.836
.3776	1.8105	20.856
.4493	2.1521	20.877
.4488	2.1520	20.855
.2856	1.3683	20.873
.2648	1.2768	20.886

Mean, 20.864, ± .0050

In the volumetric determinations the neutral chloride, prepared as before, was decomposed by means of a slight excess of potassium carbonate, and in the potassium chloride solution, after removal of the nickel or cobalt, the chlorine was measured by titration by Volhard's method with a standard solution of silver. The amount of silver thus used was comparable with the metal taken.

Nickel.

Ni.	Ag.	Ratio.
.1812	.6621260	27.366
.1662	.6079206	27.339
.2129	.7775252	27.382
.2232	.8162108	27.346
.5082	1.8556645	27.386
.1453	.5315040	27.338

Mean, 27.359, ± .0059

Cobalt.

Co.	Ag.	Ratio.
.177804	.6418284	27.702
.263538	.9514642	27.699
.245124	.8855780	27.679
.190476	.6866321	27.741
.266706	.9629146	27.696
.263538	.9503558	27.731

Mean, 27.708, ± .0064

In view of the possibility that the cobalt chloride of the foregoing experiments might contain traces of basic salt, Winkler, in a supplementary investigation,* checked them by another process. To the electrolytic cobalt, in a platinum dish, he added a quantity of neutral silver sulphate and then water. The cobalt gradually went into solution, and metallic silver was precipitated. The weights were as follows:

Co.	Ag.
.2549	.9187
.4069	1.4691

---

* Zeit. Anorg. Chem., 4, 462. 1893.

On examination of the silver it was found that traces of cobalt were retained—less than 0.5 mg. in the first determination and less than 0.2 mg. in the second. Taking these amounts as corrections, the two experiments give for the ratios $Ag_2 : Co :: 100 : x$ the subjoined values:

27.706
27.687

These figures confirm those previously found, and as they fall within the limits of the preceding series, they may fairly be included in it, when all eight values give a mean of 27.705, ± .0050.

Still another method, radically different from all of the foregoing processes, was adopted by Winkler in 1894.* The metals were thrown down electrolytically upon platinum, and so weighed. Then they were treated with a known excess of a decinormal solution of iodine in potassium iodide, which redissolved them as iodides. The excess of free iodine was then determined by titration with sodium thiosulphate, and in that way the direct ratio between metal and haloid was ascertained. The results were as follows, with the metal proportional to 100 parts of iodine given in the third column:

Cobalt.

	Wt. Co.	Wt. I.	Ratio.
First series....	.4999	2.128837	23.482
	.5084	2.166750	23.463
	.5290	2.254335	23.466
	.6822	2.908399	23.456
	.6715	2.861617	23.466
Second series..	.5185	2.209694	23.465
	.5267	2.246037	23.450
	.5319	2.268736	23.445

Mean, 23.462, ± .0027

Nickel.

	Wt. Ni.	Wt. I.	Ratio.
First series,...	.5144	2.217494	23.251
	.4983	2.148502	23.246
	.5265	2.268742	23.260
	.6889	2.970709	23.243
	.6876	2.965918	23.237
Second series..	.5120	2.205627	23.267
	.5200	2.240107	23.267
	.5246	2.259925	23.267

Mean, 23.255, ± .0091

In these experiments, as well as in some previous series, a possible source of error is to be considered in the occlusion of hydrogen by the

* Zeitsch. Anorg. Chem., 8, 1. , 1894.

metals. Accordingly, in a supplementary paper, Winkler* gives the results of some check experiments made with iron, which, however, was not absolutely pure. The conclusion is that the error, if existent, must be very small.

In 1895 Hempel and Thiele's work on cobalt appeared.† First, cobalt oxide, prepared from carefully purified materials, was reduced in hydrogen. The weights of metal and oxygen are subjoined, with the percentage of cobalt in the oxide deduced from them:

Co.	O.	Percentage.
.90068	.24429	78.664
.79159	.21445	78.686
1.31558	.35716	78.648
		Mean, 78.666, ± .0074

This mean combines with former means as follows:

Russell.	78.592, ± .0023
Zimmermann.	78.635, ± .0002
Remmler.	78.613, ± .0099
Hempel and Thiele.	78.666, ± .0074
General mean.	78.633, ± .0002

In their next series of experiments, excluding a rejected series, Hempel and Thiele weighed cobalt, converted it into anhydrous chloride, and noted the gain in weight. In four of the experiments the chloride was afterwards dissolved, precipitated with silver nitrate, and then the silver chloride was weighed. The data are as follows:

Co.	Cl Taken Up.	AgCl.
.7010	.8453	......
.3138	.3793	......
.2949	.3562	1.4340
.4691	.5657	2.2812
.5818	.7026	2.8303
.5763	.6947	......
.5096	.6142	2.4813

From these weights we get two ratios, thus:

$Cl_2 : Co :: 100 : x.$	$2AgCl : Co :: 100 : x.$
82.929	20.565
82.731	20.564
82.791	20.556
82.924	20.538
82.807	
82.957	Mean, 20.556, ± .0043
82.970	
Mean, 82.873, ± .0241	

---

* Zeitsch. Anorg. Chem., 8, 291. 1895.
† Zeitsch. Anorg. Chem., 11, 73.

NICKEL AND COBALT. 309

The second of these ratios was also studied by Winkler, and the two series combine as follows:

Winkler.................................. 20.864, ± .0050
Hempel and Thiele....................... 20.556, ± .0043

General mean..................... 20.687, ± .0033

Hempel and Thiele apply to it a correction for silver chloride retained in solution, but its amount is small and not altogether certain. For present purposes the correction may be neglected.

For the atomic weight of nickel we now have ratios as follows:

(1.) Per cent. of Ni in $NiC_2O_4.3H_2O$, 29.084, ± .006
(2.) Per cent. of CO_2 from $NiC_2O_4.2H_2O$, 44.098, ± .027
(3.) Per cent. of Ni in $NiC_2O_4.2H_2O$, 31.408, ± .0026
(4.) Per cent. of CO_2 from $NiC_2O_4.2H_2O$, 47.605, ± .053
(5.) Per cent. of Ni in brucia nickelocyanide, 5.7295, ± .0034
(6.) Per cent. of Ni in strychnia nickelocyanide, 6.595, ± .005
(7.) Per cent. of NiO in $NiSO_4$, 48.269, ± .018
(8.) Per cent. of Ni in NiO, 78.570, ± .0006
(9.) Ag_2 : $NiCl_2$: : 100 : 60.194, ± .0061
(10.) 2AgCl : Ni : : 100 : 20.590, ± .0049
(11.) Ag_2 : Ni : : 100 : 27.359, ± .0059
(12.) Au_2 : Ni_3 : : 100 : 45.209, ± .019
(13.) $BaSO_4$: $K_2Ni(SO_4)_2.6H_2O$: : 100 : 93.6505, ± .001
(14.) Ni : H_2 : : 100 : 3.4211, ± .001
(15.) I_2 : Ni : : 100 : 23.255, ± .0091

To the reduction of these ratios the following atomic and molecular weights are applicable:

O = 15.879, ± .0003   I = 125.888, ± .0069
C = 11.920, ± .0004   K = 38.817, ± .0051
N = 13.935, ± .0021   Ba = 136.392, ± .0086
S = 31.828, ± .0035   Au = 195.743, ± .0049
Ag = 107.108, ± .0031   AgCl = 142.287, ± .0037
Cl = 35.179, ± .0048

Since the proportion of water in the oxalates is not an absolutely certain quantity, the data concerning them can be best handled by employing the ratios between carbon dioxide and the metal. Accordingly, ratios (1) and (2) give a single value for Ni, and ratios (3) and (4) another. In all, there are thirteen values for the atomic weight in question:

From (1) and (2)................. Ni = 57.614, ± .0372
From (5)......................... " = 57.625, ± .0343
From (3) and (4)................. " = 57.635, ± .0644
From (6)......................... " = 57.687, ± .0439
From (8)......................... " = 58.218, ± .0020
From (7)......................... " = 58.268, ± .0428
From (13)........................ " = 58.448, ± .0206

310   THE ATOMIC WEIGHTS.

From (14).................... Ni = 58.456, ± .0316
From (15)....................... " = 58.551, ± .0231
From (9) .... " = 58.587, ± .0179
From (10)....................... " = 58.594, ± .0141
From (11)... .... " = 58.607, ± .0128
From (12)....................... " = 58.994, ± .0248

General mean............... Ni = 58.243, ± .0019

If O = 16, this becomes Ni = 58.687.

It is quite evident here that ratio (8), which includes the marvelously concordant determinations of Zimmermann, far outweighs all the other data. Whether so excessive a weight can justifiably be assigned to one set of measurements is questionable, but the general mean thus reached is not far from midway between the highest and lowest of the values, and hence it may fairly be entitled to provisional acceptance. No one of the individual values rests upon absolutely conclusive evidence, so that no one can be arbitrarily chosen to the exclusion of the others. Further investigation is evidently necessary.

For cobalt we have sixteen ratios, as follows:

(1.) Per cent. of Co in $CoC_2O_4.2H_2O$, 32.5555, ± .0149
(2.) Per cent. of CO_2 from $CoC_2O_4.2H_2O$, 47.7475, ± .0213
(3.) Per cent. of Co in CoO, 78.633, ± .0002
(4.) Per cent. of Co in purpureocobalt chloride, 23.6045, ± .0018
(5.) Per cent. of Co in phenylammonium cobalticyanide, 11.8665, ± .0124
(6.) Per cent. of Co in ammonium cobalticyanide, 21.943, ± .029
(7.) Per cent. of Co in brucia cobalticyanide, 3.7437, ± .0036
(8.) Per cent. of Co in strychnia cobalticyanide, 4.5705, ± .005
(9.) Per cent. of CoO in $CoSO_4$, 48.287, ± .0135
(10.) Ag_2 : $CoCl_2$: : 100 : 60.200, ± .0095
(11.) 2AgCl : Co : : 100 : 20.687, ± .0033
(12.) Ag_2 : Co : : 100 : 27.705, ± .0050
(13.) Au_2 : Co_3 : : 100 : 45.151, ± .025
(14.) Co : H_2 : : 100 : 3.4110, ± .0009
(15.) I_2 : Co : : 100 : 23.462, ± .0027
(16.) Cl_2 : Co : : 100 : 82.873, ± .0241

From these, using the atomic weights already cited under nickel, and combining ratios (1) and (2), we get—

From (16)....................... Co = 58.308, ± .0187
From (9)........................ " = 58.321, ± .0288
From (3)........................ " = 58.437, ± .0014
From (10)....................... " = 58.600, ± .0228
From (14)....................... " = 58.630, ± .0286
From (5)........................ " = 58.639, ± .0619
From (8)........................ " = 58.696, ± .0642
From (6)........................ " = 58.736, ± .0808
From (4)........................ " = 58.774, ± .0071
From (7)........................ " = 58.791, ± .0566

From (11).................... Co = 58.870, ± .0094
From (13).................... " = 58.920, ± .0327
From (15).................... " = 59.072, ± .0075
From (12).................... " = 59.349, ± .0108
From (1) and (2)............. " = 59.562, ± .0382

General mean................ Co = 58.487, ± .0013

If $O = 16$, this becomes $Co = 58.932$.
Here again the oxide ratio, because of Zimmermann's work, receives excessive and undue weight. The arithmetical mean of the fifteen values is $Co = 58.781$. Between this and the weighted general mean the truth probably lies, but the evidence is incomplete, and more determinations are needed.

RUTHENIUM.

The atomic weight of this metal has been determined by Claus and by Joly. Although Claus* employed several methods, we need only consider his analyses of potassium rutheniochloride, K_2RuCl_6. The salt was dried by heating to 200° in chlorine gas, but even then retained a trace of water. The percentage results of the analyses are as follows:

Ru.	*2KCl.*	*Cl₃.*
28.96	40.80	30.24
28.48	41.39	30.22
28.91	41.08	30.04
Mean, 28.78	41.09	30.17

Reckoning directly from the percentages, we get the following discordant values for Ru :

From percentage of metal................ Ru = 102.451
From percentage of KCl................. " = 106.778
From percentage of Cl₃.................. " = 96.269

These results are obviously of little importance, especially since the best of them is not in accord with the position of ruthenium in the periodic system. The work of Joly is more satisfactory.† Several compounds of ruthenium were analyzed by reduction in a stream of hydrogen with the following results :

---
* Journ. für Prakt. Chem., 34, 435. 1845.
† Compt. Rend., 108, 946.

First, reduction of RuO_2:

RuO_2.	Ru.	Per cent. Ru.
2.1387	1.6267	76.060
2.5846	1.9658	76.058
2.3682	1.8016	76.075
2.8849	2.1939	76.046

Mean, 76.060, ± .0040

Second, reduction of the salt $RuCl_3.NO.H_2O$:

Per cent. Ru.
39.78
39.66

Mean, 39.72, ±.0405

Third, reduction of $RuCl_3.NO.2NH_4Cl$:

Per cent. Ru.
29.44
29.47

Mean, 29.455, ± .0101

Computing with $O = 15.879, \pm .0003$; $N = 13.935, \pm .0021$, and $Cl = 35.179, \pm .0048$, these data give three values for ruthenium, as follows:

1. From RuO_2 .... $Ru = 100.922, \pm .0178$
2. From $RuCl_3.NO.H_2O$ .... " $= 100.967, \pm .1102$
3. From $RuCl_3.NO.2AmCl$ .... " $= 100.868, \pm .0387$

General mean .... $Ru = 100.913, \pm .0160$

If $O = 16$, $Ru = 101.682$.

RHODIUM.

Berzelius* determined the atomic weight of this metal by the analysis of sodium and potassium rhodiochlorides, Na_3RhCl_6, and K_2RhCl_5. The latter salt was dried by heating in chlorine. The compounds were analyzed by reduction in hydrogen, after the usual manner. Reduced to percentages, the analyses are as follows:

$$In\ Na_3RhCl_6.$$

Rh.	3NaCl.	Cl_3.
26.959	45.853	27.189
27.229	45.301	27.470
......	......	27.616
Mean, 27.094	Mean, 45.577	Mean, 27.425

$$In\ K_2RhCl_5.$$

Rh.	2KCl.	Cl_3.
28.989	41.450	29.561

From the analyses of the sodium salt we get the following values for Rh:

From per cent. of metal....................	Rh = 104.191
From per cent. of NaCl....................	" = 102.449
From per cent. of Cl_3...........................	" = 105.103
From ratio between Cl_3 and Rh............,....	" = 104.263
From ratio between NaCl and Rh..........	" = 103.544

These are discordant figures; but the last one fits in fairly well with the values calculated from the potassium compound, which are as follows:

From per cent. of metal....................	Rh = 103.499
From per cent. of KCl.....................	" = 103.648
From per cent. of Cl_3.......................	" = 103.485
From Rh : Cl_3 ratio.......................	" = 103.495
From Rh : KCl ratio........,...........	" = 103.540
Mean...............................	Rh = 103.533

If O = 16, this becomes Rh = 104.323.

Jörgensen's determination,† so far as I can ascertain, was published only as a preliminary note, to the effect that the atomic weight of rhodium is 103, nearly. No details are given.

---

* Poggend. Annalen, 13, 435. 1828.
† Journ. für Prakt. Chem. (2), 27, 486.

Seubert and Kobbe* determine the atomic weight by igniting rhodium pentamine chloride in hydrogen, and weighing the residual metal. Their results are given below:

$Rh(NH_3)_5Cl_3$	Rh.	Per cent. Rh.
1.8585	.6496	34.953
1.5560	.5435	34.929
1.5202	.5310	34.930
2.0111	.7031	34.961
1.8674	.6528	34.958
2.4347	.8513	34.965
2.3849	.8338	34.962
2.5393	.8881	34.974
1.4080	.4920	34.943
1.4654	.5123	34.960

Mean, 34.954, ± .0032

In the sixth experiment the ammonium chloride formed was collected in a bulb tube, and estimated by weighing as silver chloride. 3.5531 grms. of AgCl were obtained.

Computing with N = 13.935, ± .0021; Cl = 35.179, ± .0048, and AgCl = 142.287, ± .0037, we have—

From per cent. of metal............ Rh = 102.215, ± .0143
From AgCl ratio................... " = 102.287, ± .0324

General mean .... Rh = 102.227, ± .0131

If O = 16, Rh = 103.006.

In the second of these values the probable error given is only that due to the antecedent atomic weights of N, Cl, and AgCl. It is therefore lower than it should be. The two values, however, are fairly in agreement, and the result is satisfactory.

---

*Ann. d. Chem., 260, 318. 1890.

PALLADIUM.

The first work upon the atomic weight of palladium seems to have been done by Berzelius. In an early paper* he states that 100 parts of the metal united with 28.15 of sulphur. Hence $Pd = 113.06$, a result which is clearly of no present value.

In a later paper † Berzelius published two analyses of potassium palladiochloride, K_2PdCl_4. The salt was decomposed by ignition in hydrogen, as was the case with the double chlorides of potassium with platinum, osmium, and iridium. Reducing his results to percentages, we get the following composition for the substance in question:

$Pd.$	$2KCl.$	$Cl_2.$
32.726	46.044	21.229
32.655	45.741	21.604
Mean, 32.690	Mean, 45.892	Mean, 21.416

From these percentages, calculating directly, very discordant results are obtained:

From percentage of metal .... $Pd = 106.53$
From percentage of KCl........................ " $= 104.13$
From percentage of Cl_2 (loss)................ " $= 110.20$

Obviously, the only way to get satisfactory figures is to calculate from the ratio between the Pd and 2KCl, eliminating thus the influence of water in the salt. The two experiments give, as proportional to 100 parts of KCl, the following of Pd:

71.075
71.391

Mean, 71.233, ± .1066

Hence $Pd = 105.419$.

In 1847 Quintus Icilius ‡ published a determination, which need be given only for the sake of completeness. He ignited potassium palladiochloride in hydrogen, and found the following amounts of residue. His weights are here recalculated into percentages:

64.708
64.965
64.781

Mean, 64.818

From this mean, $Pd = 111.258$. This result has no present value.

---

* Poggend. Annalen, 8, 177. 1826.
† Poggend. Annalen, 13, 454. 1828.
‡ "Die Atomgewichte vom Pd, K, Cl, Ag, C, und H, nach der Methode der kleinsten Quadrate berechnet." Inaug. Diss. Göttingen, 1847. Contains no other original analyses.

In 1889 Keiser's first determinations of this constant appeared.* Finding the potassium palladiochloride to contain "water of decrepitation," he abandoned its use, and resorted to palladiammonium chloride, $Pd(NH_3Cl)_2$, as the most available compound for his purpose. This salt, heated in hydrogen, yields spongy palladium, which was allowed to cool in a current of dry air, in order to avoid gaseous occlusions. The salt itself was dried, previous to analysis, first over sulphuric acid, and then in an air bath at a temperature from 120° to 130°. Two series of experiments were made, the second series starting out from palladium produced by the first series. The data are as follows:

First Series.

$Pd(NH_3Cl)_2$.	Pd.	Per cent. Pd.
.83260	.41965	50.402
1.72635	.86992	50.391
1.40280	.70670	50.378
1.57940	.79562	50.375
1.89895	.95650	50.370
1.48065	.74570	50.363
1.56015	.78585	50.370
1.82658	.92003	50.369
2.40125	1.20970	50.378
1.10400	.55629	50.389
.93310	.47010	50.380

Mean, 50.379, ± .0008

Reduced to vacuum this becomes 50.360.

Second Series.

$Pd(NH_3Cl)_2$.	Pd.	Per cent. Pd.
2.61841	1.31900	50.374
2.23420	1.12561	50.381
1.73553	.87445	50.385
1.69160	.85210	50.372
1.72403	.86825	50.362
1.12222	.56535	50.378
1.17457	.59200	50.401
2.42760	1.22280	50.371

Mean, 50.378, ± .0028
Reduced to vacuum, 50.359

The reductions to vacuum are neglected by Keiser himself, but are here added in order to secure uniformity with later results by the same author. The mean of both series, thus corrected, gives Pd = 105.74.

Bailey and Lamb † made experiments upon several compounds of palladium, but finally settled upon palladiammonium chloride, like Keiser.

---
* Am. Chem. Journ., 11, 398. 1889.
† Journ. Chem. Soc., 61, 745. 1892.

PALLADIUM.   317

Two preliminary experiments, however, with potassium palladiochloride are given, in which the salt was reduced in hydrogen, and both Pd and KCl were weighed. The data are as follows, with the ratio (calculated as with Berzelius' experiments) given in a third column:

$2KCl.$	$Pd.$	Ratio.
1.49767	1.05627	70.528
.90484	.63738	70.441

Mean, 70.485, ± .0290

Hence Pd = 104.312.

The palladiammonium chloride was studied by two methods. First, weighed quantities of the salt were reduced in hydrogen, the ammonium chloride so formed was collected in an absorption apparatus, and then precipitated with silver nitrate. The weights found were as follows, with the $Pd(NH_4Cl)_2$ proportional to 100 parts of silver chloride given in the third column:

$Pd(NH_4Cl)_2.$	$AgCl.$	Ratio.
1.24276	1.682249	73.879
1.08722	1.468448	74.040
1.47666	2.000164	73.828
1.34887	1.837957	73.390
1.74569	2.362320	73.898

Mean, 73.807, ± .0742

Hence Pd = 105.808. Bailey and Lamb regard this as too high, and suspect loss of NH_4Cl during the operation.

The second series of data resemble Keiser's. The salt was reduced in hydrogen, and the spongy palladium was weighed in a Sprengel vacuum. The data are as follows:

$Pd(NH_4Cl)_2.$	$Pd.$	Per cent. Pd.
A. { 1.890597	.947995	50.143
{ 1.874175	.940271	50.170
⎧ 1.307076	.654687	50.088
B. ⎨ 1.340045	.633207	50.238
⎨ 1.905536	.955950	50.167
⎩ 1.685582	.846472	50.218
⎧ 1.691028	.849120	50.213
C. ⎨ 2.112530	1.059690	50.162
⎨ 2.110653	1.057910	50.122
⎩ 1.969100	.988155	50.184

Mean, 50.171, ± .0099

Hence Pd = 104.943. Bailey and Lamb's weighings are all reduced to a vacuum.

Keller and Smith,* reviewing Keiser's work, find that palladiammonium chloride, prepared as Keiser prepared it, may retain traces of foreign metals, and especially of copper. Accordingly, they prepared a quantity of the salt, after a thorough and elaborate process of purification, dried it with extreme care, and then determined the palladium by electrolysis in silver-coated platinum dishes. The precipitated palladium was dried under varying conditions, concerning which the original memoir must be consulted, and was proved to be free from occluded hydrogen. By this method two sets of experiments were made to determine the atomic weight of palladium; but for present purposes the two may fairly be treated as one. The data obtained are as follows, but the weights do not appear to have been reduced to a vacuum:

$Pd(NH_3Cl)_2$.	Pd.	Per cent. Pd.
A. 1.29960	.65630	50.504
1.05430	.53253	50.510
1.92945	.97455	50.509
B. 1.94722	.98343	50.504
1.08649	.54870	50.502
1.28423	.64858	50.503
1.68275	.85010	50.519
1.69113	.85431	50.517
1.80805	.91310	50.502

Mean, 50.508, ± .0014

Hence Pd = 106.368, a result notably higher than Keiser's.

Keller and Smith account for the difference between their determinations and Keiser's partly by the assumption that the materials used by the latter were not pure, and partly by considerations based on the process. In order to clarify the latter part of the question they made three sets of experiments by Keiser's method, slightly varying the conditions. First, the chloride was not pulverized before ignition, and slight decrepitation took place, while dark stains of palladium appeared in the reduction tube, indicating loss by volatilization. Secondly, the chloride was prepared from crude palladium exactly as described by Keiser, but was pulverized before reduction. No decrepitation ensued, but traces of palladium were volatilized. The third series, also on finely pulverized material, was like the second; but the palladiammonium chloride was purified by Keller and Smith's process. The three series, here treated as one, are as follows:

	$Pd(NH_3Cl)_2$.	Pd.	Per cent. Pd.
First series....	.62955	.31743	50.422
	.77270	.38942	50.397
	.83252	.41918	50.350
	.99055	.49895	50.371

*Amer. Chem. Journ., 14, 423. 1892.

PALLADIUM. 319

	$Pd(NH_3Cl)_2$.	Pd.	Per cent. Pd.
Second series..	1.02175	.51468	50.372
	1.10325	.55590	50.388
	.66690	.33590	50.367
	.86840	.43733	50.360
	1.41430	.71255	50.382
	1.15234	.58050	50.376
Third series...	.96229	.48502	50.403
	.97804	.49294	50.401
	.94253	.47517	50.414
	.86090	.43405	50.430

Mean, 50.388, ± .0043

The three series seem to be fairly in agreement between themselves, and with Keiser's work, but diverge seriously from the electrolytic data.

Keller and Smith also attempted to determine the atomic weight of palladium by heating the palladiammonium chloride in sulphuretted hydrogen, and so converting it into the sulphide, PdS. These data were obtained:

$Pd(NH_3Cl)_2$.	PdS.	Per cent. CdS.
.71699	.47066	65.644
1.31688	.86445	65.659

Mean, 65.651, ± .0051

Hence $Pd = 106.55$. This result, however, is affected by the work of Petrenko-Kritschenko,* who has shown the existence of the sulphide PdS to be uncertain.

Joly and Leidié,† in their determinations of this atomic weight, returned to the potassium palladiochloride, K_2PdCl_4. In their first series of experiments the salt was dried in vacuo at ordinary temperatures. It was then electrolyzed in a solution acidulated with hydrochloric acid, both the deposited palladium and the potassium chloride being weighed. The palladium was dried, ignited in a stream of hydrogen, and cooled in an atmosphere of carbon dioxide. The results were as follows, with the column added by me giving the Pd equivalent to 100 parts of KCl:

K_2PdCl_4.	Pd.	2KCl.	Ratio.
1.0255	.3919	.5520	70.996
1.2178	.3937	.5551	70.924
1.2518	.4048	.5687	71.016

Mean, 70.979, ± .0188

This series was rejected by the authors, because the salt was found to contain water—in one case 0.23 per cent. This error, however, should

---
* Zeit. Anorg. Chem., 4, 251. 1893.
† Compt. Rend., 116, 147. 1893.

not invalidate the Pd : KCl ratio. In a second series the palladiochloride was dried in vacuo at 100°, giving the following data:

K_2PdCl_4.	Pd.	$2KCl$.	Ratio.
1.3635	.4422	.6186	71.484
3.0628	.9944	1.3929	71.391
1.4845	.4816	.6782	71.011
1.7995	.5838	.8206	71.143

Mean, 71.257, ± .0736

These experiments seem to be less concordant than the preceding set. It must be noted, however, that the authors reject the KCl determinations and compute directly from the ratio between the salt and the metal. But the ratio here chosen agrees best with the determinations made by other observers, giving for this series the mean value Pd = 105.455, and is, moreover, uniform with the data given by Berzelius and by Bailey and Lamb.

Joly and Leidié also give two experiments made by reducing the K_2PdCl_4 in hydrogen, with the subjoined results:

K_2PdCl_4.	Pd.	$2KCl$.	Ratio.
2.4481	.7949	1.1168	71.177
1.8250	.5930	.8360	70.933

Mean, 71.055, ± .0823

Combining these data with previous series, we have—

Berzelius	71.233,	± .1066
Bailey and Lamb	70.485,	± .0290
Joly and Leidié, first	70.979,	± .0188
Joly and Leidié, second	71.257,	± .0736
Joly and Leidié, third	71.055,	± .0823
General mean	70.865,	± .0150

In view of the discordance among the determinations hitherto cited and because of the criticisms made by Keller and Smith, Keiser, jointly with Miss Mary B. Breed,* repeated his former work, with some variations and added precautions to ensure accuracy. His general method was the same as before, namely, the reduction of palladiammonium chloride by a stream of hydrogen. First, palladium was purified by distillation as $PdCl_2$ at low red heat in a current of chlorine. From this chloride the palladiammonium salt was then prepared. Upon heating the compound gently in a stream of hydrogen, decomposition ensued absolutely without decrepitation or loss of palladium by volatilization. Neither source of error existed. The results obtained were these:

---

*Am. Chem. Journ., 16, 20. 1894.

PALLADIUM. 321

$Pd(NH_3Cl)_2$.	Pd.	Per cent. Pd.
1.60842	.80997	50.358
2.08295	1.04920	50.371
2.02440	1.01975	50.373
2.54810	1.28360	50.375
1.75505	.88410	50.375

Mean, 50.370, ± .0023
Reduced to vacuum, 50.351

In a second series of experiments, palladium was purified as in the earlier investigation, but with special care to eliminate rhodium, iron, copper, gold, mercury, etc. The palladiammonium salt prepared from this material gave as follows:

$Pd(NH_3Cl)_2$.	Pd.	Per cent. Pd.
1.50275	.75685	50.364
1.23672	.62286	50.365
1.34470	.67739	50.375
1.49059	.75095	50.379

Mean, 50.371, ± .0026
Reduced to vacuum, 50.352

Here, again, no loss from decrepitation or volatilization occurred, although evidence of such loss was carefully sought for. The data thus obtained may now be combined with the previous series, thus:

Keiser, first series........................ 50.360, ± .0008
Keiser, second series..................... 50.359, ± .0028
Bailey and Lamb.......................... 50.171, ± .0099
Keller and Smith, electrolytic .... 50.508, ± .0014
Keller and Smith, hydrogen series .... 50.388, ± .0043
Keiser and Breed, first series.............. 50.351, ± .0023
Keiser and Breed, second series........... 50.352, ± .0026

General mean..................... 50.388, ± .00062

For palladium, ignoring the work of Quintus Icilius, the subjoined ratios are now available:

(1.) $2KCl : Pd :: 100 : 70.865, ± .0150$
(2.) Per cent. Pd in $Pd(NH_3Cl)_2$, 50.388, ± .00062
(3.) $2AgCl : Pd(NH_3Cl)_2 :: 100 : 73.807, ± .0742$
(4.) $Pd(NH_3Cl)_2 : PdS :: 100 : 65.651, ± .0051$

The antecedent data are—

$Cl = 35.179, ± .0048$ $S = 31.828, ± .0015$
$K = 38.817, ± .0051$ $AgCl = 142.287, ± .0037$
$N = 13.935, ± .0021$

Hence, for the atomic weight of palladium, we have—

From (1).................... Pd = 104.874, ± .0243
From (2).................... " = 105.858, ± .0200
From (3).................... " = 105.808, ± .2117
From (4).................... " = 106.550, ± .0491

General mean...........   Pd = 105.556, ± .0147

With O = 16, Pd = 106.364.

Taking the values separately, the second is probably the best; but in view of the work done by Bailey and Lamb on one side, and by Keller and Smith on the other, it cannot be accepted unreservedly. Until the cause of variation in the results is clearly determined, it is better to take the general mean of all the data, as given above.

OSMIUM.

The atomic weight of this metal has been determined by Berzelius, by Fremy, and by Seubert.

Berzelius* analyzed potassium osmichloride, igniting it in hydrogen like the corresponding platinum salt. 1.3165 grammes lost .3805 of chlorine, and the residue consisted of .401 grm. of potassium chloride, with .535 grm. of osmium. Calculating only from the ratio between the Os and the KCl, the data give Os = 197.523.

Fremy's determination † is based upon the composition of osmium tetroxide. No details as to weighings or methods are given; barely the final result is stated. This, if O = 16, is Os = 199.648.

When the periodic law came into general acceptance, it became clearly evident that both of the foregoing values for osmium must be several units too high. A redetermination was therefore undertaken by Seubert,‡ who adopted methods based upon that of Berzelius. First, ammonium osmichloride was reduced by heating in a stream of hydrogen. The residual osmium was weighed, and the ammonium chloride and hydrochloric acid given off were collected in a suitable apparatus, so that the total chlorine could be estimated as silver chloride. The weights were as follows:

Am_2OsCl_6.	Os.	$6AgCl$.
1.8403	7996	3.5897
2.0764	.9029	4.0460
2.1501	.9344	4.1950
2.1345	.9275	4.1614

* Poggend. Annalen, 13, 530. 1828.
† Compt. Rend., 19, 468. Journ. für Prakt. Chem., 31, 410. 1844.
‡ Berichte Deutsch. Chem. Gesell., 21, 1839. 1888.

OSMIUM.   323

Hence we have for the percentage of osmium and for the osmichloride proportional to 100 parts of AgCl—

Per cent. Os.	AgCl : Salt.
43.446	51.266
43.484	51.320
43.458	51.254
43.453	51.293
	Mean, 51.283, ± .0099

In a later paper* two more reductions are given, in which only osmium was estimated.

Salt.	Os.	Per cent. Os.
2.6687	1.1597	43.456
2.6937	1.1706	43.457

These determinations, included with the previous four as one series, give a mean percentage of Os in Am_2OsCl_6 of 43.459, ± .0036.

Secondly, potassium osmichloride was treated in the same way, but the residue weighed consisted of Os + 2KCl. From this the potassium chloride was dissolved out, recovered by evaporating the solution, and weighed separately. The volatile portion, 4HCl, was also measured by precipitation as silver chloride. In Seubert's first paper these data are given:

K_2OsCl_6	Os.	2KCl.	4AgCl.
2.5148	.....	.7796	2.9837
2.1138	.8405	.6547	2.5076

Hence, with salt proportional to 100 parts of AgCl in the last column we have—

Per cent. Os.	Per cent. KCl.	AgCl : Salt.
......	31.000	84.091
39.762	30.973	84.102
	Mean, 84.097, ± .0030	

In his second paper Seubert gives fuller data relative to the potassium osmichloride, but treats it somewhat differently. The salt was reduced by a stream of hydrogen as before, but after that the boat containing the Os + 2KCl was transferred to a platinum tube, in which, by prolonged heating in the gas, the potassium chloride was completely volatilized. The determinations of 4Cl as 4AgCl were omitted. Two series of data are given, as follows:

*Ann. d. Chem., 261, 258.

K_2OsCl_6	Os.	Per cent. Os.
1.1863	.4691	39.543
.9279	.3667	39.519
1.0946	.4330	39.558
1.6055	.6351	39.558
.4495	.1778	39.555
.8646	.3417	39.521
.7024	.2781	39.593
1.2742	.5041	39.562
1.0466	.4141	39.566

Mean, 39.553, ± .0052

K_2OsCl_6	$2KCl$	Per cent. KCl.
2.2032	.6820	30.955
2.0394	.6312	30.950
2.7596	.8544	30.961
2.4934	.7710	30.922
2.8606	.8843	30.913
2.8668	.5768	30.898
1.2227	.3778	30.899

Mean, 30.931

Earlier set, $\begin{cases} 31.000 \\ 30.973 \end{cases}$

Mean of all nine determinations, 30.941, ± .0079

The single percentage of osmium in the earlier memoir is obviously to be rejected.

The ratios to examine are now as follows:

(1.) Per cent. Os in Am_2OsCl_6, 43.459, ± .0036
(2.) $6AgCl : Am_2OsCl_6 :: 100 : 51.283$, ± .0099
(3.) $4AgCl : K_2OsCl_6 :: 100 : 84.097$, ± .0030
(4.) Per cent. Os in K_2OsCl_6, 39.553, ± .0052
(5.) Per cent. KCl in K_2OsCl_6, 30.951, ± .0079

To reduce these ratios we have—

Cl = 35.179, ± .0048   KCl = 74.025, ± .0019
K = 38.817, ± .0051   AgCl = 142.287, ± .0037
N = 13.935, ± .0021

Hence there are five independent values for osmium, as follows:

From (1)..................... Os = 190.111, ± .0300
From (2)..................... " = 190.870, ± .0901
From (3)..................... " = 189.928, ± .0371
From (4)..................... " = 188.914, ± .0243
From (5)..................... " = 189.571, ± .0928

General mean............... Os = 189.546, ± .0163

If O = 16, Os = 190.990.

These figures serve to fix the place of osmium below iridium in the periodic classification of the elements, but are not concordant enough to be fully satisfactory. More determinations are evidently needed.

IRIDIUM.

The only early determination of the atomic weight of iridium was made by Berzelius,* who analyzed potassium iridichloride by the same method employed with the platinum and the osmium salts. The result found from a single analysis was not far from $Ir = 196.7$. This is now known to be too high. I have not, therefore, thought it worth while to recalculate Berzelius' figures, but give his estimation as it is stated in Roscoe and Schorlemmer's "Treatise on Chemistry."

In 1878 the matter was taken up by Seubert,† who had at his disposal 150 grammes of pure iridium. From this he prepared the iridichlorides of ammonium and potassium $(NH_4)_2IrCl_6$ and K_2IrCl_6, which salts were made the basis of his determinations. The potassium salt was dried by gentle heating in a stream of dry chlorine.

Upon ignition of the ammonium salt in hydrogen, metallic iridium was left behind in white coherent laminæ. The results obtained were as follows:

Am_2IrCl_6.	Ir.	Per cent. Ir.
1.3164	.5755	43.725
1.7122	.7490	43.745
1.2657	.5536	43.739
1.3676	.5980	43.726
2.6496	1.1586	43.739
2.8576	1.2489	43.705
2.9088	1.2724	43.742

Mean, 43.732, ± .0035

The potassium salt was also analyzed by decomposition in hydrogen with special precautions. In the residue the iridium and the potassium chloride were separated after the usual method, and both were estimated. Eight analyses gave the following weights:

K_2IrCl_6.	Cl_6, Loss.	Ir.	KCl.
1.6316	.4779	.6507	.5030
2.2544	.6600	.8993	.6953
2.1290	.6238	.8488	.6560
1.8632	.5457	.7430	.5745
2.6898	.7878	1.0726	.8291
2.3719	.6952	.9459	.7308
2.6092	.7641	1.0406	.8040
2.5249	.7395	1.0070	.7775

* Poggend. Annalen, 13, 435. 1828.
† Ber. Deutsch. Chem. Gesell., 11, 1757. 1878.

326   THE ATOMIC WEIGHTS.

Hence we have the following percentages, reckoned on the original salt:

Ir.	$2KCl$.	Cl_4.
39.881	30.829	29.290
39.890	30.842	29.277
39.868	30.813	29.300
39.876	30.835	29.289
39.877	30.825	29.287
39.879	30.811	29.310
39.882	30.814	29.285
39.883	30.792	29.288

Mean, 39.880, ± .0015   Mean, 30.820, ± .0037   Mean, 29.291, ± .0024

Joly * studied derivatives of iridium trichloride. The salts were dried at 120°, and reduced in hydrogen. With $IrCl_3.3KCl.3H_2O$ he found as follows:

Salt.	Ir.	KCl.
1.5950	.5881	.6803
1.6386	.6037	.7000
2.6276	.9689	1.1231

These data, if the weight of the salt itself is considered, give discordant results, but the ratio $Ir : 3KCl :: 100 : x$ is satisfactory. The values of x are as follows:

$$115.677$$
$$115.952$$
$$115.915$$

Mean, 115.848, ± .0583

The ammonium salt, $IrCl_3.3NH_4Cl$, gave the subjoined data:

Wt. of Salt.	Wt. of Ir.	Per cent. Ir.
1.5772	.6627	42.017
1.6056	.6742	41.990

Mean, 42.003, ± .0094

To sum up, the ratios available for iridium are these:

(1.) Per cent. Ir in Am_2IrCl_6, 43.732, ± .0035
(2.) Per cent. Ir in K_2IrCl_6, 39.880, ± .0015
(3.) Per cent. KCl in K_2IrCl_6, 30.820, ± .0037
(4.) Per cent. Cl_4 in K_2IrCl_6, 29.291, ± .0024
(5.) Per cent. Ir in Am_2IrCl_6, 42.003, ± .0094
(6.) Ir : 3KCl :: 100 : 115.848, ± .0583

The data for computation are—

O = 15.879, ± .0003   N = 13.935, ± .0021
Cl = 35.179, ± .0048   KCl = 74.025, ± .0019
K = 38.817, ± .0051   H = 1

---

* Compt. Rend., 110, 1131. 1890.

And the six independent values for the atomic weight of iridium become—

From (1) .... Ir = 191.935, ± .0300
From (2). .... " = 191.511, ± .0221
From (3) .... " = 191.604, ± .0485
From (4) .... " = 191.641, ± .0622
From (5) .... " = 191.833, ± .0641
From (6) .... " = 191.695, ± .0966

General mean .... Ir = 191.664, ± .0154

If O = 16, Ir = 193.125.

PLATINUM.

The earliest work upon the atomic weight of this metal was done by Berzelius,* who reduced platinous chloride and found it to contain 73.3 per cent. of platinum. Hence Pt = 193.155. In a later investigation † he studied potassium chloroplatinate, K_2PtCl_6. 6.981 parts of this salt, ignited in hydrogen, lost 2.024 of chlorine. The residue consisted of 2.822 platinum and 2.135 potassium chloride. From these data we may calculate the atomic weight of platinum in four ways:

1. From loss of Cl upon ignition .... Pt = 196.637
2. From weight of Pt in residue .... " = 195.897
3. From weight of KCl in residue .... " = 195.384
4. From ratio between KCl and Pt .... " = 195.690

The last of these values is undoubtedly the best, for it is not affected by errors due to the possible presence of moisture in the salt analyzed. The work done by Andrews ‡ is even less satisfactory than the foregoing, partly for the reason that its full details seem never to have been published. Andrews dried potassium chloroplatinate at 105°, and then decomposed it by means of zinc and water. The excess of zinc having been dissolved by treatment with acetic and nitric acids, the platinum was collected upon a filter and weighed, while the chlorine in the filtrate was estimated by Pelouze's method. Three determinations gave as follows for the atomic weight of platinum:

197.86
197.68
198.12

Mean, 197.887

Unfortunately, Andrews does not state how his calculations were made.

---

* Poggend. Annalen, 8, 177. 1826.
† Poggend. Annalen, 13, 468. 1828.
‡ British Assoc. Report, 1852. Chem. Gazette, 10,

328   THE ATOMIC WEIGHTS.

In 1881 Seubert[*] published his determinations, basing them upon very pure chloroplatinates of potassium and ammonium. The ammonium salt, $(NH_4)_2PtCl_6$, was analyzed by heating in a stream of hydrogen, expelling that gas by a current of carbon dioxide, and weighing the residual metal. In three experiments the hydrochloric acid formed during such a reduction was collected in an absorption apparatus, and estimated by precipitation as silver chloride. Three series of experiments are given, representing three distinct preparations, as follows:

Series I.

Am_2PtCl_6.	Pt.	Per cent. Pt.
2.1266	.9348	43.957
1.7880	.7858	43.948
1.8057	.7938	43.960
2.6876	1.1811	43.946
4.7674	2.0959	43.963
2.0325	.8935	43.961

Mean, 43.956, ± .002

Series II.

Am_2PtCl_6.	Pt.	Per cent. Pt.
3.0460	1.3363	43.871
2.6584	1.1663	43.876
2.3334	1.0238	43.872
1.9031	.8351	43.881
3.1476	1.3810	43.875
2.7054	1.1871	43.889

Mean, 43.876, ± .001

Another portion of this preparation, recrystallized from water, of 1.4358 grm. gave 0.6311 of platinum, or 43.955 per cent.

Series III.

Am_2PtCl_6.	Pt.	Per cent. Pt.
2.5274	1.1118	43.990
3.2758	1.4409	43.986
1.9279	.8483	44.001
2.0182	.8884	44.010
1.8873	.8303	43.994
2.2270	.9798	43.996
2.4852	1.0936	44.004
2.5362	1.1166	44.026
3.0822	1.3561	43.998

Mean, 44.001, ± .003

[*] Ber. Deutsch. Chem. Gesell., 14, 865.

PLATINUM.   329

If these series are treated as independent and combined, giving each a weight as indicated by its probable error, and regarding the single experiment with preparation II as equal to one in the first series, we get a mean percentage of 43.907, ± .0009. On the other hand, if we regard the twenty-two experiments as all of equal weight in one series, the mean percentage of platinum becomes 43.953, ± .0078. Upon comparing the work with that done later by Halberstadt, the latter mean seems the fairer one to adopt.

For the chlorine estimations in the ammonium salt, Seubert gives the subjoined data. I add in the last column the weight of salt proportional to 100 parts of silver chloride.

Am_2PtCl_6.	Pt.	$6AgCl$.	Ratio.
2.7054	1.1871	5.2226	51.802
2.2748	.9958	4.3758	51.986
3.0822	1.3561	5.9496	51.805

Mean, 51.864, ± .041

The potassium salt, K_2PtCl_6, was also analyzed by ignition in hydrogen, treatment with water, and weighing both the platinum and the potassium chloride. The weights given are as follows:

K_2PtCl_6.	Pt.	$2KCl$.
5.0283	2.0173	1.5440
7.0922	2.8454	2.1793
3.5475	1.4217	1.0890
3.2296	1.2941	.9904
3.5834	1.4372	1.1001
4.2232	1.7746	1.3547
4.0993	1.6444	1.2589
4.4139	1.7713	1.3516

Hence we have these percentages, reckoned on the original salt:

Pt.	KCl.
40.119	30.706
40.120	30.728
40.076	30.698
40.070	30.666
40.107	30.700
40.120	30.627
40.114	30.710
40.130	30.621
Mean, 40.107, ± .005	Mean, 30.682, ± .009

As with the ammonium salt, three experiments were made upon the potassium compound to determine the amount of chlorine (four atoms in this case) lost upon ignition in hydrogen. In the fourth column I add the amount of K_2PtCl_6 corresponding to 100 parts of AgCl:

330   THE ATOMIC WEIGHTS.

K_2PtCl_6.	Pt.	$4AgCl$.	Ratio.
6.7771	2.7158	7.9725	85.006
3.5834	1.4372	4.2270	84.774
4.4139	1.7713	5.2144	84.648

Mean, 84.809, ± .071

Halberstadt,* like Seubert, studied the chloroplatinates of potassium and ammonium, and also the corresponding double bromides and platinic bromide as well. The metal was estimated partly by reduction in hydrogen, as usual, and partly by electrolysis. Platinic bromide gave the following results:

I. By Reduction in H.

$PtBr_4$.	Pt.	Per cent. Pt.
.6396	.2422	37.867
1.7596	.6659	37.844
.9178	.3476	37.873
1.1594	.4388	37.847
1.9608	.7420	37.842
2.0865	.7898	37.853
4.0796	1.5422	37.852
6.8673	2.5985	37.839

II. By Electrolysis.

| 1.2588 | .4763 | 37.837 |
| 1.4937 | .5649 | 37.819 |

Mean of all ten experiments, 37.847, ± .0033

The ammonium platinbromide, $(NH_4)_2PtBr_6$, was prepared in two ways, and five distinct lots were studied. With this salt, as well as with those which follow, the data are given in distinct series, with from one to several experiments in each group, but for present purposes it seems best to consolidate the material and so put it in more manageable form. The percentages of platinum and weights found are as follows:

I. By Reduction in H.

Am_2PtBr_6.	Pt.	Per cent. Pt.
.6272	.1719	27.408
1.0438	.2865	27.447
1.1724	.3215	27.422
1.4862	.4076	27.426
1.0811	.2966	27.435
1.3383	.3672	27.437

---
* Ber. Deutsch. Chem. Gesell., 17, 2962. 1884.

PLATINUM.

Am_2PtBr_6.	Pt.	Per cent. Pt.
⎧ 1.0096	.2769	27.426
⎪ 1.1935	.3269	27.390
⎨ 1.3182	.3611	27.393
⎩ 2.2476	.6159	27.402
⎧ 1.3358	.3668	27.451
⎪ 1.7859	.4899	27.431
⎪ 4.1641	1.1427	27.441
⎨ 1.1835	.3250	27.460
⎪ 2.4003	.6591	27.459
⎩ 2.5293	.6940	27.438
⎧ 1.7147	.4705	27.439
⎪ 2.3014	.6316	27.444
⎨ 3.0052	.8245	27.435
⎩ 4.8592	1.3329	27.430
⎧ 1.5337	.4210	27.449
⎨ 2.0373	.5594	27.457
⎩ 2.0939	.5751	27.465

II. By Electrolysis.

⎧ 1.5586	.4272	27.409
⎨ 1.6052	.4397	27.392
⎩ 3.1229	.8569	27.439
1.1612	.3180	27.386
⎧ 2.5817	.7081	27.427
⎨ 1.0231	.2809	27.456
⎩ 1.6744	.4591	27.418
1.6744	.4591	27.418
1.6052	.4397	27.392

Mean of all thirty-two experiments, 27.429, ± .0027

With potassium platinbromide Halberstadt found as follows:

I. By Reduction in H.

K_2PtBr_6.	Pt.	2KBr.	Per cent. Pt.	Per cent. KBr.
⎧ 2.5549	.6630	.8071	25.940	31.590
⎪ 2.6323	.6831	.8318	25.947	31.599
⎨ 2.9315	.7598	.9259	25.910	31.584
⎪ 3.4463	.8939	1.0895	25.938	31.613
⎩ 4.0081	1.0404	1.2653	25.957	31.568
3.9554	1.0266	1.2495	25.954	31.589
⎧ 2.0794	.5388	.6558	25.911	31.538
⎨ 2.1735	.5635	.6849	25.926	31.511
⎩ 2.3099	.5986	.7297	25.914	31.590
⎧ 1.4085	.3645	.4446	25.880	31.565
⎨ 2.6166	.6772	.8279	25.881	31.640
⎩ 2.6729	.6923	.8469	25.900	31.684

II. By Electrolysis.

K_2PtBr_6.	Pt.	$2KBr$.	Per cent. Pt.	Per cent. KBr.
2.2110	.5726	.6997	25.898	31.647
3.1642	.8188	.9983	25.877	31.550
1.9080	.4947	.6025	25.927	31.577
1.6754	.4341	.5286	25.915	31.550
1.3148	.3403	.4160	25.882	31.640
1.5543	.4025	.4911	25.895	31.596

Mean of eighteen experiments, 25.915, ± .0040   31.591, ± .0068

For ammonium platinchloride Halberstadt gives the following data:

I. By Reduction in H.

Am_2PtCl_6.	Pt.	Per cent. Pt.
1.0604	.4662	43.964
1.3846	.6087	43.962
1.5065	.6617	43.923
2.3266	1.0227	43.956
1.3808	.6059	43.880
1.7396	.7638	43.906
2.7420	1.2068	44.011
3.1882	1.4019	43.971
5.4644	2.4035	43.984
3.4859	1.5321	43.951

II. By Electrolysis.

.9474	.4161	43.920
1.1069	.4865	43.951
1.5101	.6634	43.930
.5345	.2347	43.910
1.6035	.7044	43.928
1.9271	.8459	43.894
1.1046	.4858	43.979
1.4179	.6233	43.959

Mean of eighteen experiments, 43.943, ± .0054
Seubert found, 43.953, ± .0078

General mean, 43.946, ± .0044

For potassium platinchloride Halberstadt's data are—

I. By Reduction in H.

K_2PtCl_6.	Pt.	$2KCl$.	Per cent. Pt.	Per cent. KCl
1.6407	.6574	.5029	40.069	30.651
1.9352	.7757	.5921	40.084	30.600
1.5793	.6334	.4836	40.106	30.621
1.6446	.6595	.5049	40.101	30.700
1.0225	.4102	.3133	40.117	30.640
2.4046	.9641	.7388	40.094	30.724
5.8344	2.3412	1.7905	40.127	30.688
7.1732	2.8776	2.1998	40.116	30.666

II. By Electrolysis.

K_2PtCl_6.	Pt.	2KCl.	Per cent. Pt.	Per cent. KCl.
1.2354	.4953	.3792	40.092	30.695
2.5754	1.0318	.7898	40.063	30.667
1.0933	.4387	.3355	40.126	30.668
1.3560	.5438	.4167	40.103	30.730
1.7345	.6956	.5298	40.104	30.545
2.0054	.8038	.6147	40.081	30.652
2.0666	.8291	.6356	40.117	30.755
1.2759	.5118	.3908	40.112	30.629
1.9376	.7763	.5927	40.065	30.589
2.3972	.9608	.7355	40.080	30.681
2.7249	1.0929	.8364	40.108	30.691

Mean of nineteen experiments, 40.098, ± .0031   30.663, ± .0080
Seubert found, 40.107, ± .0050   30.682, ± .0090

General mean, 40.101, ± .0026   30.671, ± .0060

The work of Dittmar and M'Arthur* on the atomic weight of platinum is difficult to discuss and essentially unsatisfactory. They investigated potassium platinchloride, and came to the conclusion that it contains traces of hydroxyl replacing chlorine and also hydrogen replacing potassium. It is also liable, they think, to carry small quantities of potassium chloride. In their determinations, which involve corrections indicated by the foregoing considerations, they are not sufficiently explicit, and give none of their actual weighings. They attempt, however, to fix the ratio 2KCl : Pt, and after a number of discordant, generally high results, they give the following data for the atomic weight of platinum based upon the assumption that 2KCl = 149.182:

195.54
195.48
195.60
195.37

Mean, 195.50, ± .0330.

Dittmar and M'Arthur also discuss Seubert's determinations, seeking to show that the latter also, properly treated, lead to a value nearer to 195.5 than to 195. Seubert at once replied to them,† pointing out that the concordance between his determinations by very different methods (a concordance verified by Halberstadt's investigation) precluded the existence of errors due to impurities such as Dittmar and M'Arthur assumed.

---

*Trans. Roy. Soc. Edinburgh, 33, 561. 1887.
† Ber. Deutsch. Chem. Gesell., 21, 2179. 1888.

The ratios from which to compute the atomic weight of platinum are now as follows, rejecting the work of Berzelius and of Andrews:

(1.) Percentage of Pt in ammonium platinchloride, 43.946, ± .0044
(2.) Percentage of Pt in ammonium platinbromide, 27.429, ± .0027
(3.) Percentage of Pt in potassium platinchloride, 40.101, ± .0026
(4.) Percentage of Pt in potassium platinbromide, 25.915, ± .0040
(5.) Percentage of Pt in platinic bromide, 37.847, ± .0033
(6.) Percentage of KCl in potassium platinchloride, 30.671, ± .0060
(7.) Percentage of KBr in potassium platinbromide, 31.591, ± .0068
(8.) $6AgCl : Am_2PtCl_6 :: 100 : 51.864$, ± .041
(9.) $4AgCl : K_2PtCl_6 :: 100 : 84.809$, ± .071
(10.) $2KCl : Pt :: 149.182 : 195.50$, ± .033

Computing with the subjoined atomic and molecular weights—

Cl = 35.179, ± .0048   KCl = 74.025, ± .0019
Br = 79.344, ± .0062   KBr = 118.200, ± .0073
K = 38.817, ± .0051   . AgCl = 142.287, ± .0037
N = 13.935, ± .0021

we have the following ten values for platinum:

From (1) .... Pt = 193.603, ± .0336
From (2) .... " = 193.493, ± .0248
From (3) .... " = 193.283, ± .0254
From (4) .... " = 193.684, ± .0344
From (5) .... " = 193.261, ± .0248
From (6) .... " = 193.938, ± .0746
From (7) .... " = 194.538, ± .1276
From (8) .... " = 195.836, ± .3515
From (9) .... " = 193.980, ± .4054
From (10) .... " = 194.017, ± .0331

General mean .... Pt = 193.443, ± .0114

If O = 16, Pt = 194.917.

Of these ten values the first five are obviously the most trustworthy. Their general mean is Pt = 193.414, ± .0124; or, if O = 16, Pt = 194.888. This result is preferable to the mean of all, even though the latter varies little from it. The five high values carry very little weight because of their larger probable errors.

CERIUM.

Although cerium was discovered almost at the beginning of the present century, its atomic weight was not properly determined until after the discovery of lanthanum and didymium by Mosander. In 1842 the investigation was undertaken by Beringer,* who employed several methods. His cerium salts, however, were all rose-colored, and therefore were not wholly free from didymium; and his results are further affected by a negligence on his part to fully describe his analytical processes.

First, a neutral solution of cerium chloride was prepared by dissolving the carbonate in hydrochloric acid. This gave weights of ceric oxide and silver chloride as follows. The third column shows the amount of CeO_2 proportional to 100 parts of $AgCl$:

CeO_2.	$AgCl$.	Ratio.
.5755 grm.	1.419 grm.	40.557
.6715 "	1.6595 "	40.464
1.1300 "	2.786 "	40.560
.5366 "	1.3316 "	40.297

Mean, 40.469, ± .0415

The analysis of the dry cerium sulphate gave results as follows. In a fourth column I show the amount of CeO_2 proportional to 100 parts of $BaSO_4$:

Sulphate.	CeO_2.	$BaSO_4$	Ratio.
1.379 grm.	.8495 grm.	1.711 grm.	49.649
1.276 "	.7875 "	1.580 "	49.836
1.246 "	.7690 "	1.543 "	49.838
1.553 "	.9595 "	1.921 "	49.948

Mean, 49.819, ± .042

Beringer also gives a single analysis of the formate and the results of one conversion of the sulphide into oxide. —The figures are, however, not valuable enough to cite.

The foregoing data involve one variation from Beringer's paper. Where I put CeO_2 as found he puts Ce_2O_3. The latter is plainly inadmissible, although the atomic weights calculated from it agree curiously well with some other determinations. Obviously, the presence of didymium in the salts analyzed tends to raise the apparent atomic weight of cerium.

Shortly after Beringer, Hermann † published the results of one experiment. 23.532 grm. of anhydrous cerium sulphate gave 29.160 grm. of $BaSO_4$. Hence 100 parts of the sulphate correspond to 123.926 of $BaSO_4$.

---
* Ann. Chem. Pharm., 42, 134. 1842.
† Journ. für Prakt. Chem., 30, 185. 1843.

In 1848 similar figures were published by Marignac,* who found the following amounts of $BaSO_4$ proportional to 100 of dry cerium sulphate:

122.68
122.00
122.51

Mean, 122.40, ± .138

If we give Hermann's single result the weight of one experiment in this series, and combine, we get a mean value of 122.856, ± .130.

Still another method was employed by Marignac. A definite mixture was made of solutions of cerium sulphate and barium chloride. To this were added, volumetrically, solutions of each salt successively, until equilibrium was attained. The figures published give maxima and minima for the $BaCl_2$ proportional to each lot of $Ce_2(SO_4)_3$. In another column, using the mean value for $BaCl_2$ in each case, I put the ratio between 100 parts of this salt and the equivalent quantity of sulphate. The latter compound was several times recrystallized:

$Ce_2(SO_4)_3$.		$BaCl_2$.		Ratio.
First crystallization,......	11.011 grm.	11.990 — 12.050 grm.		91.606
First crystallization,......	13.194 "	14.365 — 14.425	"	91.657
Second crystallization....	13.961 "	15.225 — 15.285	"	91.518
Second crystallization....	12.627 "	13.761 — 13.821	"	91.559
Second crystallization....	11.915 "	12.970 — 13.030	"	91.654
Third crystallization,.....	14.888 "	16.223 — 16.283	"	91.602
Third crystallization,.....	14.113 "	15.383 — 15.423	"	91.755
Fourth crystallization....	13.111 "	14.270 — 14.330	"	91.685
Fourth crystallization....	13.970 "	15.223 — 15.283	"	91.588

Mean, 91.625, ± .016

Omitting the valueless experiments of Kjerulf,† we come next to the figures published by Bunsen and Jegel ‡ in 1858. From the air-dried sulphate of cerium the metal was precipitated as oxalate, which, ignited, gave CeO_2. In the filtrate from the oxalate the sulphuric acid was estimated as $BaSO_4$:

1.5726 grm. sulphate gave .7899 grm. CeO_2 and 1.6185 grm. $BaSO_4$.
1.6967 " .8504 " 1.7500 "

Hence, for 100 parts $BaSO_4$, the CeO_2 is as follows:

48.804
48.575

Mean, 48.689, ± .077

---

* Arch. Sci. Phys. et Nat. (1), 8, 273. 1848.
† Ann. Chem. Pharm., 87, 12.
‡ Ann. Chem. Pharm., 105, 45. 1858.

One experiment was also made upon the oxalate:

.3530 grm, oxalate gave .1913 CeO_2 and .0506 H_2O.

Hence, in the dry salt, we have 63.261 per cent. of CeO_2.

In each sample of CeO_2 the excess of oxygen over Ce_2O_3 was estimated by an iodometric titration; but the data thus obtained need not be further considered.

In two papers by Rammelsberg* data are given for the atomic weight of cerium, as follows. In the earlier paper cerium sulphate was analyzed, the cerium being thrown down by caustic potash, and the acid precipitated from the filtrate as barium sulphate:

.413 grm. $Ce_2(SO_4)_3$ gave .244 grm. CeO_2 and .513 grm. $BaSO_4$.

Hence 100 $BaSO_4$ = 47.563 CeO_2, a value which may be combined with others, thus; this figure being assigned a weight equal to one experiment in Bunsen's series:

```
Beringer ........................ ............ 49.819, ± .042
Bunsen and Jegel........................ 48.689, ± .077
Rammelsberg ............................ 47.563, ± .108
         General mean.................... 49.360, ± .035
```

It should be noted here that this mean is somewhat arbitrary, since Bunsen and Rammelsberg's cerium salts were undoubtedly freer from didymium than the material studied by Beringer.

In his later paper Rammelsberg gives these figures concerning cerium oxalate. One hundred parts gave 10.43 of carbon and 21.73 of water. Hence the dry salt should yield 48.862 per cent. of CO_2, whence $Ce = 137.14$.

In all of the foregoing experiments the ceric oxide was somewhat colored, the tint ranging from one shade to another of light brown according to the amount of didymium present. Still, at the best, a color remained, which was supposed to be characteristic of the oxide itself. In 1868, however, some experiments of Dr. C. Wolff† were posthumously made public, which went to show that pure ceroso-ceric oxide is white, and that all samples previously studied were contaminated with some other earth, not necessarily didymium but possibly a new substance, the removal of which tended to lower the apparent atomic weight of cerium very perceptibly.

Cerium sulphate was recrystallized at least ten times. Even after twenty recrystallizations it still showed spectroscopic traces of didymium. The water contained in each sample of the salt was cautiously estimated, and the cerium was thrown down by boiling concentrated solutions of

---

* Poggend. Annalen, 55, 65; 108, 44.
† Amer. Journ. Science and Arts (2), 46, 53.

oxalic acid. The resulting oxalate was ignited with great care. I deduce from the weighings the percentage of CeO_2 given by the *anhydrous* sulphate:

Sulphate.	Water.	CeO_2.	Per cent. CeO_2.
1.4542 grm.	.19419 grm.	.76305 grm.	60.559
1.4104 "	.1898 "	.7377 "	60.437
1.35027 "	.1820 "	.70665 "	60.487
			Mean, 60.494

After the foregoing experiments the sulphate was further purified by solution in nitric acid and pouring into a large quantity of boiling water. The precipitate was converted into sulphate and analyzed as before:

Sulphate.	Water.	CeO_2.	Per cent. CeO_2.
1.4327 grm.	.2733 grm.	.69925 grm.	60.311
1.5056 "	.2775 "	.7405 "	60.296
1.44045 "	.2710 "	.7052 "	60.300
			Mean, 60.302

From another purification the following weights were obtained:

1.4684 grm.　　.1880 grm.　　.7717 grm.　　60.270 per cent.

A last purification gave a still lower percentage:

1.3756 grm.　　.1832 grm.　　.7186 grm.　　60.265 per cent.

The last oxide was perfectly white, and was spectroscopically free from didymium. In each case the CeO_2 was titrated iodometrically for its excess of oxygen. It will be noticed that in the successive series of determinations the percentage of CeO_2 steadily and strikingly diminishes to an extent for which no ordinary impurity of didymium can account. The death of Dr. Wolf interrupted the investigation, the results of which were edited and published by Professor F. A. Genth.

In the light of more recent evidence, little weight can be given to these observations. All the experiments, taken equally, give a mean percentage of CeO_2 from $Ce_2(SO_4)_3$ of 60.366, ± .0308. This mean has obviously little or no real significance.

The experiments of Wolf attracted little attention, except from Wing,[*] who partially verified certain aspects of them. This chemist, incidentally to other researches, purified some cerium sulphate after the method of Wolf, and made two similar analyses of it, as follows:

Sulphate.	Water.	CeO_2.	Per cent. CeO_2.
1.2885 grm.	.1707 grm.	.6732 grm.	60.225
1.4090 "	.1857 "	.7372 "	60.263
			Mean, 60 244

---

[*] Am. Journ. Sci. (2), 49, 358. 1870.

CERIUM. 339

The ceric oxide in this case was perfectly white. The cerium oxalate which yielded it was precipitated boiling by a boiling concentrated solution of oxalic acid. The precipitate stood twenty-four hours before filtering.

In 1875 Buehrig's* paper upon the atomic weight of cerium was issued. He first studied the sulphate, which, after eight crystallizations, still retained traces of free sulphuric acid. He found, furthermore, that the salt obstinately retained traces of water, which could not be wholly expelled by heat without partial decomposition of the material. These sources of error probably affect all the previously cited series of experiments, although, in the case of Wolf's work, it is doubtful whether they could have influenced the atomic weight of cerium by more than one or two tenths of a unit. Buehrig also found, as Marignac had earlier shown, that upon precipitation of cerium sulphate with barium chloride the barium sulphate invariably carried down traces of cerium. Furthermore, the ceric oxide from the filtrate always contained barium. For these reasons the sulphate was abandoned, and the atomic weight determinations of Buehrig were made with air-dried oxalate. This salt was placed in a series of platinum boats in a combustion tube behind copper oxide. It was then burned in a stream of pure, dry oxygen, and the carbonic acid and water were collected after the usual method. Ten experiments were made; in all of them the above-named products were estimated, and in five analyses the resulting ceric oxide was also weighed. By deducting the water found from the weight of the air-dried oxalate, the weight of the anhydrous oxalate is obtained, and the percentages of its constituents are easily determined. In weighing, the articles weighed were always counterpoised with similar materials. The following weights were found :

Oxalate.	Water.	CO_2.	CeO_2.
9.8541 grm.	2.1987 grm.	3.6942 grm.	......
9.5368 "	2.1269 "	3.5752 "	....
9.2956 "	2.0735 "	3.4845 "	......
10.0495 "	2.2364 "	3.7704 "	......
10.8249 "	2.4145 "	4.0586 "	......
9.3679 "	2.0907 "	3.5118 "	4.6150 grm.
9.7646 "	2.1769 "	3.6616 "	4.8133 "
9.9026 "	2.2073 "	3.7139 "	4.8824 "
9.9376 "	2.2170 "	3.7251 "	4.8971 "
9.5324 "	2.1267 "	3.5735 "	4.6974 "

These figures give us the following percentages for CO_2 and CeO_2 in the anhydrous oxalate :

---

*Journ. für Prakt. Chem., 120, 222. 1875.

340   THE ATOMIC WEIGHTS.

CO_2	CeO_2
48.256	......
48.249	......
48.248	......
48.257	......
48.257	......
48.258	63.417
48.257	63.436
48.262	63.446
48.249	63.429
48.253	63.430
Mean, 48.2546 ± .001	Mean, 63.4316 ± .0032

These results could not be appreciably affected by combination with the single oxalate experiments of Jegel and of Rammelsberg, and the latter may therefore be ignored.

Robinson's work, published in 1884,[*] was based upon pure cerium chloride, prepared by heating dry cerium oxalate in a stream of dry, gaseous hydrochloric acid. This compound was titrated with standard solutions of pure silver, prepared according to Stas, and these were weighed, not measured. In the third column I give the ratio between $CeCl_3$ and 100 parts of silver:

$CeCl_3$	Ag	Ratio
5.5361	7.26630	76.189
6.0791	7.98077	76.172
6.4761	8.50626	76.133
6.98825	9.18029	76.122
6.6873	8.78015	76.164
7.0077	9.20156	76.158
6.9600	9.13930	76.150

Mean, 76.155, ± .0065

Reduced to a vacuum this becomes 76.167.

In a later paper,[†] Robinson discusses the color of ceric oxide, and criticises the work of Wolf. He shows that the pure oxide is not white, and makes it appear probable that Wolf's materials were contaminated with compounds of lanthanum. He also urges that Wolf's cerium sulphate could not have been absolutely definite, because of defects in the method by which it was dehydrated.

Brauner,[‡] in 1885, investigated cerium sulphate with extreme care, and appears to have obtained material free from all other earths and absolutely homogeneous. The anhydrous salt was calcined with all

---
[*] Chemical News, 50, 251. Nov. 28, 1884. Proc. Roy. Soc., 37, 150.
[†] Chemical News, 54, 229. 1886.
[‡] Sitzungs. Wien. Akad., Bd. 92. July, 1885.

necessary precautions, and the data obtained, reduced to a vacuum, were as follows:

$Ce_2(SO_4)_3$.	CeO_2.	Per cent. CeO_2.
2.16769	1.31296	60.5693
2.43030	1.47205	60.5707
2.07820	1.25860	60.5620
2.21206	1.33989	60.5721
1.28448	.77845	60.6043
1.95540	1.18436	60.5687
2.46486	1.49290	60.5673
2.04181	1.23733	60.5997
2.17714	1.31878	60.5739
2.09138	1.26654	60.5605
2.21401	1.34139	60.5863
2.44947	1.48367	60.5711
2.22977	1.35073	60.5771
2.73662	1.65699	60.5486
2.62614	1.59050	60.5642
1.67544	1.01470	60.5632
1.57655	.95540	60.6007
2.72882	1.65256	60.5600
2.10455	1.27476	60.5716
2.10735	1.27698	60.5965
2.43557	1.47517	60.5692
3.01369	1.82524	60.5649
4.97694	3.01372	60.5537

Mean, 60.5729, ± .0021

This mean completely outweighs the work done by Wolf and Wing, so that upon combination the latter practically vanish. Wing's mean is arbitrarily given equal weight with Wolf's, and the combination is as follows:

Wolf............................... 60.366, ± .0308
Wing............................... 60.244, ± .0308
Brauner............................ 60.5729, ± .0021

General mean..................... 60.566, ± .0021

In 1895 several papers upon the cerite earths were published by Schutzenberger.* In the first of these a single determination of atomic weight is given. Pure CeO_2, of a yellowish white color, was converted into sulphate, which was dried in a current of dry air at 440°. This salt, dissolved in water, was poured into a hot solution of caustic soda, made from sodium, and, after filtration and washing, the filtrate, acidulated with hydrochloric acid, was precipitated with barium chloride. The trace of sulphuric acid retained by the cerium hydroxide was recovered by re-solution and a second precipitation, and added to the main amount.

---

* Compt. Rend., 120, pp. 663, 962, and 1143. 1895.

100 parts of $Ce_2(SO_4)_3$ gave 123.30 of $BaSO_4$. This may be assigned equal weight with one experiment in Marignac's series, giving the following combination:

Hermann	123.926,	± .238
Marignac	122.40,	± .138
Schutzenberger	123.30,	± .238
General mean	122.958,	± .1139

Schutzenberger, criticising Brauner's work, claims that the latter was affected by a loss of oxygen during the calcination of the cerium dioxide.

In his second and third papers Schutzenberger describes the results obtained upon the fractional crystallization of cerium sulphate. Preparations were thus made yielding oxides of various colors—canary yellow, rose, yellowish rose, reddish, and brownish red. These oxides, by synthesis of sulphates, the barium-sulphate method, etc., gave varying values for the atomic weight of cerium, ranging from 135.7 to 143.3. Schutzenberger therefore infers that cerium oxide from cerite contains small quantities of another earth of lower molecular weight; but the results as given are not sufficiently detailed to be conclusive. The third paper is essentially a continuation of the second, with reference to the didymiums.

Schutzenberger's papers were promptly followed by one from Brauner,* who claims priority in the matter of fractionation, and gives some new data, the latter tending to show that cerium oxide is a mixture of at least two earths. One of these, of a dark salmon color, he ascribes to a new element, "meta-cerium." The other he calls cerium, and gives for it a preliminary atomic weight determination. The pure oxalate, by Gibbs' method, gave 46.934 per cent. of CeO_2, and, on titration with potassium permanganate, 29.503 and 29.506 per cent. of C_2O_3. Hence $Ce = 138.799$. In mean, this ratio may be written—

$$3C_2O_3 : 2CeO_2 :: 29.5045 : 46.934,$$

and to each of its numerical terms we may roughly assign the probable error ± .001. This is derived from the average of the two titrations, and is altogether arbitrary.

The ratios, good and bad, for cerium now are—

(1.) $Ce_2(SO_4)_3 : 3BaSO_4 :: 100 : 122.958$, ± .1139
(2.) $3BaSO_4 : 2CeO_2 :: 100 : 49.360$, ± .035
(3.) $3BaCl_2 : Ce_2(SO_4)_3 :: 100 : 91.625$, ± .016
(4.) $3AgCl : CeO_2 :: 100 : 40.469$, ± .0415
(5.) Percentage CeO_2 from $Ce_2(SO_4)_3$, 60.566, ± .0021
(6.) Percentage CeO_2 from $Ce_2(C_2O_4)_3$, 63.4316, ± .0032
(7.) Percentage CO_2 from $Ce_2(C_2O_4)_3$, 48.2546, ± .001.
(8.) $3Ag : CeCl_3 :: 100 : 76.167$, ± .0065
(9.) $3C_2O_3 : 2CeO_2 :: 29.5045$, ± .001 : 46.934, ± .001

---

*Chem. News, 71, 283.

CERIUM.   343

To reduce these ratios we have—

O   = 15.879, ± .0003   C = 11.920, ± .0004
Cl = 35.179, ± .0048   S = 31.828, ± .0015
Ag = 107.108, ± .0031   Ba = 136.392, ± .0086
AgCl = 142.287, ± .0037

From the ratios, with these intermediate data, we can get two values for the molecular weight of $Ce_2(SO_4)_3$, and five for that of CeO_2. For cerium sulphate we have—

From (1).................. $Ce_2(SO_4)_3$ = 565.404, ± .1670
From (3)..................   "   = 568.304, ± .1054
                              ─────────────────────────────
General mean........ $Ce_2(SO_4)_3$ = 567.478, ± .0891

Hence Ce = 140.723, ± .0451.
For ceric oxide the values are—

From (2)...................... CeO_2 = 171.577, ± .1218
From (4)......................   "   = 172.746, ± .1772
From (5)......................   "   = 170.879, ± .0115
From (6)......................   "   = 172.125, ± .0177
From (9)......................   "   = 170.557, ± .0076
                              ─────────────────────────────
General mean............ CeO_2 = 170.827, ± .0060

And Ce = 139.069, ± .0061.
For cerium itself, four independent values are now calculable, as follows:

From molecular weight of sulphate... Ce = 140.723, ± .0451
From molecular weight of dioxide ... " = 139.069, ± .0061
From ratio (8) .... " = 139.206, ± .0263
From ratio (7) .... " = 140.516, ± .0047
                                     ─────────────────────
General mean............... Ce = 140.113, ± .0036

If O = 16, Ce = 141.181.
It must be admitted that this combination is of very questionable utility. Its component means vary too widely from each other, and involve too many uncertainties. Furthermore, Schutzenberger and Brauner both impugn the homogeneity of the supposed element, as it has hitherto been recognized. Even if no " meta-elements " are involved in the discussion, it seems clear, on chemical grounds, that the two lower values are really preferable to the two higher, and that ratio (7) receives excessive weight. The general mean obtained is probably a full unit too high. The value 139.1 is perhaps nearly correct.

LANTHANUM.

Leaving out of account the work of Mosander, and the valueless experiments of Choubine, we may consider the estimates of the atomic weight of lanthanum which are due to Hermann, Rammelsberg, Marignac, Czudnowicz, Holzmann, Zschiesche, Erk, Cleve, Brauner, Bauer, and Bettendorff.

From Rammelsberg* we have but one analysis. .700 grm. of lanthanum sulphate gave .883 grm. of barium sulphate. Hence 100 parts of $BaSO_4$ are equivalent to 79.276 of $La_2(SO_4)_3$.

Marignac,† working also with the sulphate of lanthanum, employed two methods. First, the salt in solution was mixed with a slight excess of barium chloride. The resulting barium sulphate was filtered off and weighed; but, as it contained some occluded lanthanum compounds, its weight was too high. In the filtrate the excess of barium was estimated, also as sulphate. This last weight of sulphate, deducted from the total sulphate which the whole amount of barium chloride could form, gave the sulphate actually proportional to the lanthanum compound. The following weights are given:

$La_2(SO_4)_3$.	$BaCl_2$.	1st $BaSO_4$.	2d $BaSO_4$.
4.346 grm.	4.758 grm.	5.364 grm.	.115 grm.
4.733 "	5.178 "	5.848 "	.147 "

Hence we have the following quantities of $La_2(SO_4)_3$ proportional to 100 parts of $BaSO_4$. Column A is deduced from the first $BaSO_4$ and column B from the second, after the manner above described:

A.	B.
81.022	83.281
80.934	83.662
Mean, 80.978, ± .030	Mean, 83.471, ± .128

From A... La = 138.47
From B... " = 147.13

A agrees best with other determinations, although, theoretically, it is not so good as B.

Marignac's second method, described in the same paper with the foregoing experiments, consisted in mixing solutions of $La_2(SO_4)_3$ with solutions of $BaCl_2$, titrating one with the other until equilibrium was established. The method has already been described under cerium. The weighings

---
* Poggend. Annalen. 55, 65.
† Arch. Sci. Phys. et Nat. (1), 11, 29. 1849.

give maxima and minima for $BaCl_2$. In another column I give $La_2(SO_4)_3$ proportional to 100 parts of $BaCl_2$, mean weights being taken for the latter:

$La_2(SO_4)_3$.	$BaCl_2$.	Ratio.
11.644 grm.	12.765 — 12.825 grm.	91.004
12.035 "	13.195 — 13.265 "	90.968
10.690 "	11.669 — 11.749 "	91.297
12.750 "	13.920 — 14.000 "	91.332
10.757 "	11.734 — 11.814 "	91.362
12.672 "	13.813 — 13.893 "	91.475
9.246 "	10.080 — 10.160 "	91.364
10.292 "	11.204 — 11.264 "	91.615
10.192 "	11.111 — 11.171 "	91.482

Mean, 91.322, ± .048

Hence $La = 140.2$.

Although not next in chronological order, some still more recent work of Marignac's[*] may properly be considered here. The salt studied was the sulphate of lanthanum, purified by repeated crystallizations. In two experiments the salt was calcined, and the residual oxide weighed; in two others the lanthanum was precipitated as oxalate, and converted into oxide by ignition. The following percentages are given for La_2O_3:

$\left.\begin{array}{l}57.56\\57.58\end{array}\right\}$ By calcination.

$\left.\begin{array}{l}57.50\\57.55\end{array}\right\}$ Ppt. as oxalate.

Mean, 57.5475, ± .0115

The atomic weight determinations of Holzmann [†] were made by analyses of the sulphate and iodate of lanthanum, and the double nitrate of magnesium and lanthanum. In the sulphate experiments the lanthanum was first thrown down as oxalate, which, on ignition, yielded oxide. The sulphuric acid was precipitated as $BaSO_4$ in the filtrate.

$La_2(SO_4)_3$.	La_2O_3.	$BaSO_4$.
.9663 grm.	.5157 grm.	1.1093 grm.
.6226 "	.3323 "	.7123 "
.8669 "	.4626 "	.9869 "

These results are best used by taking the ratio between the $BaSO_4$, put at 100, and the La_2O_3. The figures are then as follows:

46.489
46.652
46.873

Mean, 46.671, ± .075

---

[*] Ann. Chim. Phys. (4), 30, 68. 1873.
[†] Journ. für Prakt. Chem., 75, 321. 1858.

In the analyses of the iodate the lanthanum was thrown down as oxalate, as before. The iodic acid was also estimated volumetrically, but the figures are hardly available for present discussion. The following percentages of La_2O_3 were found:

$$23.454$$
$$23.419$$
$$23.468$$

Mean, $23.447, \pm .0216$

The formula of this salt is $La_2(IO_3)_6.3H_2O$.
The double nitrate, $La_2(NO_3)_6.3Mg(NO_3)_2.24H_2O$, gave the following analytical data:

Salt.	H_2O.	MgO.	La_2O_3.
.5327 grm.	.1569 grm.	.0417 grm.	.1131 grm.
.5931 "	.1734 "	.0467 "	.1262 "
.5662 "	.1647 "	.0442 "	.1197 "
.3757 "	......	.0297 "	.0813 "
.3263 "	......	.0256 "	.0693 "

These weighings give the subjoined percentages of La_2O_3:

$$21.231$$
$$21.278$$
$$21.141$$
$$21.640$$
$$21.238$$

Mean, $21.3056, \pm .058$

These data of Holzmann give values for the molecular weight of La_2O_3 as follows:

From sulphate................... $La_2O_3 = 322.460$
From iodate....................... " $= 320.726$
From magnesian nitrate.......... " $= 322.904$

Czudnowicz* based his determination of the atomic weight of lanthanum upon one analysis of the air-dried sulphate. The salt contained 22.741 per cent. of water.

.598 grm. gave .272 grm. La_2O_3 and .586 grm. $BaSO_4$.

The La_2O_3 was found by precipitation as oxalate and ignition. The $BaSO_4$ was thrown down from the filtrate. Reduced to the standards already adopted, these data give for the percentage of La_2O_3 in the anhydrous sulphate the figure 58.668. 79.117 parts of the salt are proportional to 100 parts of $BaSO_4$.

---
* Journ. für Prakt. Chem., 80, 33. 1860.

Hermann * studied both the sulphate and the carbonate of lanthanum. From the anhydrous sulphate, by precipitation as oxalate and ignition, the following percentages of La_2O_3 were obtained:

57.690
57.663
57.610

Mean, 57.654, ± .016

The carbonate, dried at 100°, gave the following percentages:

68.47 La_2O_3.
27.67 CO_2.
3.86 H_2O.

Reckoning from the ratio between CO_2 and La_2O_3, the molecular weight of the latter becomes 324.254.

Zschiesche's † experiments consist of six analyses of lanthanum sulphate, which salt was dehydrated at 230°, and afterwards calcined. I subjoin his percentages, and in a fourth column deduce from them the percentage of La_2O_3 in the *anhydrous* salt:

H_2O.	SO_3.	La_2O_3.	La_2O_3 in Anhydrous Salt.
22.629	33.470	43.909	56.745
22.562	33.306	44.132	56.964
22.730	33.200	44.070	57.034
22.570	33.333	44.090	56.947
22.610	33.160	44.240	57.150
22.630	33.051	44.310	57.277

Mean, 57.021, ± .051

Erk ‡ found that .474 grm. of $La_2(SO_4)_3$, by precipitation as oxalate and ignition, gave .2705 grm. of La_2O_3, or 57.068 per cent. .7045 grm. of the sulphate also gave .8815 grm. of $BaSO_4$. Hence 100 parts of $BaSO_4$ are equivalent to 79.921 of $La_2(SO_4)_3$.

From Cleve we have two separate investigations relative to the atomic weight of lanthanum. In his first series § strongly calcined La_2O_3, spectroscopically pure, was dissolved in nitric acid, and then, by evaporation with sulphuric acid, converted into sulphate:

1.9215 grm. La_2O_3 gave 3.3365 grm. sulphate.		57.590 per cent.
2.0570 "	3.5705 "	57.611 "
1.6980 "	2.9445 "	57.667 "
2.0840 "	3.6170 "	57.617 "
1.9565 "	3.3960 "	57.612 "

Mean, 57.619, ± .0085

---

* Journ. für Prakt. Chem., 82, 396. 1861.
† Journ. für Prakt. Chem., 104, 174.
‡ Jenaisches Zeitschrift, 6, 306. 1871.
§ K. Svensk. Vet. Akad. Handlingar, Bd. 2, No. 7. 1874.

From the last column, which indicates the percentage of La_2O_3 in $La_2(SO_4)_3$, we get, if $SO_3 = 80$, $La = 139.15$.

In his second paper,* published nine years later, Cleve gives results similarly obtained, but with lanthanum oxide much more completely freed from other earths. The data are as follows, lettered to correspond to different fractions of the material studied:

B.	.8390 grm. La_2O_3 gave	1.4600 sulphate.		57.466 per cent.	
	1.1861	"	2.0643	"	57.458 "
C.	.8993	"	1.5645	"	57.482 "
	.8685	"	1.5108	"	57.486 "
	.8515	"	1.4817	"	57.468 "
D.	.6486	"	1.1282	"	57.490 "
	.7329	"	1.2746	"	57.500 "
E.	1.2477	"	2.1703	"	57.490 "
F.	1.1621	".	2.0217	"	57.481 "
	1.5749	"	2.7407	"	57.463 "
G.	1.3367	"	2.3248	"	57.497 "
	1.4455	"	2.5146	"	57.484 "

Mean, 57.480, ± .0040

Hence with $SO_3 = 80$, $La = 138.22$.

From Brauner we also have two sets of determinations, both based upon the conversion of pure La_2O_3 into $La_2(SO_4)_3$.

In his first paper, Brauner† gives only two syntheses, as follows:

1.75933 grm. La_2O_3 gave	3.05707 $La_2(SO_4)_3$.		57.566 per cent.	
.92417	"	1.60589	"	57.549 "

Mean, 57.5575

This mean we may regard as of equal weight with Marignac's, and assign to it the same probable error.

In Brauner's second paper ‡ six experiments are given; but the weights are affected by a misprint in the second determination, which I am unable to correct. Only five of the syntheses, therefore, are given below.

.7850 grm. La_2O_3 gave 1.3658 $La_2(SO_4)_3$.		57.476 per cent.	
2.1052	"	3.6633 "	57.467 "
1.0010	"	1.7411 "	57.525 "
1.3807	"	2.4021 "	57.479 "
1.5275	"	2.6588 "	57.451 "

Mean, 57.480, ± .0084

Brauner's weighings are all reduced to a vacuum.

Both Bauer and Bettendorff made their determinations of the atomic

---

* K. Svensk. Vet. Akad. Handlingar, No. 2, 1883.
† Journ. Chem. Soc., Feb., 1882, p. 68.
‡ Sitzungsb. Wien. Akad., June, 1882, Bd. 85, II Abth.

LANTHANUM. 349

weight of lanthanum by the same general method as the preceding Bauer's data * are as follows:

.6431 grm. La_2O_3 gave 1.1171 sulphate.   57.569 per cent.
.7825   "   1.3613   "   57.482   "
1.0112   "   1.7571   "   57.549   "
.7325   "   1.2725   "   57.564   "

                                    Mean, 57.541, ± .0136

Bettendorff found †—

.9146 grm. La_2O_3 gave 1.5900 sulphate.   57.522 per cent.
.9395   "   1.6332   "   57.525   "
.9133   "   1.5877   "   57.523   "
1.0651   "   1.8515   "   57.526   "

                                    Mean, 57.524, ± .0006

We may now combine the similar means into general means, and deduce a value for the atomic weight of lanthanum. For the percentage of oxide in sulphate we have estimates as follows. The single experiments of Czudnowicz and of Erk are assigned the probable error and weight of a single experiment in Hermann's series:

Czudnowicz	58.668,	± .027
Erk	57.068,	± .027
Hermann	57.654,	± .016
Zschiesche	57.021,	± .051
Marignac	57.5475,	± .0115
Cleve, earlier series	57.619,	± .0085
Cleve, later series	57.480,	± .0040
Brauner, earlier series	57.5575,	± .0115
Brauner, later series	57.480,	± .0084
Bauer	57.541,	± .0136
Bettendorff	57.524,	± .0006
General mean	57.522,	± .00059

This result is practically identical with that of Bettendorff, whose work seems to receive excessive weight. The figure, however, cannot be far out of the way.

For the quantity of $La_2(SO_4)_3$ proportional to 100 parts of $BaSO_4$, we have five experiments, which may be given equal weight and averaged together:

Marignac	81.022
Marignac	80.934
Rammelsberg	79.276
Czudnowicz	79.117
Erk	79.921

                                    Mean, 80.054, ± .270

---

* Freiburg Inaugural Dissertation, 1884.
† Ann. d. Chem., 256, 168.

In all, there are six ratios from which to calculate:

(1.) Percentage of La_2O_3 in $La_2(SO_4)_3$, 57.522, ± .00059
(2.) $3BaCl_2 : La_2(SO_4)_3 :: 100 : 91.322$, ± .048—Marignac
(3.) $3BaSO_4 : La_2(SO_4)_3 :: 100 : 80.054$, ± .270
(4.) $3BaSO_4 : La_2O_3 :: 100 : 46.671$, ± .075—Holzmann
(5.) Percentage of La_2O_3 in iodate, 23.447, ± .0216—Holzmann
(6.) Percentage of La_2O_3 in magnesian nitrate, 21.3056, ± .058—Holzmann

Hermann's single experiment on the carbonate is omitted from this scheme as being unimportant.

For the reduction of these data we have—

$O = 15.879$, ± .0003 $N = 13.935$, ± .0021
$Cl = 35.179$, ± .0048 $C = 11.920$, ± .0004
$I = 125.888$, ± .0069 $Mg = 24.100$, ± .0011
$S = 31.828$, ± .0015 $Ba = 136.392$, ± .0086

For lanthanum sulphate two values are obtainable:

From (2) .... $La_2(SO_4)_3 = 566.425$, ± .2999
From (3) ....   " $= 556.542$, ± 1.8729

General mean......... $La_2(SO_4)_3 = 566.182$, ± .2961

Hence $La = 140.075$, ± .1481.

For the oxide there are four independent values, as follows:

From (1)............. $La_2O_3 = 322.825$, ± .0090
From (4)............. " $= 322.460$, ± .5215
From (5)............. " $= 320.726$, ± .3159
From (6)............. " $= 322.924$, ± .9107

A glance at these figures shows that the first alone deserves consideration, and that a combination of all would vary inappreciably from it. Taking, then, $La_2O_3 = 322.825$, ± .0090, we get—

$La = 137.594$, ± .0046;

or, with $O = 16$, $La = 138.642$.

If we take the concordant results of Cleve's and Bräuner's later series, which give the percentage of La_2O_3 in $La_2(SO_4)_3$ as 57.480, then $La = 137.316$. Possibly this value may be better than the other, but the evidence is not conclusive.

THE DIDYMIUMS.

Leaving Mosander's early experiments out of account, the atomic weight of the so-called "didymium" was determined by Marignac, Hermann, Zschiesche, Erk, Cleve, Brauner, and Bauer. All of these data now have only historical value, and may be disposed of very briefly.

Marignac* determined the ratios between didymium sulphate and barium sulphate, between silver chloride and didymia, and between didymium sulphate and didymium oxide. The other determinations all relate to the sulphate-oxide ratio. Leaving all else out of account, the earlier data for the percentage of Di_2O_3 in $Di_2(SO_4)_3$ are as follows. The atomic weight of Di in the last column is based upon $SO_3 = 80$:

	Per cent. Di_2O_3.	At. Wt. Di.
Marignac,† five experiments	58.270	143.56
Hermann,‡ one experiment	58.140	142.67
Zschiesche,§ five experiments	57.926	141.21
Erk,‖ two experiments	58.090	142.33
Cleve,¶ six experiments	58.766	147.02
Brauner,** three experiments	58.681	146.42

The discordance of the determinations is manifest, and yet up to 1883 the elementary nature of didymium seems to have been undoubted. In that year, however, Cleve and Brauner both showed, independently, that the didymia previously studied by them contained samaria, and that source of disturbance was eliminated.

In Brauner's investigation †† the didymium compounds were carefully fractionated, and the determinations of atomic weight were made by synthesis of the sulphate from the oxide in the usual way. Neglecting details, his first series gave results as follows:

Per cent. Di_2O_3.	At. Wt.
58.506	145.36
58.526	145.50
58.500	145.31
58.515	145.42
58.531	145.53

---

* Two papers: Arch. Sci. Phys. et Nat. (1), 11, 29. 1849. Ann. Chim. Phys. (3), 38, 148. 1853.
† Ann. Chim. Phys. (3), 38, 148. 1853.
‡ Journ. für Prakt. Chem., 82, 367. 1861.
§ Journ. für Prakt. Chem., 107, 74.
‖ Jenaisches Zeitschrift, 6, 306. 1871.
¶ K. Svensk. Vet. Akad. Handl., Bd. 2, No. 8. 1874.
** Berichte, 15, 109. 1882.
†† Journ. Chem. Soc., June, 1883. The values given are as computed by Brauner, with O = 16 and S = 32.07.

Another determination, with material refractionated from that used in his investigation of the previous year, gave 58.512 per cent. Di_2O_3 and $Di = 145.40$.

These determinations, although concordant among themselves, are still about a unit lower than those published in 1882, indicating that in the earlier research some earth of higher molecular weight was present. Accordingly, another series of fractionations was carried out, and the several fractions of " didymia " obtained gave the following values:

Fraction.	Per cent. Di_2O_3.	At. Wt. "Di."
1	58.355	144.32
2	58.479	145.16
3	58.510	145.39
4	58.755	147.10
5	59.071	149.35
	59.086	149.46

The last fraction is evidently near samaria (Sm = 150), and this earth was proved to be present by a study of the absorption spectra of the material investigated.

Similar results, but in some respects more explicit, were obtained by Cleve,* who also found that his earlier research had been vitiated by the presence of samaria. He gives two series of syntheses of sulphate from oxide, with two different lots of material, after eliminating samaria, and obtains, computing with $SO_3 = 80$, values for Di as follows:

First Series.

Per cent. Di_2O_3.	At. Wt. Di.
58.088	142.31
58.113	142.49
58.047	142.03
58.099	142.39
58.104	142.42
58.098	142.38
58.104	142.42
58.103	142.42
58.070	142.19
58.079	142.25

Second Series.

Per cent. Di_2O_3.	At. Wt. Di.
58.125	142.57
58.093	142.35
58.088	142.31
58.111	142.47
58.056	142.10
58.097	142.38
58.057	142.10

In short, the atomic weight of this " didymium " is not far from 142.

---

* Bull. Soc. Chim., 39, 289. 1883. Öfv. K. Vet. Akad. Förhaudl., No. 2, 1883.

THE DIDYMIUMS.

Bauer's little known determinations* were also made by the synthesis of the sulphate. They have corroborative value and are as follows:

Per cent. Di_2O_3.	At. Wt. Di.
58.285	143.56
58.100	142.40
58.133	142.64
58.098	142.38

In 1885 all of the foregoing determinations were practically brushed aside by Auer von Welsbach,† who by the most laborious fractionations proved that the so-called "didymia" was really a mixture of oxides, whose metals he names neodidymium and praseodidymium, names which are now commonly shortened into neodymium and praseodymium. One of these metals gives deep rose-colored salts, the other forms green compounds, and the difference of color is almost as strongly marked as in the cases of cobalt and nickel. Their atomic weights, determined by the sulphate method, are given by Welsbach a:—

$$Pr = 143.6$$
$$Nd = 140.8$$

No further details as to these determinations are cited, and whether they rest upon $O = 16$, $SO_3 = 80$, or $O = 15.96$ is uncertain. Fuller determinations are evidently needed.

---
* Freiburg Inaugural Dissertation, 1884.
† Monatsh. Chem., 6, 490. 1885.

SCANDIUM.

Cleve,[*] who was the first to make accurate experiments on the atomic weight of this metal, obtained the following data: 1.451 grm. of sulphate, ignited, gave .5293 grm. of Sc_2O_3. .4479 grm. of Sc_2O_3, converted into sulphate, yielded 1.2255 grm. of the latter, which, upon ignition, gave .4479 grm. of Sc_2O_3. Hence, for the percentage of Sc_2O_3 in $Sc_2(SO_4)_3$ we have:

$$36.478$$
$$36.556$$
$$36.556$$

Mean, $36.530, \pm .0175$

Hence, if $SO_3 = 79.465$, $Sc = 44.882$.

Later results are those of Nilson,[†] who converted scandium oxide into the sulphate. I give in a third column the percentage of oxide in sulphate:

.3379 grm. Sc_2O_3 gave .9343 grm. $Sc_2(SO_4)_3$.			36.166 per cent.
.3015 " .8330 "			36.194 "
.2998 " .8257 "			36.187 "
.3192 " .8823 "			36.178 "

Mean, $36.181, \pm .004$

Hence $Sc = 43.758$.
Combining the two series, we have—

Cleve .... $36.530, \pm .0175$
Nilson .... $36.181, \pm .0040$

General mean .... $36.190, \pm .0039$

Hence, with $SO_3 = 79.465, \pm .00175$,

$$Sc = 43.784, \pm .0085.$$

If $O = 16$, $Sc = 44.118$.

As between the two values found, the presumption is in favor of the lower. The most obvious source of error would be the presence in the scandia of earths of higher molecular weight.

---

[*] Compt. Rend., 89, 419.
[†] Compt. Rend., 91, 118.

YTTRIUM.

All the regular determinations of the atomic weight of yttrium depend upon analyses or syntheses of the sulphate. A series of analyses of the oxalate, however, by Berlin,* is sometimes cited, and the data are as follows. In three experiments upon the salt $Yt_2(C_2O_4)_3\, 3H_2O$ the subjoined percentages of oxide were found:

$$45.70$$
$$45.65$$
$$45.72$$

Mean, $45.69, \pm .0141$

Hence with $O = 15.879$ and $C = 11.920$,

$$Yt = 88.943.$$

Ignoring the early work of Berzelius,† the determinations to be considered are those of Popp, Delafontaine, Bahr and Bunsen, Cleve, and Jones.

Popp‡ evidently worked with material not wholly free from earths of higher molecular weight than yttria. The yttrium sulphate was dehydrated at 200°; the sulphuric acid was then estimated as barium sulphate, and after the excess of barium in the filtrate had been removed the yttrium was thrown down as oxalate and ignited to yield oxide. The following are the weights given by Popp:

Sulphate.	$BaSO_4$.	Yt_2O_3.	H_2O.
1.1805 grm.	1.3145 grm.	.4742 grm.	.255 grm.
1.4295 "	1.593 "	.5745 "	.308 "
.8455 "	.9407 "	.3392 "	.1825 "
1.045 "	1.1635 "	.4195 "	.2258 "

Eliminating water, these figures give us for the percentages of Yt_2O_3 in $Yt_2(SO_4)_3$ the values in column A. In column B I put the quantities of Yt_2O_3 proportional to 100 parts of $BaSO_4$:

A.	B.
51.237	36.075
51.226	36.064
51.161	36.058
51.209	36.055
Mean, $51.208, \pm .011$	Mean, $36.063, \pm .003$

From B, $Yt = 101.54$. The values in A will be combined with similar data from other experimenters.

---

* Forhandlingar ved de Skandinaviske Naturforskeres, 8, 452. 1860.
† Lehrbuch, V Aufl., 3, 1225.
‡ Ann. Chem. Pharm., 131, 179. 1854.

In 1865 Delafontaine* published some results obtained from yttrium sulphate, the yttrium being thrown down as oxalate and weighed as oxide. In the fourth column I give the percentages of Yt_2O_3 reckoned from the anhydrous sulphate:

Sulphate.	Yt_2O_3.	H_2O.	Per cent. Yt_2O_3.
.9545 grm.	.371 grm.	.216 grm.	50.237
2.485 "	.9585 "	.565 "	49.922
2.153 "	.827 "	.4935 "	49.834

Mean, 49.998, ± .081

In another paper † Delafontaine gives the following percentages of Yt_2O_3 in dry sulphate. The mode of estimation was the same as before:

48.23
48.09
48.37

Mean, 48.23, ± .055

Bahr and Bunsen, ‡ and likewise Cleve, adopted the method of converting dry yttrium oxide into anhydrous sulphate, and noting the gain in weight. Bahr and Bunsen give us the two following results. I add the usual percentage column:

Yt_2O_3.	$Yt_2(SO_4)_3$.	Per cent. Yt_2O_3.
.7266 grm.	1.4737 grm.	49.304
.7856 "	1.5956 "	49.235

Mean, 49.2695, ± .0233

Cleve's first results are published in a joint memoir by Cleve and Hoeglund,§ and are as follows:

Yt_2O_3.	$Yt_2(SO_4)_3$.	Per cent. Yt_2O_3.
1.4060 grm.	2.8925 grm.	48.608
1.0930 "	2.2515 "	48.545
1.4540 "	2.9895 "	48.637
1.3285 "	2.7320 "	48.627
2.3500 "	4.8330 "	48.624
2.5780 "	5.3055 "	48.591

Mean, 48.605, ± .0096

In a later paper Cleve ‖ gives syntheses of yttrium sulphate made with yttria, which was carefully freed from terbia. The weights and percentages are as follows:

---

* Ann. Chem. Pharm., 134, 108. 1865.
† Arch. Sci. Phys. et Nat. (2), 25, 119. 1866.
‡ Ann. Chem. Pharm., 137, 21. 1866.
§ K. Svenska Vet. Akad. Handlingar, Bd. 1, No. 8. 1873.
‖ K. Svenska Vet. Akad. Handlingar, No. 9, 1882. See also Bull. Soc. Chim., 39, 120. 1883.

YTTRIUM.

Yt_2O_3.	$Yt_2(SO_4)_3$.	Per cent. Yt_2O_3.
.8786	1.8113	48.507
.8363	1.7234	48.526
.8906	1.8364	48.497
.7102	1.4645	48.494
.7372	1.5194	48.519
.9724	2.0047	48.506
.9308	1.9197	48.487
.8341	1.7204	48.483
1.0224	2.1073	48.517
.9384	1.9341	48.519
.9744	2.0093	48.494
1.5314	3.1586	48.484

Mean, 48.503, ± .0029

Hence Yt = 88.449.

The yttria studied by Jones* had been purified by Rowland's method—that is, by precipitation with potassium ferrocyanide—and certainly contained less than one-half of one per cent. of other rare earths as possible impurities. Two series of determinations were made—one by ignition of the sulphate, the other by its synthesis. The results were as follows, with the usual percentage column added:

First Series. Syntheses.

Yt_2O_3.	$Yt_2(SO_4)_3$.	Per cent. Yt_2O_3.
.2415	.4984	48.455
.4112	.8485	48.462
.2238	.4617	48.473
.3334	.6879	48.466
.3408	.7033	48.457
.3418	.7049	48.489
.2810	.5798	48.465
.3781	.7803	48.456
.4379	.9032	48.483
.4798	.9901	48.460

Mean, 48.467, ± .0025

Second Series. Analyses.

$Yt_2(SO_4)_3$.	Yt_2O_3.	Per cent. Yt_2O_3.
.5906	.2862	48.459
.4918	.2383	48.455
.5579	.2705	48.485
.6430	.3117	48.478
.6953	.3369	48.454
1.4192	.6880	48.478
.8307	.4027	48.477
.7980	.3869	48.484
.8538	.4139	48.477
1.1890	.5763	48.469

Mean, 48.472, ± .0024

---

* Amer. Chem. Journ., 17, 154. 1895.

From syntheses.................. Yt = 88.287
From analyses...................... " = 88.309

These data of Jones were briefly criticised by Delafontaine,* who regards a lower value as more probable. In a brief rejoinder† Jones defended his own work; but neither the attack nor the reply needs farther consideration here. They are referred to merely as part of the record.

For the percentage of yttria in the sulphate we now have eight series of determinations, to be combined in the usual way:

Popp........................... 51.208, ± .0110
Delafontaine, first............. 49.998, ± .0810
Delafontaine, second............ 48.230, ± .0550
Bahr and Bunsen................. 49.2695, ± .0233
Cleve, earlier.................. 48.605, ± .0096
Cleve, later.................... 48.503, ± .0029
Jones, syntheses................ 48.467, ± .0025
Jones, analyses................. 48.472, ± .0024

General mean.................... 48.532, ± .0015

Hence, if O = 15.879, ± .0003, and S = 31.828, ± .0015,

$$Yt = 88.580, \pm .0053.$$

If O = 16, Yt = 89.255.

If only the four series by Cleve and by Jones are considered, the mean percentage of yttria in the sulphate becomes 48.481. Hence Yt = 88.350, or, with O = 16, 89.023.

This result is preferable to that derived from all the data, for it throws out determinations which are certainly erroneous. Cleve's early series might also be rejected, but its influence is insignificant.

---

* Chem. News, 71, 243.
† Chem. News, 71, 305.

SAMARIUM, GADOLINIUM, ERBIUM, AND YTTERBIUM.

The data relative to the atomic weights of these rare elements are rather scanty, and all depend upon analyses or syntheses of the sulphates.

SAMARIUM.

Atomic weight given by Marignac,[*] without details, as 149.4, and by Brauner,[†] as 150.7 in maximum. The first regular series of determinations was by Cleve,[‡] who effected the synthesis of the sulphate from the oxide. Data as follows:

Sm_2O_3.	$Sm_2(SO_4)_3$.	Per cent. Sm_2O_3.
1.6735	2.8278	59.180
1.9706	3.3301	59.175
1.1122	1.8787	59.201
1.0634	1.7966	59.190
.8547	1.4440	59.190
.7447	1.2583	59.183

Mean, 59.1865, ± .0025

Hence Sm = 149.038.

Another set of determinations by Bettendorff,[§] after the same general method, gave as follows:

Sm_2O_3.	$Sm_2(SO_4)_3$.	Per cent. Sm_2O_3.
1.0467	1.7675	59.219
1.0555	1.7818	59.238
1.0195	1.7210	59.225

Mean, 59.227, ± .0038

Hence Sm = 149.328.

Combining the two series, we have—

Cleve..................................... 59.1865, = .0025
Bettendorff............................. 59.227, ± .0038

General mean..................... 59.199, ± .0021

Hence, if $SO_3 = 79.465$, ± .00175,

$$Sm = 149.127, ± .0115.$$

If O = 16, Sm = 150.263.

According to Demarçay,[||] samaria contains an admixed earth whose properties are yet to be described.

---

[*] Arch. Sci. Phys. et Nat. (3), 3, 435. 1880.
[†] Journ. Chem. Soc., June, 1883.
[‡] Journ. Chem. Soc., August, 1883. Compt. Rend., 97, 94.
[§] Ann. Chem. Pharm., 263, 164. 1891.
[||] Compt. Rend., 122, 728. 1896.

GADOLINIUM.

This element, discovered by Marignac, must not be confounded with the mixture of metals from the gadolinite earths to which Nordenskiöld gave the same name. Several determinations of its atomic weight have been made, but Bettendorff's only were published with proper details.* He effected the synthesis of the sulphate from the oxide, and his weights were as follows. The percentage of Gd_2O_3 in $Gd_2(SO_4)_3$ is given in the third column:

Gd_2O_3.	$Gd_2(SO_4)_3$.	Per cent. Gd_2O_3.
1.0682	1.7779	60.082
1.0580	1.7611	60.076
1.0796	1.7969	60.081

Mean, 60.080, ± .0013

Hence, with $SO_3 = 79.465$, Gd = 155.575.
If O = 16, Gd = 156.761.
Boisbaudran † found Gd = 155.33, 156.06, 155.76, and 156.12. The last he considers the best, but gives no details as to antecedent values. He also quotes Marignac, who found Gd = 156.75, and Cleve, who found 154.15, 155.28, 155.1, and 154.77. Probably these all depend upon $SO_3 = 80$.

ERBIUM.

Since the earth which was formerly regarded as the oxide of this metal is now known to be a mixture of two or three different oxides, the older determinations of its molecular weight have little more than historical interest. Nevertheless the work done by several investigators may properly be cited, since it sheds some light upon certain important problems.

First, Delafontaine's ‡ early investigations may be considered. A sulphate, regarded as erbium sulphate, gave the following data. An oxalate was thrown down from it, which, upon ignition, gave oxide. The percentages in the fourth column refer to the anhydrous sulphate. In the last experiment water was not estimated, and I assume for its water the mean percentage of the four preceding experiments:

Sulphate.	Er_2O_3.	H_2O.	Per cent. Er_2O_3.
.827 grm.	.353 grm.	.177 grm.	54.308
1.0485 "	.4475 "	.226 "	54.407
.803 "	.3415 "	.171 "	54.035
1.232 "	.523 "	.264 "	54.028
1.1505 "	.495 "	....	54.760

Mean, 54.308, ± .0915

Hence Er = 117.86.

---

* Ann. Chem. Pharm., 270, 376. 1892.
† Compt. Rend., 111, 409. 1890.
‡ Ann. Chem. Pharm., 134, 108. 1865.

Bahr and Bunsen * give a series of results, representing successive purifications of the earth which was studied. The final result, obtained by the conversion of oxide into sulphate, was as follows:

.7870 grm. oxide gave 1.2765 grm. sulphate. 61.653 per cent. oxide.

Hence Er = 167.82.

Hoeglund,† following the method of Bahr and Bunsen, gives these results:

Er_2O_3.	$Er_2(SO_4)_3$.	Per cent. Er_2O_3.
1.8760 grm.	3.0360 grm.	61.792
1.7990 "	2.9100 "	61.821
2.8410 "	4.5935 "	61.848
1.2850 "	2.0775 "	61.853
1.1300 "	1.827 "	61.850
.8475 "	1.370 "	61.861

Mean, 61.8375, ± .0063

Hence Er = 169.33.

According to Thalén,‡ spectroscopic evidence shows that the "erbia" studied by Hoeglund was largely ytterbia.

Humpidge and Burney § give data as follows:

1.9596 grm. $Er_2(SO_4)_3$ gave 1.2147 grm. Er_2O_3.   61.987 per cent.
1.9011   "   1.1781   "   61.965   "

Mean, 61.976, ± .0074

Hence Er = 170.46.

The foregoing data were all published before the composite nature of the supposed erbia was fully recognized. It will be seen, however, that three sets of results were fairly comparable, while Delafontaine evidently studied an earth widely different from that investigated by the others. Since the discovery of ytterbium, some light has been thrown on the matter. The old erbia is a mixture of several earths, to one of which, a rose-colored body, the name erbia is now restricted. For the atomic weight of the true erbium Cleve || gives three determinations, based on syntheses of the sulphate after the usual method. His weights were as follows, with the percentage ratio added:

Er_2O_3.	$Er_2(SO_4)_3$.	Per cent. Er_2O_3.
1.0692	1.7436	61.321
1.2153	1.9820	61.317
.7850	1.2808	61.290

Mean, 61.309, ± .0068

Hence, with SO_3 = 79.465, Er = 165.059.
If O = 16, Er = 166.316.

* Ann. Chem. Pharm., 137, 21. 1866.
† K. Svenska Vet. Akad. Handlingar, Bd. 1, No. 6.
‡ Wiedemann's Beiblätter, 5, 122. 1881.
§ Journ. Chem. Soc., Feb., 1879, p. 116.
|| K. Svensk. Vet. Akad. Handlingar, No. 7, 1880. Abstract in Compt. Rend., 91, 382.

It is not worth while to combine this result with the earlier determinations, for they are now worthless.

YTTERBIUM.

For ytterbium we have one very good set of determinations by Nilson.* The oxide was converted into the sulphate after the usual manner:

Yb_2O_3.	$Yb_2(SO_4)_3$.	Per cent. Yb_2O_3.
1.0063 grm.	1.6186 grm.	62.171
1.0139 "	1.6314 "	62.149
.8509 "	1.3690 "	62.155
.7371 "	1.1861 "	62.145
1.0005 "	1.6099 "	62.147
.8090 "	1.3022 "	62.126
1.0059 "	1.6189 "	62.134

Mean, 62.147, ± .0036

Hence, with $SO_3 = 79.465$, Yb = 171.880.
If O = 16, Yb = 173.190.

TERBIUM, THULIUM, HOLMIUM, DYSPROSIUM, ETC.

For these elements the data are both scanty and vague. Concerning the atomic weights of holmium and dysprosium, practically nothing has been determined. To thulium, Cleve † assigns a value of Tm = 170.7, approximately, but with no details as to weighings. Probably the value was computed with $SO_3 = 80$.

For terbium, ignoring older determinations, Lecoq de Boisbaudran has published two separate estimates.‡ First, for two preparations, one with a lighter and one with a darker earth, he gives Tb = 161.4 and 163.1 respectively. In his second paper he gives Tb = 159.01 to 159.95. These values probably are all referred to $SO_3 = 80$.

---

*Compt. Rend., 91, 56. 1880. Berichte, 13. 1430.
† Compt. Rend., 91, 329. 1880.
‡ Compt. Rend., 102, 396, and 111, 474.

ARGON AND HELIUM.

The true atomic weights of these remarkable gases are still in doubt, and so far can only be inferred from their specific gravities.

For argon, the discoverers, Rayleigh and Ramsay,* give various determinations of density, ranging, with hydrogen taken as unity, from 19.48 to 20.6. In an addendum to the same paper, Ramsay alone gives for the density of argon prepared by the magnesium method the mean value of 19.941. In a later communication † Rayleigh gives determinations made with argon prepared by the oxygen method, and puts the density at 19.940.

For the density of helium, Ramsay ‡ gets 2.18, while Langlet § finds the somewhat lower value 2.00.

From one set of physical data both gases appear to be monatomic, but from other considerations they are supposably diatomic. Upon this question controversy has been most active, and no final settlement has yet been reached. If diatomic, argon and helium have approximately the atomic weights two and twenty respectively; if monatomic, these values must be doubled. In either case helium is an element lying between hydrogen and lithium, but argon is most difficult to classify. With the atomic weight 20, argon falls in the eighth column of the periodic system between fluorine and sodium, but if it is 40 the position of the gas is anomalous. A slightly lower value would place it between chlorine and potassium, and again in the eighth column of Mendelejeff's table; but for the number 40 no opening can be found.

It must be noted that neither gas, so far, has been proved to be absolutely homogeneous, and it is quite possible that both may contain admixtures of other things. This consideration has been repeatedly urged by various writers. If argon is monatomic, a small impurity of greater density, say of an unknown element falling between bromine and rubidium, would account for the abnormality of its atomic weight, and tend towards the reduction of the latter. If the element is diatomic, its classification is easy enough on the basis of existing data. Its resemblances to nitrogen, as regards density, boiling point, difficulty of liquefaction, etc., lead me personally to favor the lower figure for its atomic weight, and the same considerations may apply to helium also. Until further evidence is furnished, therefore, I shall assume the values two and twenty as approximately true for the atomic weights of helium and argon.

---

* Phil. Trans., 186, pp. 220 to 223, and 238. 1895.
† Chem. News, 73, 75. 1896.
‡ Journ. Chem. Soc., 1895, p. 684.
§ Zeitsch. Anorg. Chem., 10, 289. 1895.

TABLE OF ATOMIC WEIGHTS.

The following table contains the values for the various atomic weights found or adopted in the preceding calculations. As the table is intended for practical use, the figures are given only to the second decimal, the third being rarely, if ever, significant. In most cases even the first decimal is uncertain, and in some instances whole units may be in doubt.

	$H = 1.$	$O = 16.$
Aluminum	26.91	27.11
Antimony	119.52	120.43
Argon	?	?
Arsenic	74.44	75.01
Barium	136.39	137.43
Bismuth	206.54	208.11
Boron	10.86	10.95
Bromine	79.34	79.95
Cadmium	111.10	111.95
Cæsium	131.89	132.89
Calcium	39.76	40.07
Carbon	11.92	12.01
Cerium	139.10	140.20
Chlorine	35.18	35.45
Chromium	51.74	52.14
Cobalt	58.49	58.93
Columbium	93.02	93.73
Copper	63.12	63.60
Erbium	165.06	166.32
Fluorine	18.91	19.06
Gadolinium	155.57	156.76
Gallium	69.38	69.91
Germanium	71.93	72.48
Glucinum	9.01	9.08
Gold	195.74	197.23
Helium	?	?
Hydrogen	1.000	1.008
Indium	112.99	113.85
Iodine	125.89	126.85
Iridium	191.66	193.12
Iron	55.60	56.02
Lanthanum	137.59	138.64
Lead	205.36	206.92
Lithium	6.97	7.03
Magnesium	24.10	24.28
Manganese	54.57	54.99
Mercury	198.49	200.00
Molybdenum	95.26	95.99
Neodymium	139.70	140.80
Nickel	58.24	58.69

TABLE OF ATOMIC WEIGHTS.

	H = 1.	O = 16.
Nitrogen	13.93	14.04
Osmium	189.55	190.99
Oxygen	15.88	16.00
Palladium	105.56	106.36
Phosphorus	30.79	31.02
Platinum	193.41	194.89
Potassium	38.82	39.11
Praseodymium	142.50	143.60
Rhodium	102.23	103.01
Rubidium	84.78	85.43
Ruthenium	100.91	101.68
Samarium	149.13	150.26
Scandium	43.78	44.12
Selenium	78.42	79.02
Silicon	28.18	28.40
Silver	107.11	107.92
Sodium	22.88	23.05
Strontium	86.95	87.61
Sulphur	31.83	32.07
Tantalum	181.45	182.84
Tellurium	126.52	127.49
Terbium	158.80	160.00
Thallium	202.61	204.15
Thorium	230.87	232.63
Thulium	169.40	170.70
Tin	118.15	119.05
Titanium	47.79	48.15
Tungsten	183.43	184.83
Uranium	237.77	239.59
Vanadium	50.99	51.38
Ytterbium	171.88	173.19
Yttrium	88.35	89.02
Zinc	64.91	65.41
Zirconium	89.72	90.40

INDEX TO AUTHORITIES.

A

Agamennone.................. 14, 25
Allen.......................... 89
Allen and Pepys................ 24
Alibegoff................. 266, 300
Anderson...................... 130
Andrews.................. 118, 327
Arago................... 24, 58, 72
Arfvedson........... 84, 263, 282
Aston..................... 52, 172
Awdejew...................... 132

B

Bahr.......................... 137
Bahr and Bunsen........... 356, 361
Bailey.................... 197, 231
Bailey and Lamb............... 316
Balard......................... 44
Baubigny......... 93, 148, 180, 244, 300
Bauer.................... 349, 353
Becker.......................... 1
Beringer...................... 335
Berlin................ 238, 251, 355
Bernoulli..................... 257
Berzelius.. 5, 8, 24, 34, 38, 43, 44, 50, 58,
    72, 82, 84, 91, 101, 110, 112, 121, 123,
    127, 132, 135, 146, 171, 176, 188, 196,
    204, 209, 211, 213, 216, 236, 238, 250,
    255, 263, 268, 271, 277, 282, 287, 313,
    315, 322, 325, 327, 355
Bettendorff............. 349, 359, 360
Biot and Arago............. 24, 58, 72
Blomstrand.............. 234, 236
Boisbaudran........... 181, 360, 362
Bongartz...................... 226
Bongartz and Classen......... 200, 201
Borch, von.................... 256
Boussingault............... 24, 58
Brauner..272, 274, 340, 342, 348, 351, 359
Breed......................... 320
Bucher........................ 160
Buehrig....................... 339
Buff...................... 24, 72
Bunsen............. 87, 89, 356, 361
Bunsen and Jegel............. 336
Burney........................ 361

Burton........................ 151
Burton and Vorce............. 142

C

Capitaine..................... 287
Cavendish..................... 24
Chikashigé.................... 275
Choubine..................... 344
Christensen................... 280
Chydenius.................... 204
Clark........................... 9
Clarke....................... 159
Classen............. 200, 201, 231, 232
Claus......................... 311
Cleve.. 206, 347, 348, 351, 352, 354, 356,
    359, 360, 361, 362
Cleve and Hoeglund........... 356
Commaille..................... 91
Cooke........ 27, 81, 157, 221, 222, 224
Cooke and Richards............ 13
Crafts.................... 25, 58
Crookes...................... 185
Czudnowicz............... 211, 346

D

Davy.......................... 24
Debray................... 133, 251
Delafontaine....... 205, 356, 358, 360
De Luca....................... 278
Demarçay..................... 359
Demoly....................... 191
De Saussure............... 24, 72
Desi.......................... 262
Deville....................... 291
Deville and Troost............ 235
Dewar and Scott.............. 283
Dexter........................ 217
Diehl.......................... 84
Dittmar........................ 85
Dittmar and Henderson...... 12, 19
Dittmar and M'Arthur......... 333
Döbereiner............... 127, 287
Dulong and Berzelius..... 8, 24, 58, 72
Dumas.. 9, 39, 45, 50, 51, 72, 80, 91, 110,
    112, 113, 119, 129, 140, 156, 176, 188,
    199, 201, 209, 213, 217, 229, 251, 256,
    269, 278, 279, 282, 289, 294

(367)

THE ATOMIC WEIGHTS.

Dumas and Boussingault... ..... 24, 58
Dumas and Stas.................. 76

E

Ebelmen........................... 264
Ekman and Pettersson............ 269
Erdmann........................... 146
Erdmann and Marchand... 11, 76, 110, 111, 166, 268, 288, 291
Erk............................ 347, 351
Ewan and Hartog 171

F

Faget............................. 37
Favre............................. 147
Fourcroy.......................... 24
Fownes............................ 72
Fremy............................. 322
Friedel........................... 77

G

Gay-Lussac... .... 32, 135, 146
Genth.............................. 338
Gerhardt........................... 36
Gibbs......................... 298, 342
Gladstone and Hibbard............ 152
Gmelin............................ 84
Godeffroy....................... 87, 90
Gooch and Howland 274
Gray............................... 98

H

Hagen.............................. 84
Halberstadt....................... 330
Hampe.............................. 92
Hardin 34, 63, 74, 163, 167
Hartog............................ 171
Hauer, von.............. 156, 271, 283
Hebberling........................ 184
Hempel and Thiele................ 308
Henry.............................. 6
Hermann 84, 196, 206, 234, 236, 335, 347, 351
Heycock............................ 88
Hibbard........................... 152
Hibbs...................... 67, 68, 215
Hinrichs........................... 6
Hoeglund..................... 356, 361
Holzmann.......................... 345
Hoskyns-Abrahall................. 171
Howland........................... 274
Humboldt and Gay-Lussac......... 32
Humpidge and Burney.............. 361
Huntington.................... 46, 157

I

Isnard............................ 176

J

Jacquelain............. 136, 146, 238
Jegel............................. 336
Johnson........................... 17
Johnson and Allen................ 89
Jolly............................. 59
Joly......................... 311, 326
Joly and Leidié................... 319
Jones.............. 159, 357, 358
Jörgensen......................... 313

K

Keiser............... 15, 150, 316
Keiser and Breed................. 320
Keller and Smith................. 318
Kemp.............................. 252
Kessler..... 214, 216, 218, 224, 241, 242
Kirwan............................ 24
Kjerulf........................... 336
Klatzo............................ 132
Kobbe............................. 314
Kralovanzky...................... 84
Krüss............................. 102
Krüss and Alibegoff......... 266, 300
Krüss and Moraht................. 133
Krüss and Nilson 207
Krüss and Schmidt................ 303

L

Lagerhjelm........................ 229
Lamb.............................. 316
Lamy.............................. 184
Langer............................ 301
Langlet........................... 363
Laurent....................... 34, 171
Laurie............................ 103
Lavoisier................. 24, 58, 72
Le Conte.......................... 25
Leduc............. 20, 27, 32, 59, 78
Lee............................... 298
Lefort............................ 240
Leidié............................ 319
Lenssen........................... 156
Lepierre.......................... 186
Levol............................. 102
Liebig............................ 44
Liebig and Redtenbacher.......... 72
Liechti and Kemp................. 252
Longchamp.................. 127, 135

Lorimer and Smith............ 159
Louyet................ 277, 279, 280
Löwe.......................... 231
Löwig......................... 44

M

Maas........................... 252
M'Arthur....................... 333
Macdonnell..................... 136
Malaguti....................... 255
Mallet............... 84, 105, 150, 177
Marchand......... 11, 72, 76, 110, 111,
  · 166, 256, 263, 268, 288, 291
Marchand and Scheerer.......... 138
Marignac.. 34, 35, 36, 38, 39, 41, 43, 44,
  45, 47, 48, 49, 60, 62, 65, 74, 110, 114,
  115, 118, 121, 122, 123, 129, 141, 148,
  196, 230, 235, 236, 284, 292, 336, 344,
  345, 351, 359, 360.
Mather......................... 176
Maumené........ 34, 36, 39, 43, 75, 288
Meineke........................ 244
Meyer.......................... 252
Meyer and Seubert........... 1, 5, 6
Millon...................... 48, 167
Millon and Commaille........... 91
Mitscherlich 72
Mitscherlich and Nitzsch....... 268
Moberg......................... 239
Moissan.............. 278, 279, 280
Mond, Langer, and Quincke...... 301
Moraht......................... 133
Morley................ 12, 21, 27, 32
Morse and Burton............... 151
Morse and Jones................ 159
Morse and Keiser............... 150
Mosander........... 190, 335, 344, 351
Mulder......................... 6
Mulder and Vlaanderen.......... 199

N

Nilson................ 207, 354, 362
Nilson and Pettersson.......... 133
Nitzsch........................ 268
Nordenfeldt.................... 137
Nordenskiöld 360
Norlin......................... 287
Noyes...................... 16, 17

O

Ostwald........... 1, 6, 57, 71, 83, 131
Oudemans....................... 6

P

Parker......................... 142
Partridge 157
Peligot................. 238, 264, 265
Pelouze................. 35, 51, 60,
  113, 118, 188, 209, 213
Penfield....................... 186
Pennington and Smith........... 258
Penny........ 35, 39, 50, 62, 64, 66, 67
Pepys.......................... 24
Persoz......................... 257
Petrenko-Kritschenko 319
· Pettersson............... 133, 269
Pfeifer........................ 225
Piccard 87
Pierre......................... 191
Pollard 252
Popp........................... 355
Popper......................... 225

Q

Quincke........................ 301
Quintus Icilius................ 315

R

Rammelsberg... 234, 252, 263, 337, 344
Ramsay................... 149, 363
Ramsay and Aston 52, 172
Rawack......................... 283
Rawson......................... 244
Rayleigh.. 14, 16, 25, 26, 58, 59, 98, 363
Rayleigh and Ramsay 363
Rayleigh and Sidgwick.......... 98
Redtenbacher................. 72
Regnault................ 24, 25, 72
Reich and Richter 182
Remmler........................ 302
Reynolds and Ramsay............ 149
Richards..... 13, 46, 82, 92, 93, 94, 96,
  97, 115, 119, 121, 123, 124, 154
Richards and Parker............ 142
Richards and Rogers..... 141, 152, 153
Riche.......................... 259
Richter........................ 182
Rimbach........................ 174
Rivot.......................... 289
Robinson 340
Rogers 141, 152, 153
Roscoe............. 77, 211, 257, 262
Rose.............. 190, 217, 234, 236
Rothhoff....................... 291
Russell.................. 294, 295

S

Sacc.................................. 268
Salvétat................. 110, 113, 118
Scheerer............ 135, 136, 138, 139
Scheibler........................... 260
Schiel............................... 188
Schmidt........................ 203, 303
Schneider...... 216, 224, 229, 232, 255, 258, 282, 291, 292, 297
Schrötter........................... 209
Schutzenberger.............. 301, 341
Scott............................ 32, 283
Sebelien............................ 1, 7
Sefström............................ 165
Seubert........ 1, 322, 323, 325, 328, 333
Seubert and Kobbe................ 314
Seubert and Pollard............... 252
Shaw................................. 98
Shinn................................ 259
Sidgewick............................ 98
Siewert............................. 243
Smith............... 159, 258, 318
Smith and Desi.................... 262
Smith and Maas.................... 252
Sommaruga.......................... 297
Spring................................ 6
Stas.. 6, 37, 38, 40, 41, 42, 44, 45, 47, 48, 49, 51, 52, 57, 61, 62, 64, 65, 66, 71, 73, 76, 78, 80, 82, 83, 85, 128, 130, 131
Staudenmaier...................... 274
Strecker............................. 73
Stromeyer.............. 84, 156, 287
Struve................. 81, 82, 123, 250
Svanberg...................... 130, 167
Svanberg and Nordenfeldt........ 137
Svanberg and Norlin.............. 287
Svanberg and Struve....... 82, 250

T

Terreil.............................. 177
Thalén.............................. 361
Thiele.............................. 308
Thomsen.. 13, 22, 30, 57, 69, 71, 83, 131
Thomson........................ 24, 58
Thorpe.............................. 192
Thorpe and Laurie................. 103
Thorpe and Young................. 189
Tissier.............................. 176
Torrey..................... 151, 180, 289
Troost........................... 84, 235
Turner.. 38, 64, 121, 122, 123, 128, 166, 167, 282

U

Unger............................... 221

V

Van der Plaats... 6, 57, 71, 77, 83, 131, 149, 200, 210
Van Geuns............................ 5
Vanni................................ 98
Vauquelin........................... 24
Vlaanderen......................... 199
Vogel................................. 6
Vorce............................... 142

W

Wackenroder....................... 287
Waddell............................ 258
Wallace....................... 44, 213
Warrington.......................... 33
Weber............................... 217
Weeren........................ 132, 285
Weibull............................ 196
Wells and Penfield................ 186
Welsbach........................... 353
Wertheim........................... 265
Werther............................ 184
Weselsky........................... 298
Wildenstein........................ 241
Wills............................... 271
Wing................................ 338
Winkler.... 182, 195, 297, 305, 306, 307
Wolf........................... 337, 340
Woskresensky....................... 72
Wrede........................... 24, 72

Y

Young.............................. 189

Z

Zettnow............................ 260
Zimmermann.................. 266, 300
Zschiesche.................... 347, 351

www.ingramcontent.com/pod-product-compliance
Lightning Source LLC
Chambersburg PA
CBHW032042220426
43664CB00008B/816